High Speed Flows

High Speed Flows

Editor

Olga A. Azarova

MDPI

Basel • Beijing • Wuhan • Barcelona • Belgrade • Novi Sad • Cluj • Manchester

Editor
Olga A. Azarova
Federal Research Center
"Computer Science and
Control" of the Russian
Academy of Sciences
Moscow
Russia

Editorial Office
MDPI
St. Alban-Anlage 66
4052 Basel, Switzerland

This is a reprint of articles from the Special Issue published online in the open access journal *Fluids* (ISSN 2311-5521) (available at: https://www.mdpi.com/journal/fluids/special_issues/high_speed_flows).

For citation purposes, cite each article independently as indicated on the article page online and as indicated below:

Lastname, A.A.; Lastname, B.B. Article Title. *Journal Name* **Year**, *Volume Number*, Page Range.

ISBN 978-3-7258-0685-0 (Hbk)
ISBN 978-3-7258-0686-7 (PDF)
doi.org/10.3390/books978-3-7258-0686-7

Cover image courtesy of Olga Azarova

Contents

About the Editor

Olga A. Azarova

Olga A. Azarova, Ph.D., DSci, AIAA Senior Member, is a Full Member of the Russian Academy of Natural Sciences and a Leading Research Scientist in the Department of Mathematical Modeling of Computer-Aided Design Systems, Federal Research Center "Computing Science and Control" of the Russian Academy of Sciences (FRS CSC RAS), Moscow, Russia. Olga Azarova graduated from the Lomonosov Moscow State University. Olga Azarova received her Ph.D. and DSci degrees in Mechanics of Fluids, Gases, and Plasma at the Dorodnicyn Computing Centre of RAS. Her research interests include fluid mechanics, computational fluid dynamics, numerical simulation, aerodynamics, flow control, CFD coding, Richtmyer–Meshkov instability, vortex dynamics, compressible turbulence, and shock waves. Olga Azarova collaborated with Rutgers University, NJ, USA. She has participated in 68 international conferences, and a full list of her publications includes more than 150 names.

Preface

The presented reprint collection of papers covers a wide variety of modern areas of experimental and numerical research into the high-speed and complex flows of a solid medium, causing problems of internal and external gas dynamics, as well as some new practical problems. The considered topics include the following areas: aircraft concepts, developing new hypersonic vehicles, flows around high-speed vehicles, supersonic/hypersonic flows, flow control, supersonic nozzles, synthetic jets, shock waves, vortices and vortex structures, turbulence and boundary layers, construction of numerical methods, and modeling and application problems. The authors of the articles are professional researchers in the fields of CFD, physics, aerospace and general engineering. The Guest Editor expresses his deep gratitude to the authors, the editors of *Fluids*, and numerous anonymous reviewers.

Olga A. Azarova
Editor

fluids

Editorial

High Speed Flows

Olga A. Azarova

Federal Research Center "Computer Science and Control" of the Russian Academy of Sciences, Vavilova Str. 44, 119333 Moscow, Russia; olgazarov@gmail.com

High speed gas flows occur during the movement of aircrafts, rockets, and descent vehicles, as well as in combustion chambers, nozzles, and many other technological applications. High speed flows are characterized by a complex shock–vortex structure and the presence of large gradients of gas parameters due to the emerging shock waves, areas of shear deformations and the possible development of gas-dynamic instabilities. This Special Issue of *Fluids*, entitled "High Speed Flows", is focused on recent advances in the numerical and experimental modeling of high speed flows. The topics considered by the Issue include the following areas: aircraft concepts, developing new hypersonic vehicles, flows around high speed vehicles, supersonic/hypersonic flows, flow control, supersonic nozzles, synthetic jets, shock waves, vortices and vortex structures, turbulence, and boundary layers. Together with these topics devoted to modeling the problems of application, attention has been paid to the construction of numerical methods and the application of the apparatus of solid and fluid media to new technological problems.

Topics concerning aircraft concepts, developing new hypersonic vehicles, and investigation of high speed flows around vehicles are considered in [1–4]. The development of hypersonic aircraft requires the simulation of hypersonic flows using new computational methods. In particular, this concerns the modeling of heat fluxes to the boundary of a streamlined body. The Reynolds-averaged Navier–Stokes (RANS) method proved to be insufficiently accurate for solving such problems. Therefore, new methods were explored, including large eddy simulation (LES). In [1], Knight and Kianvashrad proposed a new LES method for predicting the dynamics of a boundary layer over a flat plate, using a new recirculation-scaling approach. The method is based on the calculation of total enthalpy and static pressure, along with velocity components, to obtain the best results for hard wall and the parameters of turbulence. The results of the Law of the Wall, Reynolds Analogy Factor, turbulent stresses, and energy spectrum were compared with the previous methods. It was shown that the new recycling–rescaling method improves the prediction of the Strong Reynolds Analogy and turbulent Prandtl number.

The development of new CFD approaches for simulating high speed flows is presented by Struchkov et al. in [2], where the features of the implementation of a flow limiter in solving 3D aerodynamic problems using the system of Navier–Stokes equations on unstructured grids are presented. The paper describes the discretization of the system of Navier–Stokes equations by the finite volume method, a mathematical model which includes the Spalart–Allmaras turbulence model, and the calculation scheme of the splitting method using a second-order approximation scheme. To monotonize the method, the Venkatakrishnan limiter function was chosen by the authors. It was shown that when calculating on unstructured grids, the Venkatakrishnan limiter can lead to the appearance of areas of its accidental operation, which affects the accuracy of the result. A modified variant of the Venkatakrishnan limiter for unstructured grids was proposed, which was free from this shortcoming. To study the applicability of the limiting function, the problems with supersonic flow in a channel with a wedge and a transonic flow around the NACA0012 airfoil were simulated on an unstructured grid. The analysis of the flow field around the NACA0012 airfoil revealed the absence of areas of accidental triggering in the case of using the modified version of the limiter function.

Citation: Azarova, O.A. High Speed Flows. *Fluids* **2023**, *8*, 109. https://doi.org/10.3390/fluids8040109

Received: 21 March 2023
Accepted: 22 March 2023
Published: 24 March 2023

In [3], Zhang and Wu simulated the expansion problem and used the characteristic directions of wave propagation and determined three zones—the U-zone, the M-zone, and the D-zone—within which the characteristics of pressure fluctuations exhibit different behavior in the boundary layer. The D-zone was defined as being located downstream of the first family characteristic line passing through the corner. The U-zone was defined as being located upstream of the second family characteristic line passing through the corner. The middle zone (M-zone) was defined as the zone between the U-zone and D-zone. The results of numerical analysis through the Detached Eddy Simulation (DES) made it possible to reveal this difference in behavior of pressure fluctuations. In addition, it was found that in the M-zone, the spatial distributions of fluctuation properties can differ at different levels, which, in the opinion of the authors, is explained by the action of the feedback mechanism.

The study by Gueraiche et al. [4] aimed to research the flight stability of an aircraft with a light box wing and a pusher propeller in the rear fuselage. New solutions were proposed, which included the use of a propeller in a fairing and several configurations of small vertical stabilizers in combination with vortex generators on the surface of the fuselage. Experiments were carried out in a wind tunnel, the results of which were confirmed by CFD modeling. Thus, the dynamics of the flow were explained for each of the proposed solutions. It was shown that effect of the expansion angle on pressure fluctuations is an important issue in supersonic flow around high speed vehicles.

Currently, non-mechanical control of highspeed flows is a widely studied research topic, both experimentally and numerically. The studies in [5,6] are devoted to this topic. The research in [5] by Azarova and Kravchenko focuses on the investigation, based on the Navier–Stokes equations, of the effect of the thermally stratified energy deposition in front of the bow shock wave (SW) in the supersonic flow created by an aerodynamic (AD) body in air. A new multi-vortex mechanism for the impact of a stratified energy source on a supersonic flow/flight is described, which is due to the multiple manifestation of the Richtmyer–Meshkov instabilities. Flow regimes are obtained for which almost complete destruction of the bow SW in the density field occurs due to the multiple generation of this instability. It is also shown that, by changing the temperature in the layers of a stratified energy source, it is possible to influence the drag forces of the AD body and ensure the emergence and change in the magnitude of the lift (pitch) force. Thus, the basic principles for controlling non-stationary high speed flows using stratified energy sources were established.

The research in [6] by Znamenskaya et al. is devoted to the experimental and numerical investigation of the influence of a high-energy plasma formation (plasmoid) on the supersonic flow past a blunt body. The experimentally obtained series of Schlieren patterns of the unsteady interaction of the bow SW with explosive waves is compared with the results of modeling the flow dynamics based on the Euler equations. A qualitative agreement between the calculated flow patterns and the experimental ones is shown. Based on this comparison, the dynamics of the shock-wave structure caused by the interaction of the bow SW and the blast flow were studied, and a scheme was constructed for the initiation and dynamics of the generated SWs and contact discontinuities. A significant drop in drag force and stagnation pressure (up to 80%) was obtained, and the dynamics of the zone of low density and high gas temperature was studied. The dynamics of the drag forces of the front surface were also considered for various values of the plasmoid energy.

The construction of a supersonic wind tunnel facility, as well as the development of supersonic nozzles and synthetic jets are studied in [7–9]. The aim of the study in [7] by Andrews et al. is to characterize the flow in the SBR-50 facility (University of Notre Dame, Notre Dame, IN, USA) containing a supersonic shock tube to study the dynamics of gas temperature. Using thermocouple measurements and laser spark velocimetry, a detailed set of gas parameters along the entire length of the pipe was obtained, which was compared with 3D modeling based on the Navier–Stokes equations. This study proved that the original scheme of the experimental setup allows longer operation with a constant stagnation temperature.

The application of the Shock Vector Control (SVC) approach to an axisymmetric supersonic nozzle was numerically investigated by Resta et al. [8]. In the SVC method, the injection of a secondary air stream creates an asymmetrical pressure distribution on the wall. Forcing the SVC axisymmetric nozzle created fully three-dimensional flows that interact with the external flow. Experimental data on a nozzle designed and tested for a passenger supersonic aircraft were used to validate numerical software at various flight Mach numbers and nozzle pressures. Then, as a result of the fully 3D flow simulations, the optimal position of the slot was found at the Mach number M = 0.9 for various values of SVC forcing.

In [9], Pellessier et al. investigated several methods for visualization of pulsed synthetic jets for cooling applications. The visualization techniques under consideration include smoke, Schlieren imaging, and thermography. The Schlieren images were analyzed using Proper Orthogonal Decomposition (POD) and numerical methods for processing the videos. The results showed that for the particular nozzle under study, the optimal cooling occurs at a frequency of 80 Hz. It was shown that the combination of Schlieren visualization and POD is a unique method for optimizing synthetic jets.

For a long time, a particularly wide research area in the study of high speed flows has been the dynamics of the interaction of SWs with other SWs, barriers, boundary layers, and inhomogeneities of the medium (vortices, simple waves, and rarefaction waves). In this Special Issue, this topic is touched upon in [10,11], where the Mach reflection of SWs and shock–vortex interactions are studied. The Mach stem arises as an additional SW in the SW reflection of the Mach type in a steady supersonic flow. In [10], Bai and Wu showed that the normalized length of a Mach stem is almost linear with respect to the normalized wedge trailing edge height, which was justified by the theoretical analysis. This result gives the possibility to obtain analytical models for expression of the length of a Mach stem through the flow parameters.

In [11], Skews presented the results of experimental studies of shock-vortex interactions accompanied by the numerical simulations. In the experiments, the vortex is formed due to flow separation from the corner and is accompanied by the appearance of a shear layer; next, the SW is diffracted at the edge. A review of the results of various experiments is presented, in which two independent SW reach the corner at different times, the diffracting SW is reflected from different surfaces back into the vortex, and the flow around bends is studied, where the SW is reflected from the far wall back into the vortex. In most cases it was obtained that the vortex retained its integrity after passing through a SW. Some studies with curved SWs showed signs of the decay of a vortex and development of turbulent spots, as well as a significant change in the vortex shape.

The papers [12,13] are devoted to new directions of using the concept of solid media and fluid description for modeling the gas dynamics of solid and fluid substances. A paper [12] by Sposobin and Reviznikov is devoted to numerical simulation of the gas-dynamic interaction of solid particles with the shock layer; in particular, the heat transfer by high-inertia particles. The particles rebound from the surface and destroy the bow SW front, which changes the structure of the whole flow. It is shown that by the successive action of particles, the impact jet flowing onto the surface is generated. In the impact jet, the values of pressure and heat flux are increased, which is the reason for the effect obtained.

The flooding of railway ballasts has been the subject of several experimental investigations. In [13], Alrdadi and Meylan presented the results of numerical simulation of two experiments on the flooding of railway ballast. The fluid flow is modelled by Darcy's law, which the authors extend to the free fluid flowing above the ballast. The equations are solved using the finite element method. The results of numerical calculations were compared with the experimental ones reported in the literature and a good agreement was demonstrated. The method was then extended, taking into account the realistic railway ballast.

Thus, the Special Issue "High Speed Flows" of the *Fluids* journal covers a wide variety of modern areas of experimental and numerical research into high speed flows of a solid medium, arising both in problems of internal and external gas dynamics as well as in some new practical problems.

Acknowledgments: Throughout the editing of this Special Issue, the Guest Editor was strongly encouraged and supported by the Editors of *Fluids*. For this, the Guest Editor expresses his deep gratitude to them. It is also important to note the work of the anonymous Reviewers on the above articles. Without their contributions, this Special Issue would not have been possible, and the Guest Editor extends his gratitude to them.

Conflicts of Interest: The author declares no conflict of interest.

References

1. Kianvashrad, N.; Knight, D. Large Eddy Simulation of Hypersonic Turbulent Boundary Layers. *Fluids* **2021**, *6*, 449. [CrossRef]
2. Struchkov, A.V.; Kozelkov, A.S.; Zhuchkov, R.N.; Volkov, K.N.; Strelets, D.Y. Implementation of Flux Limiters in Simulation of External Aerodynamic Problems on Unstructured Meshes. *Fluids* **2023**, *8*, 31. [CrossRef]
3. Zhang, L.; Wu, Z.-N. Numerical Simulation of Pressure Fluctuation near an Expansion Corner in a Supersonic Flow of M = 3.01. *Fluids* **2021**, *6*, 268. [CrossRef]
4. Gueraiche, D.; Karpovich, E.; Pikulev, M.; Kuznetsov, A.; Popov, S. Experimental and CFD Investigation of Directional Stability of a Box-Wing Aircraft Concept. *Fluids* **2022**, *7*, 340. [CrossRef]
5. Azarova, O.A.; Kravchenko, O.V. Principles of Unsteady High-Speed Flow Control Using a Time-Limited Thermally Stratified Energy Source. *Fluids* **2022**, *7*, 326. [CrossRef]
6. Znamenskaya, I.; Chernikov, V.; Azarova, O. Dynamics of Shock Structure and Frontal Drag Force in a Supersonic Flow Past a Blunt Cone under the Action of Plasma Formation. *Fluids* **2021**, *6*, 399. [CrossRef]
7. Andrews, P.; Lax, P.; Elliott, S.; Firsov, A.; Leonov, S. Flow Characterization at Heated Air Supersonic Facility SBR-50. *Fluids* **2022**, *7*, 168. [CrossRef]
8. Resta, E.; Marsilio, R.; Ferlauto, M. Thrust Vectoring of a Fixed Axisymmetric Supersonic Nozzle Using the Shock-Vector Control Method. *Fluids* **2021**, *6*, 441. [CrossRef]
9. Pellessier, J.E.; Dillon, H.E.; Stoltzfus, W. Schlieren Flow Visualization and Analysis of Synthetic Jets. *Fluids* **2021**, *6*, 413. [CrossRef]
10. Bai, C.-Y.; Wu, Z.-N. A Study of the Dependence of the Mach Stem Height on the Trailing Edge Height. *Fluids* **2021**, *6*, 313. [CrossRef]
11. Skews, B. Experiments in Shock-Vortex Interactions. *Fluids* **2021**, *6*, 303. [CrossRef]
12. Sposobin, A.; Reviznikov, D. Impact of High Inertia Particles on the Shock Layer and Heat Transfer in a Heterogeneous Supersonic Flow around a Blunt Body. *Fluids* **2021**, *6*, 406. [CrossRef]
13. Alrdadi, R.; Meylan, M.H. Modelling Experimental Measurements of Fluid Flow through Railway Ballast. *Fluids* **2022**, *7*, 118. [CrossRef]

fluids

MDPI

Article

Large Eddy Simulation of Hypersonic Turbulent Boundary Layers

Nadia Kianvashrad *,† **and Doyle Knight** *,†

Department of Mechanical and Aerospace Engineering, Rutgers—The State University of New Jersey, New Brunswick, NJ 08903, USA
* Correspondence: nadiakianvashrad@gmail.com (N.K.); doyleknight@gmail.com (D.K.)
† These authors contributed equally to this work.

Abstract: The recent revival of interest in developing new hypersonic vehicles brings attention to the need for accurate prediction of hypersonic flows by computational methods. One of the challenges is prediction of aerothermodynamic loading over the surface of the vehicles. Reynolds Average Navier-Stokes (RANS) methods have not shown consistent accuracy in prediction of such flows. Therefore, new methods including Large Eddy Simulations (LES) should be investigated. In this paper, the LES method is used for prediction of the boundary layer over a flat plate. A new recycling-rescaling method is tested. The method uses total enthalpy and static pressure along with the velocity components to produce the best results for the Law of the Wall, turbulent statistics and turbulent Prandtl number.

Keywords: large eddy simulation (LES); turbulent boundary layer; hypersonic flow

Citation: Kianvashrad, N.; Knight, D. Large Eddy Simulation of Hypersonic Turbulent Boundary Layers. *Fluids* **2021**, *6*, 449. https://doi.org/10.3390/fluids6120449

Academic Editor: Giuliano De Stefano

Received: 30 September 2021
Accepted: 7 December 2021
Published: 11 December 2021

1. Introduction

There is a recent revival of interest in developing hypersonic vehicles. Examples includes Boeing Hypersonic Airliner [1], SpaceLiner [2], and LAPCAT A2 [3]. This recent interest focuses attention on several fields in hypersonic flow physics including prediction of aerothermodynamic loading over the vehicle's surface. There is no consistently accurate prediction of aerothermodynamic loading over the entire body of a vehicle. One example of the problematic prediction of aerothermodynamic loading is the prediction in the regions with shock wave boundary layer interactions. The inaccurate prediction of aerothermodynamic loading in the regions of shock wave boundary layer interactions can result in disastrous failure in the structure of the vehicle [4,5].

Reynolds Average Navier-Stokes (RANS) methods have not shown consistent accuracy in prediction of aerothermodynamic loading in the shock wave boundary layer interaction regions in hypersonic flows [6]. As an example, Kumar et al. [7] use different RANS models to predict the flowfield over a hollow cylinder flare at Mach 9.6. The RANS models include Spalart-Allmaras (SA) RC-QCR2013, SST, Goldberg-Rt, $k-\epsilon$-Rt, SA-Catris, and SA-Neg turbulent models. Their results show a significant change in prediction of the separation region size from no separation to a very large separation region.

An alternate computational method is Large Eddy Simulation (LES). It is proven that LES can provide accurate prediction of shock wave boundary layer interaction in supersonic flows. Examples are Loginov et al. [8], Touber and Sandham [9], Morgan et al. [10], Ritos et al. [11], Ritos et al. [12], and Hadjadj [13]. However, LES is rarely used for prediction of hypersonic shock wave boundary layer interactions. Examples of the usage of LES in hypersonic shock wave boundary layer interactions are discussed below.

Edwards et al. [14] performed a Wall Modeled LES (hybrid RANS/LES) of a shock wave boundary layer interaction over a compression ramp at Mach 5 with $T_w/T_{aw} = 1.0$, where T_w is wall temperature and T_{aw} is the adiabatic wall temperature. The RANS model uses the SST turbulent model of Menter [15]. In general, their results are in reasonable agreement with the experimental data. However, the plateau pressure in the separated region is underestimated and the pressure recovery downstream of the interaction region is overestimated.

Fang et al. [16] performed an LES of a hypersonic flow past a single fin at Mach 5 and $T_w/T_{aw} = 1.0$ for fin and $T_w/T_{aw} = 0.81$ for the plate. The subgrid scale stresses and heat transfer are modeled by the dynamic Smagorinsky model [17,18]. The instantaneous time-dependent inflow condition is generated by a separate LES simulation using a wall blowing and suction technique. The inflow condition shows a good agreement with the Law of the Wall and incompressible and low speed compressible density scaled turbulent fluctuations. The surface pressure over the fin is in agreement with the experimental data; however, the peak skin friction coefficient in the vicinity of the first reattachment point is underpredicted by a factor of four.

Fu et al. [19] performed a Wall Modeled LES of a hypersonic turbulent flow past two parallel fins at Mach 8.23. The inflow condition upstream of the leading edge of the flat plate over which two sharp fins are installed is generated as uniform flow with turbulent fluctuations created by a synthetic turbulence method. The subgrid scale stresses are modeled by a static coefficient Vreman model. The LES results are in general agreement with experimental normalized mean surface pressure and mean heat transfer over the flat plate and fins.

One possible explanation for the rarity of LES applications in hypersonic flows is the existence of cold walls in hypersonic experiments. The existence of a cold wall makes it difficult to create a model to generate the instantaneous time-dependent inflow condition in the boundary layer. To generate the inflow condition, five variables are needed to be recycled for the inflow condition, namely, three components of velocity and two state parameters. One method was introduced by Sheng Xu and Pino Martin [20] in which the boundary layer is divided into three sections namely, viscous sublayer, logarithmic region, and outer layer. In their model, each of these three sections is modeled separately and then blended together using three weighting functions.

The objective of this paper is to evaluate a new recycling-rescaling method for generation of the instantaneous time-dependent inflow boundary layer to be used in hypersonic turbulent LES. The new method recycles total enthalpy and static pressure in addition to the three components of velocity. This new method does not assume a constant pressure inside the boundary layer. The new recycling-rescaling method is compared to the more traditional ways that assume constant pressure in the boundary layer and recycles the three components of velocity and generates the inflow temperature either using Walz's equation [21] or recycling the mean temperature. The new recycling-rescaling method is tested for different wall temperatures and using different grids. This method will enable researchers to create a dynamic inflow condition to be used for solving hypersonic turbulent shock wave boundary layer interactions with a cold wall.

2. Materials and Methods

2.1. Governing Equations

The compressible Large Eddy Simulation governing equations are obtained by spacial filtering of the time-dependent compressible Navier-Stokes equations. The Favre averaging is the common spatial averaging used for this purpose. The Favre averaging of an arbitrary variable $\mathcal{F}(x_i, t)$ is defined as

$$\widetilde{\mathcal{F}} = \frac{\overline{\rho\mathcal{F}}}{\overline{\rho}}, \tag{1}$$

where $\overline{\rho}$ is the mean density. Therefore, the instantaneous expression of an arbitrary variable $\mathcal{F}$ is

$$\mathcal{F}(x_i, t) = \widetilde{\mathcal{F}} + \mathcal{F}''. \tag{2}$$

The conventional spatial average of a variable $\mathcal{G}(x_i, t)$ is denoted $\overline{\mathcal{G}}$ and thus

$$\mathcal{G}(x_i, t) = \overline{\mathcal{G}} + \mathcal{G}'. \tag{3}$$

The Favre-averaged governing equations for a perfect gas are

$$\frac{\partial \bar{\rho}}{\partial t} + \frac{\partial \bar{\rho}\tilde{u}_i}{\partial x_i} = 0, \tag{4}$$

$$\frac{\partial \bar{\rho}\tilde{u}_i}{\partial t} + \frac{\partial \bar{\rho}\tilde{u}_i\tilde{u}_j}{\partial x_j} = -\frac{\partial \bar{p}}{\partial x_i} + \frac{\partial \mathcal{T}_{ij}}{\partial x_j}, \tag{5}$$

$$\frac{\partial \bar{\rho}\tilde{e}}{\partial t} + \frac{\partial}{\partial x_j}(\bar{\rho}\tilde{e} + \bar{p})\tilde{u}_j = \frac{\partial \mathcal{H}_j}{\partial x_j}, \tag{6}$$

$$\bar{p} = \bar{\rho}R\tilde{T}, \tag{7}$$

where $\tilde{u}_i$ are the Favre-averaged velocity components in the Cartesian coordinates, x_i are the Cartesian coordinates ($i = 1,2,3$), $\bar{p}$ is the average pressure, $\mathcal{T}_{ij}$ is the total stress, $\mathcal{H}_j$ is the energy flux (due to heat transfer and work done by the total stress), and $\tilde{e}$ is the Favre-averaged total energy per unit mass

$$\bar{\rho}\tilde{e} = \bar{\rho}c_v\tilde{T} + \frac{1}{2}\bar{\rho}\tilde{u}_i\tilde{u}_i + \bar{\rho}k, \tag{8}$$

where $\bar{\rho}k$ is the subgrid scale turbulence kinetic energy per unit volume

$$\bar{\rho}k = \frac{1}{2}\bar{\rho}(\widetilde{u_i u_i} - \tilde{u}_i\tilde{u}_i). \tag{9}$$

The total stress is

$$\mathcal{T}_{ij} = \tau_{ij} + \bar{\sigma}_{ij}, \tag{10}$$

where τ_{ij} is the subgrid scale stress tensor

$$\tau_{ij} = -\bar{\rho}(\widetilde{u_i u_j} - \tilde{u}_i\tilde{u}_j), \tag{11}$$

and therefore $\tau_{ii} = -2\bar{\rho}k$. The molecular viscous stress tensor $\bar{\sigma}_{ij}$ can be approximated as [18]

$$\bar{\sigma}_{ij} = \mu(\tilde{T})\left(-\frac{2}{3}\frac{\partial \tilde{u}_k}{\partial x_k}\delta_{ij} + \frac{\partial \tilde{u}_i}{\partial x_j} + \frac{\partial \tilde{u}_j}{\partial x_i}\right), \tag{12}$$

where $\mu(\tilde{T})$ is the molecular viscosity based on the Favre-averaged static temperature $\tilde{T}$.

The energy flux $\mathcal{H}_j$ is

$$\mathcal{H}_j = \mathcal{Q}_j + \mathcal{T}_{ij}\tilde{u}_i, \tag{13}$$

where the total heat transfer $\mathcal{Q}_j$ is

$$\mathcal{Q}_j = Q_j + \bar{q}_j. \tag{14}$$

The subgrid scale heat flux Q_j is

$$Q_j = -c_p\bar{\rho}\left(\widetilde{u_j T} - \tilde{u}_j\tilde{T}\right), \tag{15}$$

and $\bar{q}_j$ is the molecular heat flux

$$\bar{q}_j = \kappa(\tilde{T})\frac{\partial \tilde{T}}{\partial x_j}, \tag{16}$$

where $\kappa(\tilde{T})$ is the molecular thermal conductivity based on the Favre-averaged static temperature. The transport properties and thermodynamic data are obtained from Gupta, Yos, Thompson and Lee [22] (NASA-RP-1232) database. Since in this database, air consists of different species, Wilke's Rule [23] is used to calculate the mixture viscosity μ and thermal conductivity κ. The molecular Prandtl number $Pr = 0.74$. Hereafter, $\bar{}$ and $\tilde{}$ are dropped for simplicity.

To have a closed system of equations for Equations (4)–(7), it is required to model the subgrid scale stress τ_{ij} and the heat flux Q_j in addition to having appropriate initial and boundary conditions. In this paper, an implicit SubGrid Scale (SGS) model is implemented. The implicit SGS models are also known as Monotone Integrated Large Eddy Simulation (MILES) or Implicit Large Eddy Simulation (ILES). These methods are based on the concept that dissipation of energy from the resolved scale to the subgrid scales is achievable through the inviscid flux algorithm. Therefore, the subgrid scale stresses and heat flux are zero, i.e., $\tau_{ij} = 0$ and $Q_j = 0$. Grinstein et al. [24] provided a detailed review of this model.

2.2. Numerical Algorithm

2.2.1. Finite Volume Code

The governing equations are solved using a finite volume C^{++} code developed by the authors for a block structured grid. The code is capable of solving non-equilibrium hypersonic flows; however, the low stagnation enthalpy of the simulations presented in this paper, make the calculations to be thermally perfect. The inviscid fluxes are calculated using the Roe's method with second-order Monotone Upstream Scheme for Conservation Laws (MUSCL) [25], and viscous fluxes are determined using a second-order central differencing method. The second-order Data Parallel Line Relaxation (DPLR) method [26] is used for time integration to achieve computational efficiency. To parallelize the code, the Message Passing Interface (MPI) is used.

2.2.2. Boundary Conditions

Figure 1 shows the schematic of the computational domain. The turbulent boundary layer at the inflow is calculated using the recycling-rescaling method, which is described in Section 2.2.4. The recycling-rescaling method provides time-dependent values of ρ, ρu_i, and ρe at the inflow boundary. The outflow boundary is the zero gradient boundary condition. The spanwise boundaries are periodic boundary conditions. The fixed condition at the freestream flow is applied at the top boundary and the bottom boundary is the no-slip isothermal wall boundary condition.

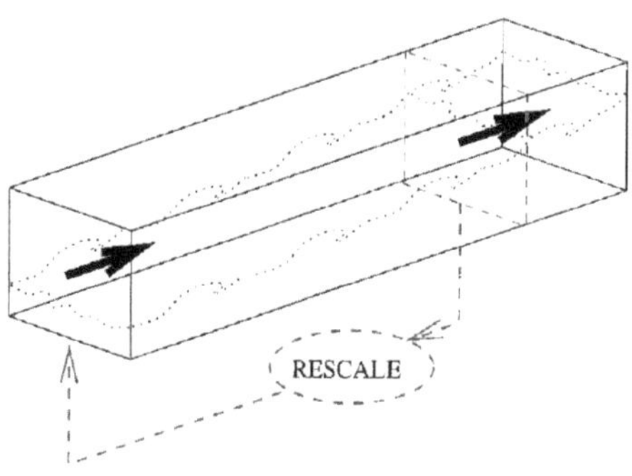

Figure 1. Computational domain with recycling-rescaling [27].

2.2.3. Initial Condition

The initial condition of the streamwise velocity is calculated in two parts, namely, the streamwise velocity of the viscous sublayer and the streamwise velocity of the Law of Wall and Wake. The initial condition of the mean streamwise velocity in the viscous sublayer is

$$u = \frac{\tau_w y}{\mu_w} \quad \text{for } y^+ \lesssim 10, \tag{17}$$

where τ_w is the mean shear stress at the wall, μ_w is the molecular viscosity at the wall, $y^+ = yu_\tau/\nu_w$, ν_w is the kinematic molecular viscosity at the wall, and

$$u_\tau = \sqrt{\frac{\tau_w}{\rho_w}}. \tag{18}$$

The initial condition of the mean streamwise velocity of the Law of Wall and Wake region is

$$\frac{u_{vd}}{u_\tau} = \frac{1}{\kappa}\log\left(\frac{yu_\tau}{\nu_w}\right) + C + \frac{2\Pi}{\kappa}\sin^2\left(\frac{\pi}{2}\frac{y}{\delta}\right), \tag{19}$$

where the Von Karman's constant is $\kappa = 0.4 \pm 0.01$ and u_{vd} is the Van Driest transformed velocity

$$u_{vd} = \frac{U_\infty}{A}\left[\sin^{-1}\left(\frac{2A^2(u/U_\infty) - B}{\sqrt{B^2 + 4A^2}}\right) + \sin^{-1}\left(\frac{B}{\sqrt{B^2 + 4A^2}}\right)\right], \tag{20}$$

where U_∞ is the freestream velocity and

$$A = \sqrt{\frac{(\gamma-1)}{2} Pr_t\, M_\infty^2 \frac{T_\infty}{T_w}}, \tag{21}$$

$$B = \left[1 + \frac{(\gamma-1)}{2} Pr_t M_\infty^2\right]\frac{T_\infty}{T_w} - 1, \tag{22}$$

where the turbulent Prandtl number is $Pr_t \approx 0.89$ and T_w is the wall temperature. Since the Reynolds number is high, $\Pi = 0.55$ is appropriate for this problem. The value of C ranges from 5.2 to 6.82 for an adiabatic wall. However, for isothermal cold wall cases, the value of C can be higher than 6.82. At the y^+ location that the mean streamwise velocity profiles of the viscous sublayer (Equation (17)) and the Law of the Wall and Wake (Equation (19)) provide the same value, the transition from viscous sublayer to Law of Wall and Wake occurs. The initial mean streamwise velocity u is the mean streamwise velocity by combining the mean streamwise velocity of viscous sublayer and Law of Wall and Wake profiles.

The value of u_τ at the inflow boundary is calculated by applying the Law of Wall and Wake at the edge of the boundary layer

$$\frac{u_{vd,\infty}}{u_\tau} = \frac{1}{\kappa}\log\left(\frac{\delta u_\tau}{\nu_w}\right) + C + \frac{2\Pi}{\kappa}, \tag{23}$$

where $u_{vd,\infty}$ is obtained from Equation (20). The value of u_τ at the inflow boundary depends upon the Reynolds number based upon the boundary layer thickness Re_δ at the inflow boundary, freestream Mach number M_∞, and the ratio of the wall temperature to the adiabatic wall temperature (obtained from Equation (25)) T_w/T_{aw}. The initial condition of the mean spanwise and mean wall normal velocities is zero. The mean static temperature initial condition is obtained from Walz's expression [21]

$$T = T_w + (T_{aw} - T_w)\left(\frac{u}{U_\infty}\right) + (T_\infty - T_{aw})\left(\frac{u}{U_\infty}\right)^2, \tag{24}$$

where

$$T_{aw} = T_\infty\left(1 + \frac{(\gamma-1)}{2} Pr_t M_\infty^2\right). \tag{25}$$

The mean density (ρ) initial condition is calculated from the equation of state assuming uniform mean static pressure (p) across the boundary layer. Random perturbations are added to three component of velocities, i.e., streamwise, wall normal, and spanwise velocities to initiate the turbulence.

2.2.4. Recycling-Rescaling Method

The turbulent inflow at each time step is calculated by a recycling-rescaling method. The method using here is a two layer method. A cross-stream plane ("recycling plane") located at a distance L_r from the inflow boundary is used to achieve the mean and fluctuating velocities, total enthalpy, and total and static pressure (Figure 1). The mean values are averaged over a period of time t_{aver} at the streamwise position L_r and each spanwise position, and then averaged in the spanwise direction to obtain the profiles in the wall normal direction. This method is called time and spanwise averaging method. For the simulations presented in this paper, $L_r = 10\delta$ and $t_{aver} = 20\delta/U_\infty$. The time and spanwise average of the Favre-averaged variables are denoted as $\overline{\mathcal{F}}$.

To fully define the inflow boundary, instantaneous velocities (i.e., streamwise velocity $u(y,z,t)$, wall normal velocity $v(y,z,t)$, and spanwise velocity $w(y,z,t)$), static temperature $T(y,z,t)$, and density $\rho(y,z,t)$ should be calculated. The mean velocity at the inflow plane is obtained separately for the inner layer and the outer layer. The mean velocity at the inflow in the inner layer $\bar{u}^{\text{inner}}_{\text{inflow}}$ is obtained from the mean velocity at the recycle plane in the inner layer $\bar{u}^{\text{inner}}_{\text{recycle}}$ according to

$$\bar{u}^{\text{inner}}_{\text{inflow}}\left(y^+_{\text{inflow}}\right) = \beta \bar{u}^{\text{inner}}_{\text{recycle}}\left(y^+_{\text{inflow}}\right), \tag{26}$$

where

$$\beta = \frac{u_\tau^{\text{inflow}}}{u_\tau^{\text{recycle}}}, \tag{27}$$

where u_τ^{inflow} is fixed and determined from Equation (23) and u_τ^{recycle} is obtained from $u^{\text{inner}}_{\text{recycle}}$ using Equation (18) based upon the computed mean shear stress at the recycle station and assuming the mean static pressure is constant across the boundary layer.

The mean velocity at the inflow plane in the outer layer $\bar{u}^{\text{outer}}_{\text{inflow}}$ is based upon the Van Driest transformation. Several other recent transformations have been proposed for cold wall hypersonic turbulent boundary layers including Trettle and Larsson [28] and Griffin et al. [29]. However, a substantial amount of experimental data for hypersonic cold wall turbulent boundary layers have shown that the Van Driest transformation provides an accurate conversion of the mean compressible streamwise velocity profile to the incompressible Law of the Wall and Wake. Examples include Hill [30], Winkler and Cha [31], Danberg [32], Young [33], Samuels et al. [34], Horstman and Owen [35], Owen and Horstman [36], and Keener and Hopkins [37]. Our future research will examine the alternate transformations. First, the Van Driest transformation of the outer layer mean velocity at the recycle station is calculated, and then the Van Driest velocity is rescaled to the inflow boundary according to

$$\bar{u}^{\text{outer}}_{\text{vd, inflow}}(\eta_{\text{inflow}}) = \beta \bar{u}^{\text{outer}}_{\text{vd, recycle}}(\eta_{\text{inflow}}) + (1-\beta)\bar{u}_{\text{vd},\infty}. \tag{28}$$

Then, the Van Driest transformed velocity $\bar{u}^{\text{outer}}_{\text{vd, inflow}}$ is inverted to obtain the mean velocity in the outer layer $\bar{u}^{\text{outer}}_{\text{inflow}}$.

The mean wall normal velocity at the inflow boundary is calculated by scaling the recycle values as

$$\bar{v}^{\text{inner}}_{\text{inflow}}\left(y^+_{\text{inflow}}\right) = \bar{v}^{\text{inner}}_{\text{recycle}}\left(y^+_{\text{inflow}}\right), \tag{29}$$

$$\bar{v}^{\text{outer}}_{\text{inflow}}(\eta_{\text{inflow}}) = \bar{v}^{\text{outer}}_{\text{recycle}}(\eta_{\text{inflow}}). \tag{30}$$

The mean spanwise velocity at the inflow boundary is set to zero.

The mean temperature at the inflow boundary is calculated from the mean total enthalpy $\bar{H}_t = c_p\bar{T} + \bar{u}_i\bar{u}_i/2$. The selection of total enthalpy instead of the temperature itself is due to large temperature gradient near the wall especially at cold wall conditions. On the other hand, the total enthalpy does not have a large gradient in the boundary layer. The created turbulent boundary layer is an equilibrium turbulent boundary layer. Therefore, the viscous heating and the wall heat flux are at equilibrium near the wall due to (1) rescaling according to y^+, and (2) negligible streamwise variation in the shape factor. The mean total enthalpy and static pressure in the inner and outer regions is scaled as

$$\bar{X}^{\text{inner}}_{\text{inflow}}(y^+_{\text{inflow}}) = \bar{X}^{\text{inner}}_{\text{recycle}}(y^+_{\text{inflow}}), \tag{31}$$

$$\bar{X}^{\text{outer}}_{\text{inflow}}(\eta_{\text{inflow}}) = \bar{X}^{\text{outer}}_{\text{recycle}}(\eta_{\text{inflow}}), \tag{32}$$

with $X = \tilde{H}_t, \bar{p}$. The mean density $\bar{\rho}$ is obtained from the equation of state. This overall method is denoted $H_t\&p$.

For comparison purposes, we also used two other methods for recycling-rescaling for calculation of the mean inflow temperature. In other words, velocity components are recycled by Equations (26)–(30) while the mean temperature and mean pressure are recycled differently. The first alternative method is called Walz's method in which the mean temperature is calculated from Equation (24) knowing the scaled mean velocity at the inflow. The second alternative method is denoted the T method and interpolates mean temperature in the inner and outer region using Equations (31) and (32) with $X = \tilde{T}$. The mean density $\bar{\rho}$ for both Walz and T methods is obtained from the equation of state assuming uniform static pressure across the boundary layer.

The fluctuating velocities are rescaled in the inner region according to

$$u''_{\text{inflow}}(y^+_{\text{inflow}}, z^+_{\text{inflow}}, t) = \beta u''_{\text{recycle}}(y^+_{\text{inflow}}, z^+_{\text{inflow}}, t), \tag{33}$$

$$v''_{\text{inflow}}(y^+_{\text{inflow}}, z^+_{\text{inflow}}, t) = \beta v''_{\text{recycle}}(y^+_{\text{inflow}}, z^+_{\text{inflow}}, t), \tag{34}$$

$$w''_{\text{inflow}}(y^+_{\text{inflow}}, z^+_{\text{inflow}}, t) = \beta w''_{\text{recycle}}(y^+_{\text{inflow}}, z^+_{\text{inflow}}, t), \tag{35}$$

and in the outer region

$$u''_{\text{inflow}}(\eta_{\text{inflow}}, \zeta_{\text{inflow}}, t) = \beta u''_{\text{recycle}}(\eta_{\text{inflow}}, \zeta_{\text{inflow}}, t), \tag{36}$$

$$v''_{\text{inflow}}(\eta_{\text{inflow}}, \zeta_{\text{inflow}}, t) = \beta v''_{\text{recycle}}(\eta_{\text{inflow}}, \zeta_{\text{inflow}}, t), \tag{37}$$

$$w''_{\text{inflow}}(\eta_{\text{inflow}}, \zeta_{\text{inflow}}, t) = \beta w''_{\text{recycle}}(\eta_{\text{inflow}}, \zeta_{\text{inflow}}, t), \tag{38}$$

where $\eta = y/\delta, \zeta = z/\delta$. A similar expression is used for the inner and outer fluctuations of other parameters.

The instantaneous streamwise velocity combines the inner and outer region using the Lund et al. blending function [38] ($W(\eta)$) according to

$$\begin{aligned} u_{\text{inflow}}(y, z, t) = & \left[\tilde{u}^{\text{inner}}_{\text{inflow}}(y^+_{\text{inflow}}) + u''^{\text{inner}}_{\text{inflow}}(y^+_{\text{inflow}}, z^+_{\text{inflow}}, t)\right][1 - W(\eta_{\text{inflow}})] + \\ & \left[\tilde{u}^{\text{outer}}_{\text{inflow}}(\eta_{\text{inflow}}) + u''^{\text{outer}}_{\text{inflow}}(\eta_{\text{inflow}}, \zeta_{\text{inflow}}, t)\right] W(\eta_{\text{inflow}}), \end{aligned} \tag{39}$$

where

$$W(\eta) = \frac{1}{2}\left(1 + [\tanh(4)]^{-1} \tanh\left[\frac{4(\eta - B)}{(1 - 2B)\eta + B}\right]\right), \tag{40}$$

where $B = 0.2$ to provide a smooth transition at $\eta = 0.2$. Similar equations are used for wall normal velocity $v_{\text{inflow}}(y, z, t)$, spanwise velocity $w_{\text{inflow}}(y, z, t)$, total enthalpy $H_{t,\text{inflow}}(y, z, t)$, and static pressure $p_{\text{inflow}}(y, z, t)$.

2.3. Description of Problem

Turbulent flow over a cold wall flat plate at hypersonic speed is considered. The freestream gas is dry air with N_2 and O_2 mass fractions of 0.765 and 0.235, respectively. The freestream conditions are presented in Table 1. The friction Reynolds number $\mathrm{Re}_\tau = u_\tau \delta / \nu_w$ is also provided in the table.

Table 1. Flow conditions.

Condition No.	Mach Number	Temperature (K)	Pressure (kPa)	Re_δ	Re_τ	T_w/T_{aw}
1	6.0	223.3	26.5	10^5	204	1.0
2	6.0	223.3	26.5	10^5	263	0.79
3	6.0	223.3	26.5	10^5	435	0.54

3. Results and Discussion

In this section, we evaluate the effect of the three recycling-rescaling methods, the effect of wall temperature, and the effect of number of cells in the boundary layer. In Table 2, the grid properties of the four grids used in this paper are present, where L_x, L_y, and L_z are respectively the dimensions of the computational region in the streamwise, wall normal, and spanwise directions. Grid 1 is used for freestream Condition 1, Grids 2 and 2(b) are used for Condition 2, and Grid 3 is used for Condition 3.

Table 2. Grid Properties for the Slightly Cold Wall Flat Plate.

Grid	Δx (μm)	Δy_{min} (μm)	Δz (μm)	Δx^+	Δy^+	Δz^+	L_x/δ	L_y/δ	L_z/δ	Cells
1	96.5	4.82	96.5	10	0.5	10	29.5	4.00	4.22	6.6 M
2	74.8	3.74	74.8	10	0.5	10	22.8	2.85	3.27	5.8 M
2(b)	74.8	3.74	74.8	10	0.5	10	22.8	2.95	3.27	8.7 M
3	45.15	2.26	45.15	10	0.5	10	22.8	3.16	3.26	16.9 M

3.1. *Effect of the Recycling-Rescaling Method*

In this section the effect of the three different recycling-rescaling methods on the prediction of turbulent properties is examined. To do so, the freestream properties are the same as Condition 2 of Table 1 and the grid is Grid 2. The calculated velocity profile, Reynolds Analogy Factor, Strong Reynolds Analogy, turbulent Prandtl number, dimensionless turbulent shear stress, turbulent normal stresses in streamwise, wall normal, and spanwise directions, and energy spectra are examined to evaluate the different recycling-rescaling models and the performance of the proposed method of $H_t\&p$.

Figure 2 shows the calculated mean streamwise velocity and the Van Driest transformed Law of the Wall at $x/\delta = 20$. The continuous lines are calculated mean streamwise velocity of the the three recycling-rescaling methods, namely $H_t\&p$, Walz, and T. The dashed-dotted line is is the Van Driest transformed velocity in the viscous sublayer, and the dashed line is the Van Driest transformed velocity of the Law of the Wall. The calculated mean streamwise velocity of all three recycling-rescaling methods agree well with the Van Driest transformed velocity of viscous sublayer and Law of the Wall.

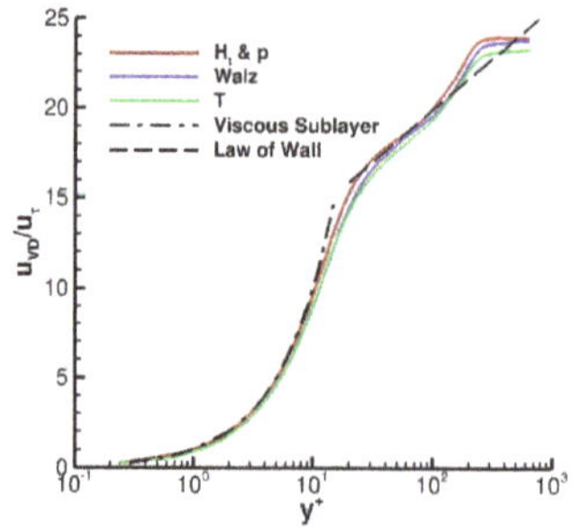

Figure 2. Effect of recycling-rescaling method on comparison of velocity profile with Law of Wall at $x/\delta = 20$ for Condition No. 2.

Table 3 presents the Reynolds Analogy Factor of each recycling-rescaling method. The conventional value ($2St/C_f = Pr_t^{-1} = 1.12$ with $Pr_t = 0.89$) is also provided. Figure 3 shows the experimental scattering of Reynolds Analogy Factor versus Mach number [39]. The calculated Reynolds Analogy Factors by the three methods are within the experimental uncertainty of experimental data.

Table 3. Effect of recycling-rescaling method on Reynolds Analogy Factor ($2St/C_f$) for condition No. 2.

	$H_t\&p$	Walz	T	Conventional
$2St/C_f$	1.21	1.33	0.95	1.12

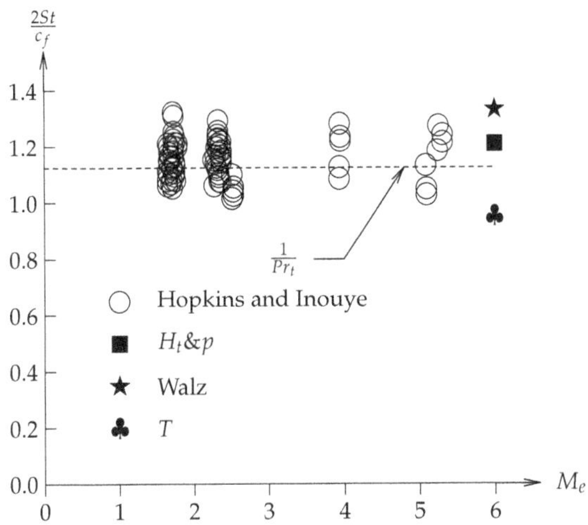

Figure 3. Scatter in Reynolds Analogy Factor (Hopkins and Inouye [39]).

To examine the effect of the recycling-rescaling method on the calculated Strong Reynolds Analogy, the Strong Reynolds Analogies proposed by Morkovin [40] and Huang [41] are selected. The Morkovin Strong Reynolds Analogy [40] is

$$\frac{\sqrt{\overline{T'^2}}/\overline{T}}{(\gamma-1)\,\overline{M}^2\,\sqrt{\overline{u'^2}}/\overline{u}} = 1. \tag{41}$$

The Huang Strong Reynolds Analogy [41] is

$$\frac{\sqrt{\overline{T'^2}}/\overline{T}}{(\gamma-1)\,\overline{M}^2\,\sqrt{\overline{u'^2}}/\overline{u}} = \frac{1}{Pr_t}\frac{1}{\left|d\overline{T_t}/d\overline{T}-1\right|}, \tag{42}$$

where,

$$\overline{T_t} \approx \overline{T} + \frac{\overline{u}^2}{2c_p}, \tag{43}$$

and $Pr_t = 0.89$. Figure 4 presents the two Strong Reynolds Analogies evaluated at $x/\delta = 20$. In each graph, the blue lines are the MSRA, where MSRA is calculated by

$$\text{MSRA} = \frac{\sqrt{\overline{T'^2}}/\overline{T}}{(\gamma-1)\,\overline{M}^2\,\sqrt{\overline{u'^2}}/\overline{u}}. \tag{44}$$

The red lines are the HSRA, where HSRA is calculated by

$$\text{HSRA} = \left(\frac{\sqrt{\overline{T'^2}}/\overline{T}}{(\gamma - 1)\,\overline{M}^2\,\sqrt{\overline{u'^2}}/\overline{u}} \right) \Big/ \left(\frac{1}{Pr_t} \frac{1}{|d\overline{T_t}/d\overline{T} - 1|} \right). \tag{45}$$

The black line is the line of constant value one. The closer the MSRA and HSRA are to the line of constant one (i.e., the black line), the better is the prediction. In general, Huang Strong Reynolds Analogy is a better estimate than Morkovin Strong Reynolds Analogy which is in agreement with the DNS results of Duan et al. [42]. Importantly, the best agreement of the Morkovin and Huang Reynolds Analogy is for the $H_t\&p$ recycling-rescaling method.

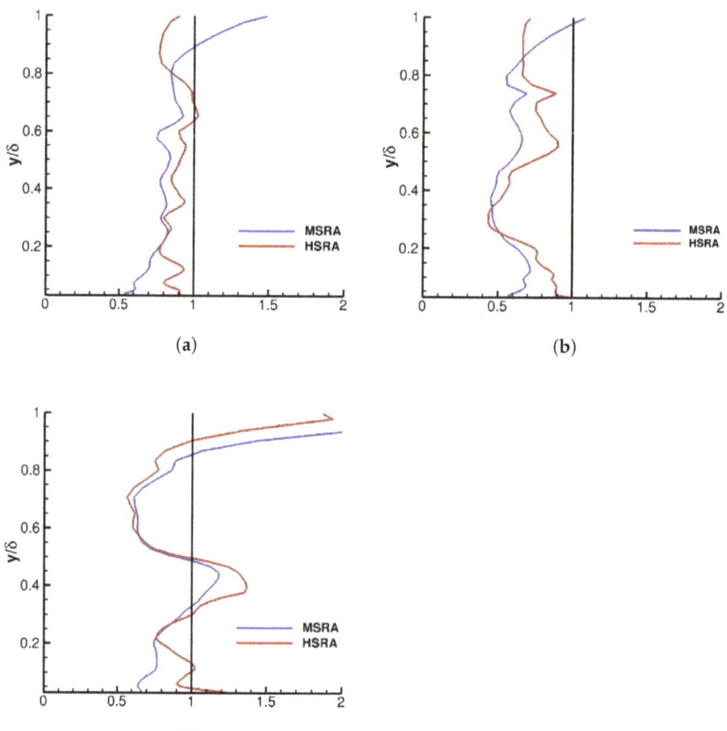

Figure 4. Effect of recycling-rescaling method on Strong Reynolds Analogy for Condition No. 2: (**a**) $H_t\&p$ recycling-rescaling method. (**b**) Walz recycling-rescaling method. (**c**) T recycling-rescaling method.

Figure 5 shows the calculated turbulent Prandtl number by the three recycling-rescaling methods evaluated at $x/\delta = 20$. The Prandtl number in a turbulent boundary layer is defined as

$$Pr_t = \frac{\overline{\rho u'v'}\partial\overline{T}/\partial y}{\overline{\rho v'T'}\partial\overline{u}/\partial y}. \tag{46}$$

The black line in the figure is the constant Prandtl number of 0.89. The calculated turbulent Prandtl number of $H_t\&p$ and T methods are closer to this line in comparison to the Walz method. According to DNS results of Zhang et al. [43], although the turbulent Prandtl number is not constant in the boundary layer, it stays close to the conventional value of 0.89.

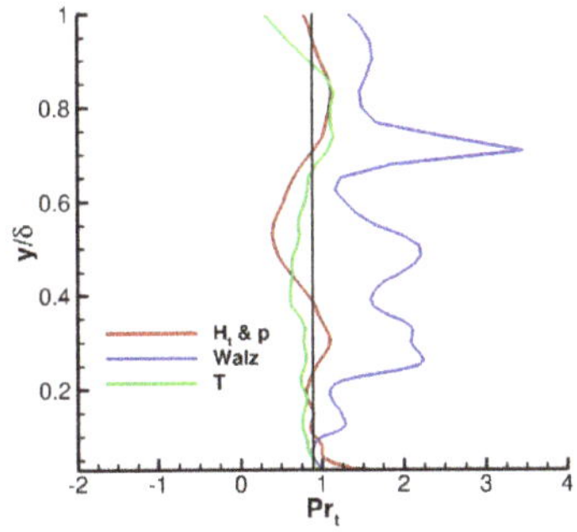

Figure 5. Effectof recycling-rescaling method on turbulent Prandtl number for Condition No. 2.

Figure 6 presents the calculated Morkovin scaled dimensionless turbulent shear stress ($-\overline{\rho u''v''}/\tau_w$) by the three recycling-rescaling methods at $x/\delta = 20$. Additionally, for comparison purposes, the incompressible shear stress data of Klebanoff [44] is also presented. The predicted Morkovin scaled dimensionless turbulent shear stresses are in general agreement with the incompressible data although the region with non-zero turbulent shear stress is smaller in the incompressible data. The Morkovin scaled dimensionless turbulent shear stress is zero at the wall, increases to its maximum value of about one at some distance near the wall and then reduces to zero at the edge of the boundary layer. The Walz and H_t&p methods predict the same maximum turbulent shear stress while the T method predicts a larger maximum shear stress.

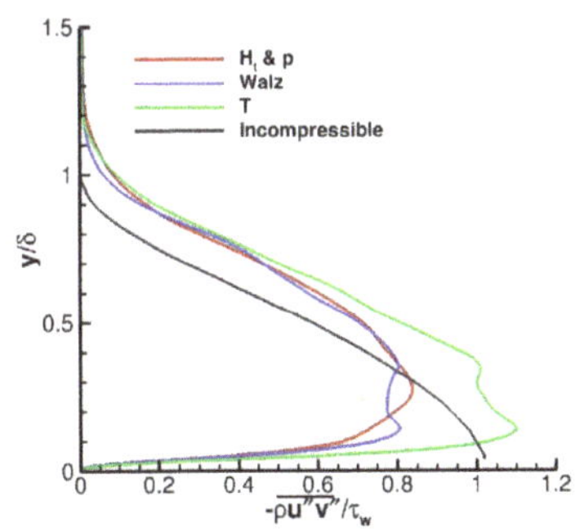

Figure 6. Effect of the recycling-rescaling method on Morkovin scaled dimensionless turbulent shear stress at $x/\delta = 20$ for condition No. 2.

Figure 7 shows the predicted Morkovin scaled dimensionless turbulent normal stresses ($\overline{\rho u_i'' u_i''}/\tau_w$, $i = 1, 2, 3$) respectively in streamwise, wall normal, and spanwise directions by the three recycling-rescaling methods at $x/\delta = 20$. All the methods predict the same behavior for the turbulent normal stresses. In the outer part of boundary layer, the Morkovin scaled dimensionless turbulent normal stresses are essentially the same for all three methods. However, the T method predicts the largest peak normal stresses for each of the Morkovin scaled dimensionless turbulent normal stresses while H_t&p method has the smallest peaks.

The comparison of the predicted Morkovin scaled dimensionless turbulent streamwise stress with the DNS results of Duan et al. [42] shows a general agreement; however, the maximum stress is smaller for Duan et al. Further investigation is required to understand the reason behind this. It should be mentioned here that T_w/T_{aw} for the presented result of Duan et al. is 0.68 while in our case, T_w/T_{aw} is 0.79.

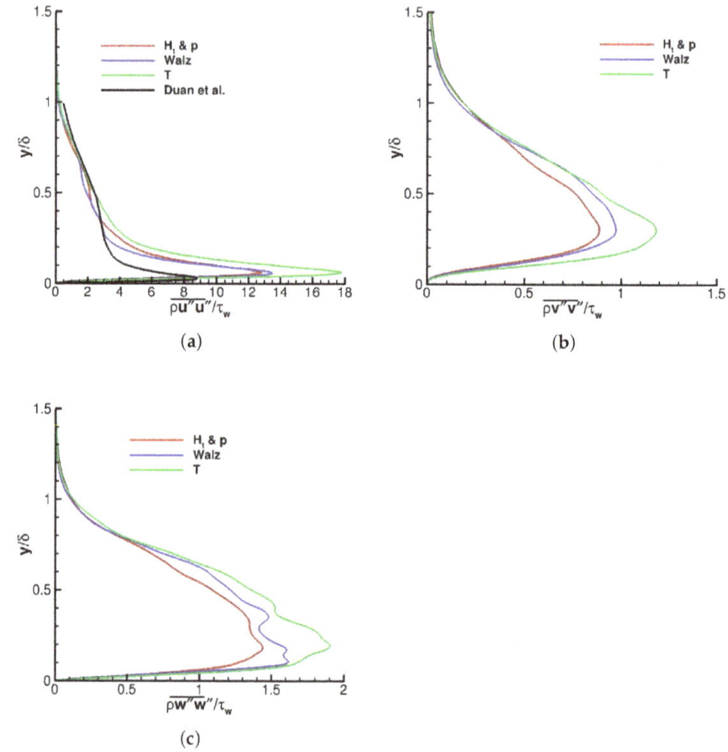

Figure 7. Effect of recycling-rescaling method on Morkovin scaled dimensionless turbulent normal stresses of condition No. 2 at $x/\delta = 20$ for: (**a**) Streamwise direction. (**b**) Wall normal direction. (**c**) Spanwise direction.

Figure 8 presents the energy spectrum of the total energy per unit mass of the three recycling-rescaling methods at $x/\delta = 20$ and $y/\delta = 0.7$ for the dimensionless time interval $\Delta t U_\infty/\delta = 100$. The inertial subrange for this problem is between dimensionless frequencies ($f = f^*\delta/U_\infty$) of 1 and 10. From these graphs, all three methods predict the slope of the energy spectrum of $-5/3$ in the inertial subrange. The slope $-5/3$ is the theoretical prediction for the energy spectrum slope at the inertial subrange.

In summary, the introduced H_t&p recycling-rescaling method has a better prediction of the turbulent properties. The H_t&p method has the best prediction for Strong Reynolds Analogy and turbulent Prandtl number compared to the Walz and T methods and has comparable results for Law of the Wall, Reynolds Analogy Factor, turbulent shear and normal stresses, and energy spectrum. Therefore, the H_t&p method is used for the recycling-rescaling in the rest of the paper.

3.2. *Effect of Wall Temperature*

In this section, the proposed H_t&p recycling-rescaling method is used to calculate the turbulent properties at different wall temperatures. Conditions 1 to 3 of Table 1 are considered. The Grids 1, 2, and 3 of Table 2 are used respectively for Conditions 1, 2, and 3. Table 2 shows that the LES calculation becomes computationally expensive as the wall temperature decreases. Decreasing the wall temperature from adiabatic wall (T_{aw}) to $0.54T_{aw}$, the grid spacing normal to the wall decreases by a factor of 2.13 to keep the same Δy^+.

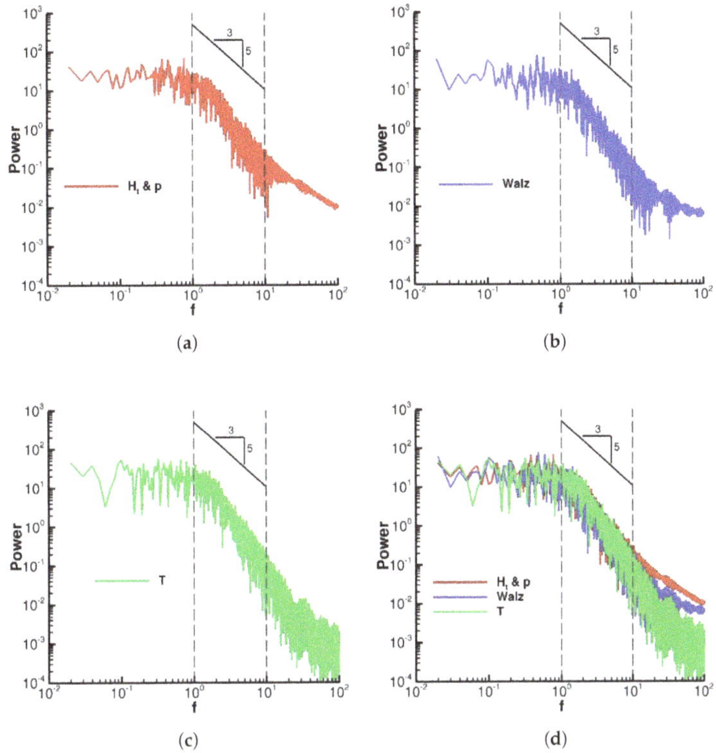

Figure 8. Effect of recycling-rescaling method on energy spectra for condition No. 2: (**a**) $H_t\&p$ recycling-rescaling method. (**b**) Walz recycling-rescaling method. (**c**) T recycling-rescaling method. (**d**) Combined.

Figure 9 shows the calculated Law of the Wall at $x/\delta = 20$ for the three wall temperatures. The continuous lines are the Van Driest transformed mean streamwise velocity at T_w/T_{aw} of 0.54, 0.79, and 1.0. Additionally, the dashed lines show the Law of the Wall for each wall temperature and the dashed-dotted line is the Van Driest transformed velocity at viscous sublayer. The predicted Van Driest transformed velocity agrees well with the Van Driest transformed velocities of the viscous sublayer and Law of the Wall for all three wall temperature. Increasing the wall temperature increases the constant C in the logarithmic region. The constant C in the Law of the Wall is 7.2, 8.4, and 9.5 respectively for T_w/T_{aw} of 0.54, 0.79, and 1.0. This trend is in opposition with the trend reported by Danberg [32] as shown in Figure 10. Table 4 presents the values of the compressible turbulent boundary layer displacement (δ^*) and momentum (θ) thicknesses, as well as the shape factor $H = \delta^*/\theta$ for all three wall temperature. Decreasing the wall temperature decreases the displacement thickness while increasing the momentum thickness and decreasing the shape factor.

Table 5 presents the Reynolds Analogy Factor for T_w/T_{aw} of 0.54 and 0.79. Again, the conventional value ($2St/C_f = Pr_t^{-1} = 1.12$) for Reynolds Analogy Factor is also presented in the table. The predictions are within the experimental uncertainty of available experimental data in the literature. Figure 11 shows a sample of the Reynolds Analogy Factor from the literature where the open symbols are from Keener and Polek [45].

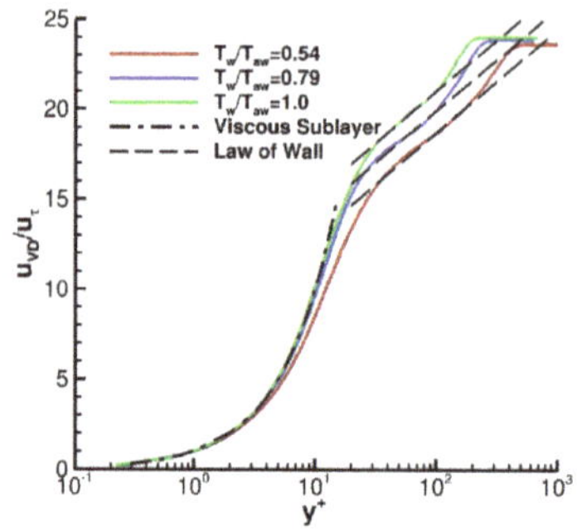

Figure 9. Van Driest transformed mean streamwise velocity for three wall temperature at $x/\delta = 20$ and comparison with the Law of Wall.

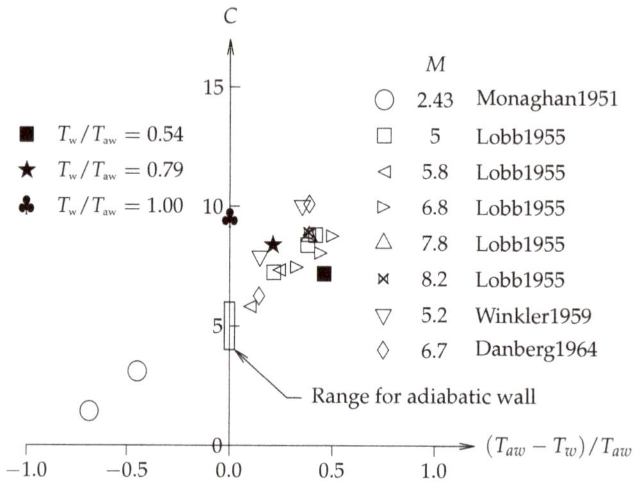

Figure 10. Relation of constant C in the logarithmic region with wall temperature (Danberg [32]).

Table 4. Effect of wall temperature on boundary layer displacement and momentum thickness.

	$T_w/T_{aw} = 0.54$	$T_w/T_{aw} = 0.79$	$T_w/T_{aw} = 1.00$
δ^*/δ	0.4312 ± 0.00004	0.4658 ± 0.00027	0.5344 ± 0.0018
θ/δ	0.0428 ± 0.00028	0.0357 ± 0.00027	0.0322 ± 0.00046
H	10.07 ± 0.07	13.1 ± 0.1	16.8 ± 0.6

Table 5. Effect of wall temperature on Reynolds Analogy Factor ($2St/C_f$).

	$T_w/T_{aw} = 0.54$	$T_w/T_{aw} = 0.79$	**Conventional**
$2St/C_f$	1.02	0.98	1.12

Figure 12 shows the Morkovin and Huang Strong Reynolds Analogies for the three wall temperature T_w/T_{aw} of 0.54, 0.79, and 1.0 evaluated at $x/\delta = 20$. In each graph, the blue lines are the MSRA calculated by Equation (44), the red lines are the HSRA calculated by Equation (45), and the black lines are the constant value of one. The turbulent Prandtl number (Pr_t) is 0.89. In general, the Huang Strong Reynolds Analogy is a better estimate compared to the Morkovin Strong Reynolds Analogy. Moreover, the Huang Strong Reynolds Analogy is more accurate for the cold wall condition. In other words, the accuracy of Huang Reynolds Analogy is reduced by increasing the wall temperature, especially in the inner region.

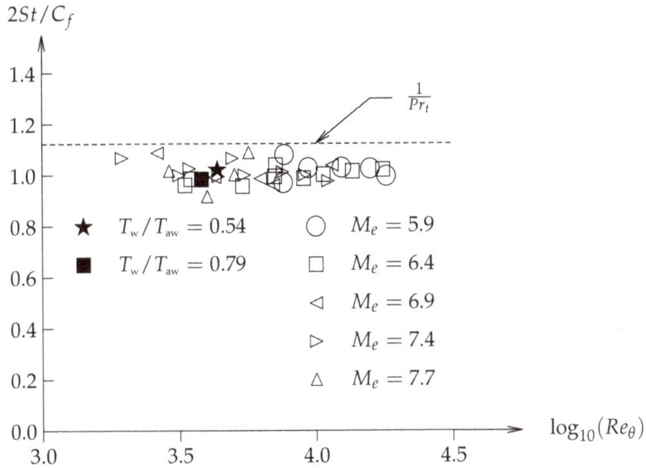

Figure 11. Scatter in Reynolds Analogy Factor (Keener and Polek [45]).

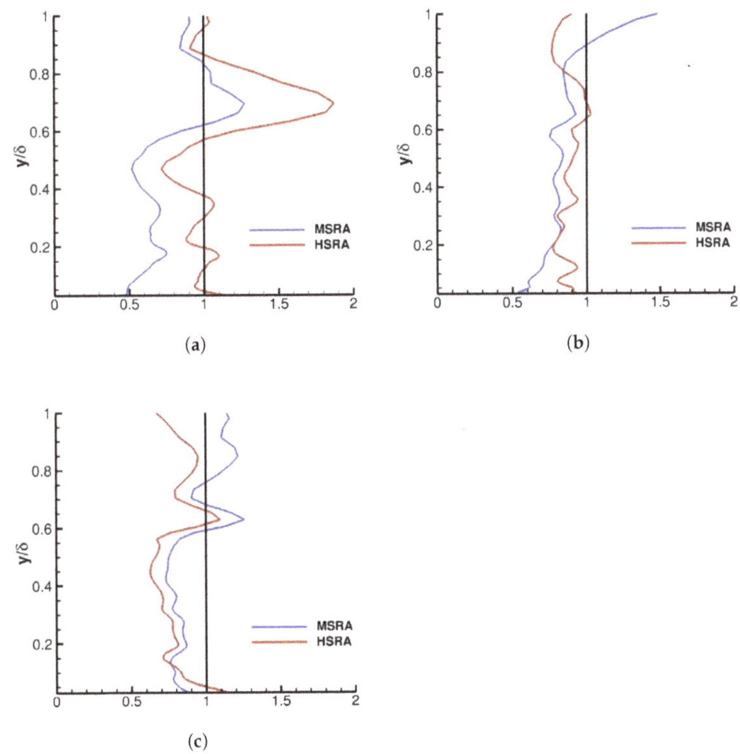

Figure 12. Strong Reynolds Analogy for: (**a**) $T_w/T_{aw} = 0.54$. (**b**) $T_w/T_{aw} = 0.79$. (**c**) $T_w/T_{aw} = 1.0$.

Figure 13 shows the calculated turbulent Prandtl number (Equation (46)) for the three wall temperatures. The black line represents the constant turbulent Prandtl number of

0.89. The calculated turbulent Prandtl number remains close to the constant line of 0.89, especially in the inner region.

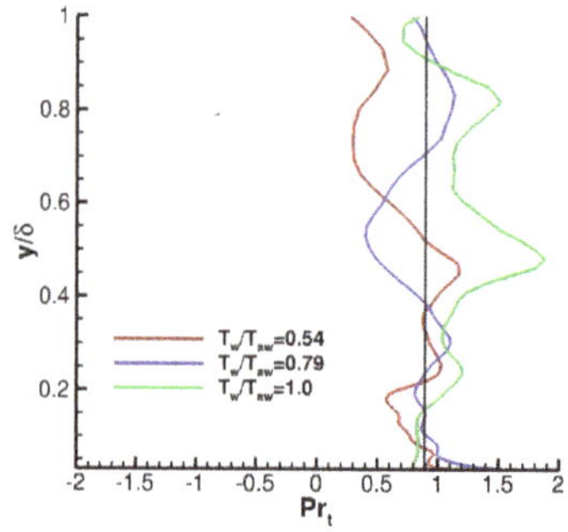

Figure 13. Effect of wall temperature on turbulent Prandtl number.

Figure 14 presents the Morkovin scaled dimensionless turbulent shear stress $(-\overline{\rho u''v''}/\tau_w)$ at $x/\delta = 20$ for three wall temperature T_w/T_{aw} of 0.54, 0.79, and 1.0. The black line in the figure is the incompressible turbulent shear stress from Klebanoff [44]. For all wall temperatures, the Morkovin scaled dimensionless turbulent shear stress is zero at the wall and increases by increasing distance from the wall until it reaches its maximum value and then starts decreasing until reaching to zero at the edge of the boundary layer. The results have the same behavior as the incompressible data; however, the region with non-zero turbulent stress is larger in the compressible results.

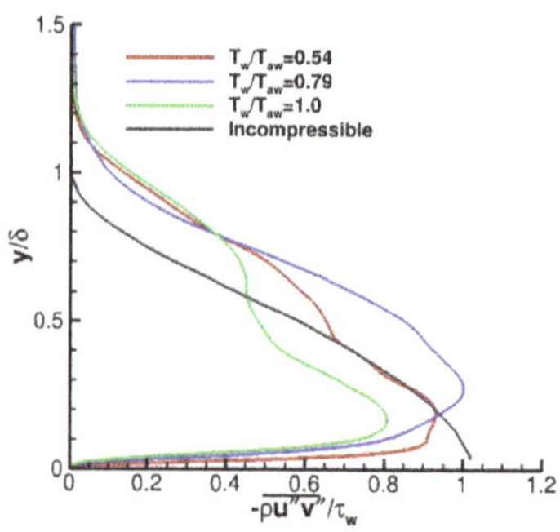

Figure 14. Morkovin scaled dimensionless turbulent shear stress at $x/\delta = 20$ for three wall temperature.

Figure 15 shows the Morkovin scaled dimensionless turbulent normal stresses $(\overline{\rho u_i'' u_i''}/\tau_w)$ in streamwise, wall normal, and spanwise directions. Additionally, the DNS results of Duan et al. [42] for the Morkovin scaled dimensionless turbulent streamwise stress for three wall temperatures are also presented. From the figure, in the outer region, the Morkovin scaled dimensionless turbulent normal stresses are independent of wall temperature. However, in the inner region, where the peak normal stresses are located, increasing the wall temperature decreases the maximum wall normal and spanwise turbulent stresses while increasing the streamwise turbulent stress.

Additionally, increasing the wall temperature moves the location of maximum turbulent streamwise stress away from the wall. It should be mentioned here that the Morkovin scaled dimensionless turbulent streamwise stress yields almost the same peak value and peak location for the T_w/T_{aw} of 0.79 and 1.0. However, the DNS results of Duan et al. shows no significant dependence to the wall temperature.

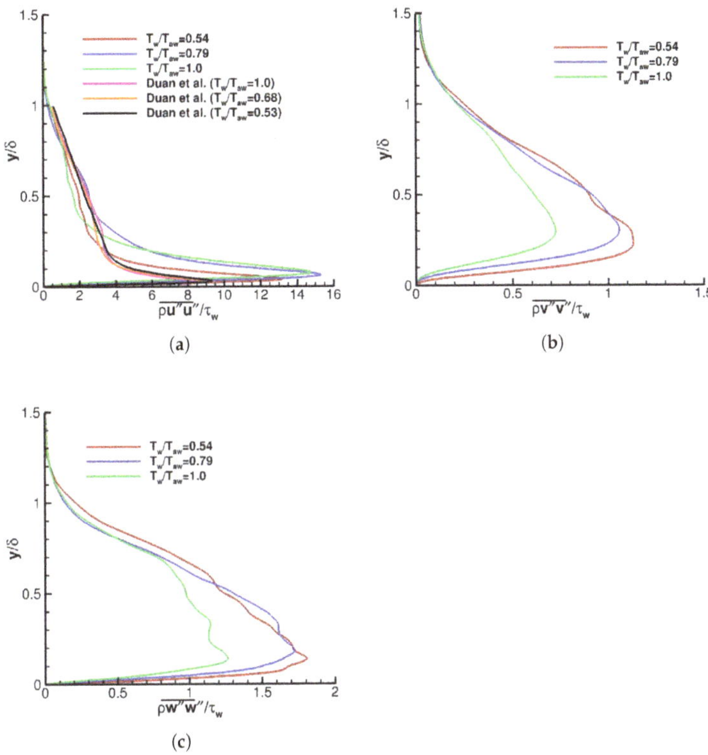

Figure 15. Effect of wall temperature on Morkovin scaled dimensionless turbulent normal stresses at $x/\delta = 20$ for: (**a**) Streamwise direction. (**b**) Wall normal direction. (**c**) Spanwise direction.

Figure 16 represents the energy spectrum of the total energy per unit mass at $x/\delta = 20$ and $y/\delta = 0.7$ over the dimensionless time interval of 100 for three wall temperatures. The range of dimensionless frequencies for the inertial subrange is 1 to 10. It can be seen that the slope of the energy spectrum in the inertial subrange for all the wall temperatures is $-5/3$. This slope of $-5/3$ is what is expected for the inertial subrange in the theory.

3.3. Effect of The Number of Cells in the Boundary Layer

In this section, the effect of the grid is examined by changing the number of cells in the wall normal direction in the boundary layer. For this purpose, two grids, namely Grids 2 and 2(b) of Table 2 are used for Condition 2 of Table 1. The grids have the same y^+ near the wall and then are stretched using geometric stretching until reaching the edge of the boundary layer. Grid 2 and 2(b) respectively have 80 and 120 cells in the wall normal direction in the boundary layer. The different number of cells in the wall normal direction means that the stretching factor is smaller in Grid 2(b) and thus, the grid cells are smaller in size in the boundary layer in comparison to Grid 2. Both calculations are performed using $H_t\&p$ recycling-rescaling method.

Figure 17 shows the comparison of the calculated Van Driest transformed mean streamwise velocity with the Law of the Wall at $x/\delta = 20$. Additionally, the dashed line in the figure represents the Law of Wall, and the dashed dotted line represents the viscous sublayer. The calculated Van Driest transformed streamwise velocity of both grids are in good agreement with the theoretical values for Law of Wall and viscous sublayer. It is worth mentioning here that the u_τ for both grids is essentially the same and the difference is less than a percent.

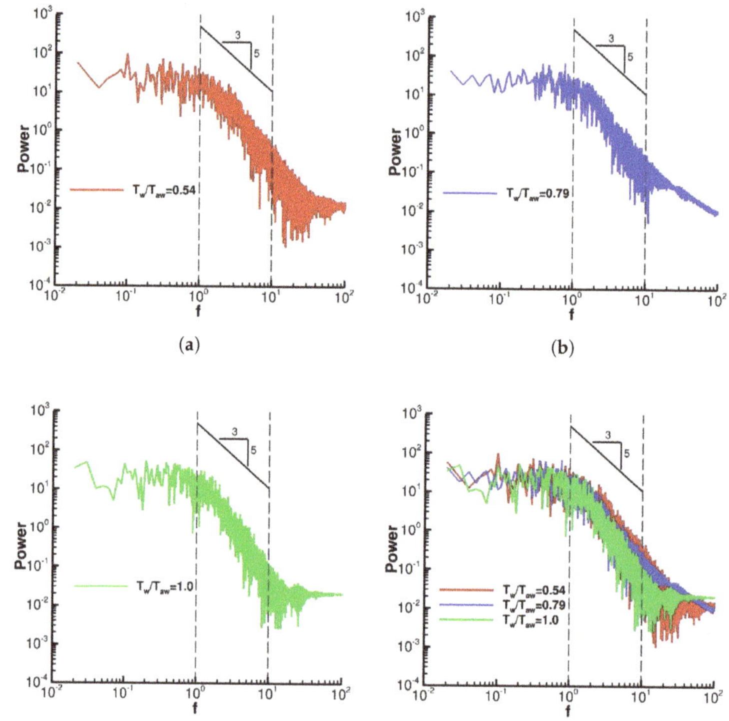

Figure 16. Energy Spectra for: (**a**) $T_w/T_{aw} = 0.54$. (**b**) $T_w/T_{aw} = 0.79$. (**c**) $T_w/T_{aw} = 1.00$. (**d**) Combined.

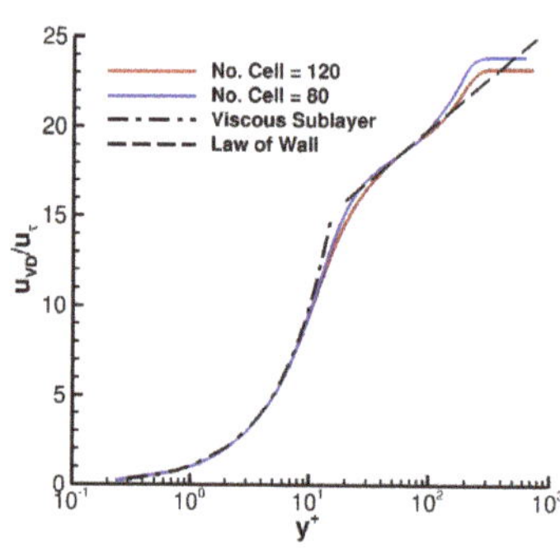

Figure 17. Van Driest transformed mean streamwise velocity at $x/\delta = 20$ for condition No. 2. for two grids with different number of cells in the wall normal direction.

Table 6 shows the Reynolds Analogy Factor for the two grids with 80 and 120 cells in the wall normal direction in the boundary layer. Additionally, the conventional value of Reynolds Analogy Factor ($2St/C_f = Pr_t^{-1} = 1.12$) is also presented. Both grids have the Reynolds Analogy Factor within the range of the experimental uncertainty in the literature (see Figure 11).

Table 6. Effect of Number of Cells in the Boundary Layer on Reynolds Analogy Factor ($2St/C_f$) for Condition No. 2.

	$j = 120$	$j = 80$	**Conventional**
$2St/C_f$	0.98	1.02	1.12

Figure 18 shows the Morkovin and Huang Strong Reynolds Analogies for the two grids with 80 and 120 cells in the wall normal direction inside the boundary layer. In each graph, the blue line is the MSRA calculated by Equation (44), the red line is the HSRA calculated by Equation (45), and the black line is the line of constant value one. The turbulent Prandtl number in the Strong Reynolds Analogy is $Pr_t = 0.89$. The Huang Strong Reynolds Analogy has a better result as expected. Surprisingly, the grid with 80 cells in the wall normal direction inside the boundary layer has better prediction of the Strong Reynolds Analogy compared to having a finer grid with 120 cells in the wall normal direction inside the boundary layer.

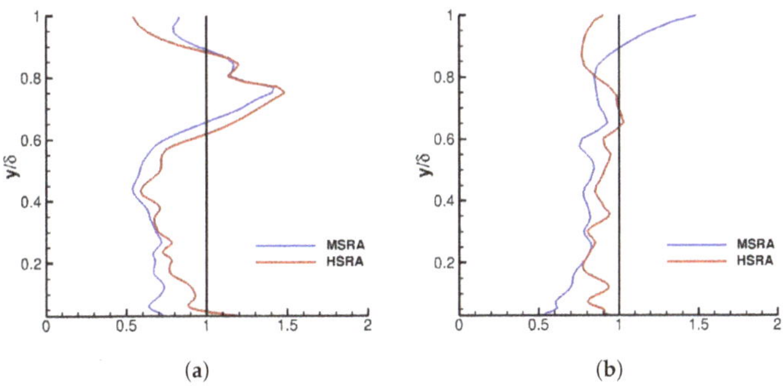

Figure 18. Effect of number of cells in the boundary layer on Strong Reynolds Analogy for condition No. 2: (**a**) 120 cells in the wall normal direction inside the boundary layer. (**b**) 80 cells in the wall normal direction inside the boundary layer.

Figure 19 presents the calculated turbulent Prandtl number for the two grids with 80 and 120 cells in the wall normal direction inside the boundary layer. Additionally, the constant turbulent Prandtl number of 0.89 is also presented by a black line. In general, both grids have the Prandtl number close to the constant turbulent Prandtl number of 0.89. In the inner region, both grids have good agreement with each other; however, in the outer region the grid with 80 cells in wall normal direction inside the boundary layer stays closer to the constant line of 0.89.

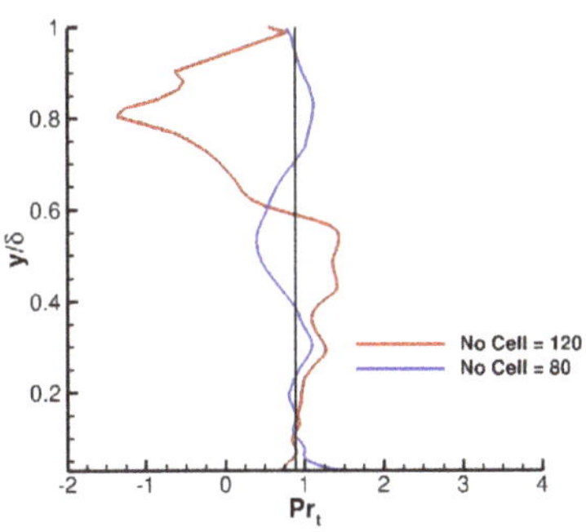

Figure 19. Turbulent Prandtl number of condition No. 2 for two grids with 80 and 120 cells in the wall normal direction inside the boundary layer.

Figure 20 presents the Morkovin scaled dimensionless turbulent shear stress ($-\overline{\rho u''v''}/\tau_w$) for the two grids with 80 and 120 cells in the wall normal direction inside the boundary layer. For comparison purposes, the incompressible dimensionless shear stress values of Klebanoff [44] are also presented. Once again, the compressible data have a larger region with non-zero shear stresses. However, the compressible and incompressible data have the same trends. Moreover, the maximum shear stress is larger for the grid with 80 cells compared to the grid with 120 cells in the wall normal direction inside the boundary layer. It is worth mentioning here that the u_τ for both grids is essentially the same.

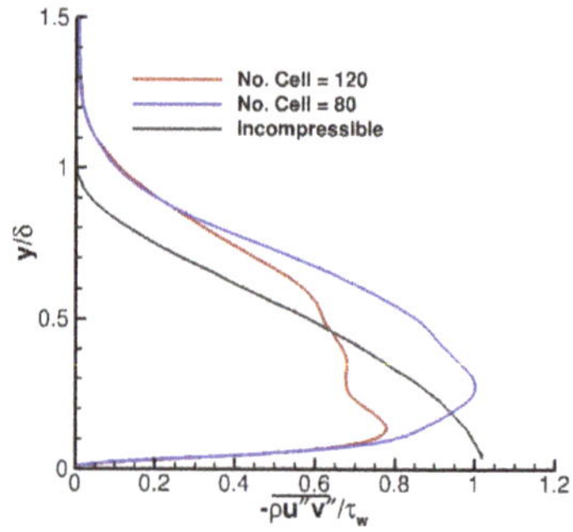

Figure 20. Morkovin scaled dimensionless turbulent shear stress at $x/\delta = 20$ for condition No. 2 and two grids with 80 and 120 cells in wall normal direction inside the boundary layer.

Figure 21 shows the Morkovin scaled dimensionless turbulent normal stresses ($\overline{\rho u_i''u_i''}/\tau_w$) in streamwise, wall normal, and spanwise directions for two grids with 80 and 120 cells in the wall normal direction inside the boundary layer. It can be seen that the thickness of the layer normal to the wall with non-zero normal stresses are the same for both grids. Moreover, the maximum values of the turbulent normal stresses decreases by increasing the number of cells in the wall normal direction. Comparison of the results with the DNS results of Duan et al. [42] shows similar trend; however, the peak value of the wall normal stress is smaller in Duan et al. calculations.

Figure 22 presents the energy spectrum of the total energy per unit mass for two grids with 80 and 120 cells in the wall normal direction inside the boundary layer at $x/\delta = 20$ and $y/\delta = 0.7$ over the dimensionless time interval of 100. The inertial subrange for this problem is bounded by dimensionless frequencies ($f = f^*\delta/U_\infty$) of 1 and 10. From the figure it can be seen that both grid predict the slope of $-5/3$ in the inertial subrange. This slope of $-5/3$ is the theoretical value for the slope of the inertial subrange.

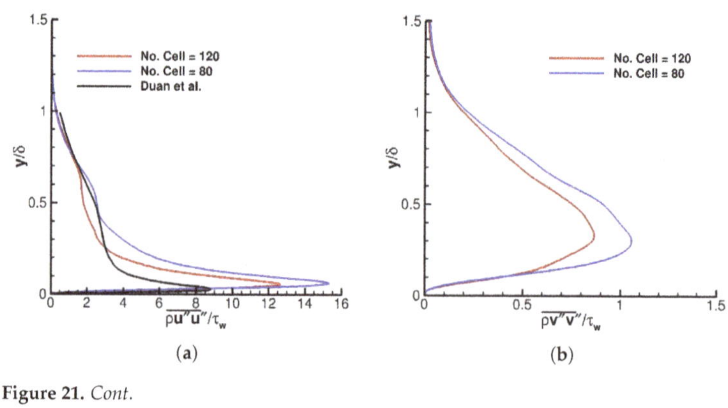

Figure 21. *Cont.*

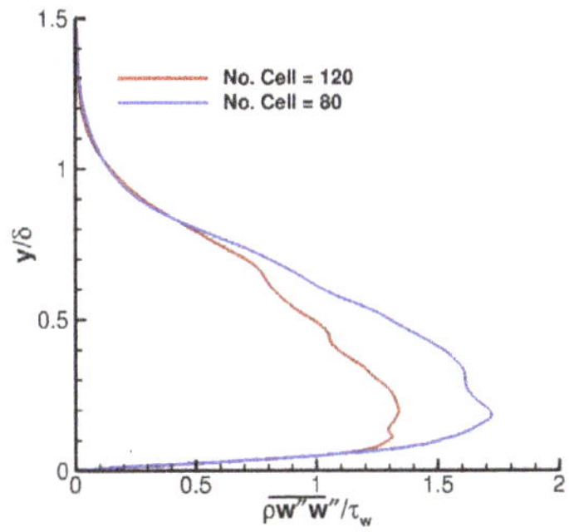

(c)

Figure 21. Morkovin scaled dimensionless turbulent normal stress at $x/\delta = 20$ for condition No. 2 for two grids with 80 and 120 cells in the wall normal direction inside the boundary layer: (**a**) Streamwise stress. (**b**) Wall normal stress. (**c**) Spanwise stress.

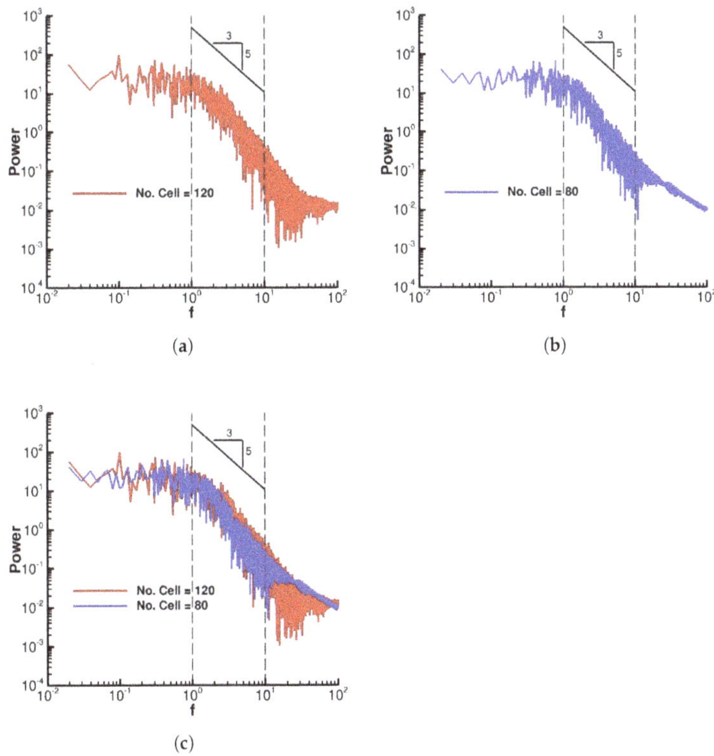

Figure 22. Energy spectra for condition No. 2 for two grids: (**a**) 120 cells in the wall normal direction inside the boundary layer. (**b**) 80 cells in the wall normal direction inside the boundary layer. (**c**) Combined.

4. Conclusions

A new recycling-rescaling method is evaluated for generation of the instantaneous time-dependent inflow boundary. The new method considers the change in the pressure along the boundary layer and recycles total enthalpy, static pressure, and three components of velocity. This method is compared with two more traditional ways of recycling-rescaling by comparing the Law of the Wall, Reynolds Analogy Factor, Morkovin and Huang Strong Reynolds Analogy, turbulent Prandtl number, turbulent shear stress, turbulent normal stresses, and energy spectrum.

The new recycling-rescaling method improves the prediction of the Strong Reynolds Analogy and turbulent Prandtl number. The results of Law of the Wall, Reynolds Analogy Factor, turbulent stresses, and energy spectrum are comparable with the previous methods.

Author Contributions: Methodology, N.K. and D.K.; Software: N.K. and D.K.; Validation, N.K. and D.K.; Formal Analysis, N.K. and D.K.; Data Curation, N.K. and D.K.; Writing—Original Draft Preparation, N.K. and D.K.; Writing—Review & Editing, N.K. and D.K.; Visualization, N.K. and D.K. All authors have read and agreed to the published version of the manuscript.

Funding: This research received no external funding.

Acknowledgments: The computational resources are provided partially by a grant from the Emil Buehler Foundation and partially by Rutgers Office of Advanced Research Computing for the provision of computational time on Caliburn Supercomputer.

Conflicts of Interest: The authors declare no conflict of interest.

Nomenclature

C	Constant in the Law of the Wall
e	Total energy per unit mass
$\widetilde{\mathcal{F}}$	Favre averaging of variable $\mathcal{F}$
$\overline{\mathcal{F}}$	Conventional spatial averaging of variable $\mathcal{F}$
$\mathcal{F}'$	Conventional spatial fluctuation of variable $\mathcal{F}$
$\mathcal{F}''$	Favre fluctuation of variable $\mathcal{F}$
$\mathcal{H}_j$	Energy flux
H_t	Total enthalpy per unit mass
L_r	Distance of the recycling plan from inflow plan
M	Mach number
p	pressure or mean pressure
Pr	Molecular Prandtl number
Pr_t	Turbulent Prandtl number
Q_j	Subgrid scale heat flux
$\mathcal{Q}_j$	Total heat transfer
$\bar{q}_j$	Molecular heat flux
Re_δ	Reynolds number based upon the boundary layer thickness
T	Static temperature
t_{aver}	Averaging time period
T_{w}	Wall temperature
T_{aw}	Adiabatic wall temperature
$\mathcal{T}_{ij}$	Total stress
u_i	Velocity components in the Cartesian coordinates
u	Streamwise velocity
u_{vd}	Van Driest transformed velocity
U_∞	Freestream velocity
v	Wall normal velocity
w	Spanwise velocity
$W(\eta)$	Lund et al. blending function
x_i	Cartesian coordinate
y^+	yu_τ/ν_{w}

Greek Symbols

η	y/δ
κ	Von Karman's constant
$\kappa(\tilde{T})$	Molecular thermal conductivity
μ	Molecular viscosity
μ_w	Molecular viscosity at the wall
ν_w	Kinematic molecular viscosity at the wall
ρ	Density or mean density
$\bar{\rho}$	Mean density
$\bar{\rho}k$	Subgrid scale turbulence kinetic energy per unit volume
$\tilde{\sigma}_{ij}$	Molecular viscous stress tensor
τ_{ij}	Subgrid scale stress tensor
τ_w	Mean shear stress at the wall

Subscripts

aw	Adiabatic wall
inflow	Value at the inflow plane
recycle	Value at the recycling plane
vd	Van Driest Transformed
w	Wall
∞	Freestream condition

Superscripts

inflow	Value at the inflow plane
inner	Inner layer
recycle	Value at the recycling plane

Abbreviations

DPLR	Data Parallel Line Relaxation
ILES	Implicit Large Eddy Simulation
LES	Large Eddy Simulation
MILES	Monotone Integrated Large Eddy Simulation
MPI	Message Passing Interface
MUSCL	Monotone Upstream Scheme for Conservation Laws
RANS	Reynolds Average Navier-Stokes
SA	Spalart Allmaras
SGS	SubGrid Scale

References

1. Malik, T. Boeing's Hypersonic Vision: A Sleek Passenger Plane That Can Hit Mach 5. 2018. Available online: https://www.space.com/41042-boeing-hypersonic-passenger-plane-concept.html (accessed on)29 June 2018.
2. Artist's Impression of the SpaceLiner 7 Full Configuration during the Ascent Phase. 2013. ToSch1983 Available online: https://commons.wikimedia.org/w/index.php?curid=25664959 (accessed on)1 April 2013.
3. Reaction Engines Limited. LAPCAT. 2006. Available online: https://web.archive.org/web/20090705025020/http://www.reactionengines.co.uk/lapcat.html (accessed on)5 July 2009.
4. Bertin, J.; Cummings, R. Fifty Years of Hypersonics: Where We've Been, Where We're Going. *Prog. Aerosp. Sci.* **2003**, *39*, 511–536. [CrossRef]
5. Kostoff, R.; Cummings, R. Highly Cited Literature of High Speed Compressible Flow Research. *Aerosp. Sci. Technol.* **2013**, *26*, 216–234. [CrossRef]
6. Gaitonde, D. Progress in Shock Wave/Boundary Layer Interactions. *Prog. Aerosp. Sci.* **2015**, *72*, 80–99. [CrossRef]
7. Kumar, R.; Wheaton, B.; Marineau, E.; Kurits, I.; Lafferty, J. *Computational Analyses to Support Design of Hollow Cylinder Flare Experiments at AEDC Hypervelocity Tunnel 9*; Technical Report 2020-2221; AIAA: Reston, VA, USA, 2020.
8. Loginov, M.; Adams, N.; Zheltovodov, A. Large-Eddy Simulation of Shock-Wave/Turbulent-Boundary-Layer Interaction. *J. Fluid Mech.* **2006**, *565*, 135–169. [CrossRef]
9. Touber, E.; Sandham, N. Large-Eddy Simulation of Low-Frequency Unsteadiness in a Turbulent Shock-Induced Separation Bubble. *Theor. Comput. Fluid Dyn.* **2009**, *23*, 79–107. [CrossRef]
10. Morgan, B.; Duraisamy, K.; Nguyen, N.; Kawai, S.; Lele, S. Flow Physics and RANS Modelling of Oblique Shock/Turbulent Boundary Layer Interaction. *J. Fluid Mech.* **2013**, *729*, 231–284. [CrossRef]
11. Ritos, K.; Kokkinakis, I.; Drikakis, D.; Spottswood, M. Implicit Large Eddy Simulation of Acoustic Loading in Supersonic Turbulent Boundary Layers. *Phys. Fluids* **2017**, *29*, 046101. [CrossRef]
12. Ritos, K.; Kokkinakis, I.; Drikakis, D. Physical Insight into the Accuracy of Finely-Resolved ILES in Turbulent Boundary Layers. *Comput. Fluids* **2018**, *169*, 309–316. [CrossRef]

13. Hadjadj, A. Large-Eddy Simulation of Shock/Boundary Layer Interaction. *AIAA J.* **2012**, *50*, 2919–2927. [CrossRef]
14. Edwards, J.; Choi, J.I.; Boles, J. Large-Eddy/Reynolds-Averaged Navier-Stokes Simulation of a Mach 5 Compression-Corner Interaction. *AIAA J.* **2008**, *46*, 977–991. [CrossRef]
15. Menter, F. Two Equation Eddy Viscosity Turbulence Models for Engineering Applications. *AIAA J.* **1994**, *32*, 1598–1605. [CrossRef]
16. Fang, J.; Lu, L.; Yao, Y.; Zheltovodov, A. *Large Eddy Simulation of a Three Dimensional Hypersonic Shock Wave/Turbulent Boundary Layer Interaction of a Single Fin*; AIAA Paper 2015-1062; AIAA: Reston, VA, USA, 2015.
17. Ghosal, S.; Lund, T.; Mon, P.; Akselvoll, K. A Dynamic Localization Model for Large Eddy Simulation o Turbulent Flows. *J. Fluid Mech.* **1995**, *286*, 229–255. [CrossRef]
18. Moin, P.; Squires, K.; Cabot, W.; Lee, S. A Dynamic Subgrid-scale Model for Compressible Turbulence and Scalar Transport. *Phys. Fluids A* **1991**, *11*, 2746–2757. [CrossRef]
19. Fu, L.; Bose, S.; Moin, P. *Wall-Modeled LES of Three-Dimensional Intersecting Shock Wave Turbulent Boundary Layer Interactions*; Annual Research Brief; Center for Turbulence Research, Stanford University: Stanford, CA, USA, 2020.
20. Xu, S.; Martin, M. Assessment of Inflow Boundary Conditions for Compressible Turbulent Boundary Layers. *Phys. Fluids* **2004**, *16*, 2623–2639. [CrossRef]
21. Walz, A. *Boundary Layers of Flow and Temperature*; The MIT Press: Cambridge, MA, USA, 1969.
22. Gupta, R.; Yos, J.; Thompson, R.; Lee, K. *A Review of Reaction Rates and Thermodynamic and Transport Properties for an 11-Species Air Model for Chemical and Thermal Nonequilibrium Calculations to 30000 K*; Reference Report 1232; NASA: Washington, DC, USA, 1990.
23. Wilke, C. A Viscosity Equation for Gas Mixtures. *J. Chem. Phys.* **1950**, *18*, 517–519. [CrossRef]
24. Grinstein, F.; Margolin, L.; Rider, W. (Eds.) *Implicit Large Eddy Simulation—Computing Turbulent Fluid Dynamics*; Cambridge University Press: Cambridge, UK, 2007.
25. Knight, D. *Elements of Numerical Methods for Compressible Flows*; Cambridge University Press: New York, NY, USA, 2006.
26. Wright, M.; Candler, G.; Bose, D. Data-Parallel Line Relaxation Method for the Navier-Stokes Equations. *AIAA J.* **1998**, *36*, 1603–1609. [CrossRef]
27. Kianvashrad, N.; Knight, D.. *Large Eddy Simulation of Hypersonic Cold Wall Flat Plate-Part II*; AIAA Paper 2021-2882; AIAA: Reston, VA, USA, 2021.
28. Trettel, A.; Larsson, J. Mean Velocity Scaling for Compressible Wall Turbulence with Heat Transfer. *Phys. Fluids* **2016**, *28*, 026102. [CrossRef]
29. Griffin, K.; Fu, L.; Moin, P. Velocity Transformation for Compressible Wall-Bounded Turbulent Flows With and Without Heat Transfer. *Proc. Natl. Acad. Sci. USA* **2021**, *118*, e2111144118. [CrossRef]
30. Hill, F. Boundary-Layer Measurements in Hypersonic Flow. *J. Aeronaut. Sci.* **1956**, *23*, 35–42. [CrossRef]
31. Winkler, E.; Cha, M. *Investigation of Flat Plate Hypersonic Turbulent Boundary Layer with Heat Transfer at a Mach Number of 5.2*; NAVORD Report 6631; United States Naval Ordnance Laboratory: White Oak, MD, USA, 1959.
32. Danberg, J. *Characteristics of the Turbulent Boundary Layer with Heat and Mass Transfer at M = 6.7*; Technical Report NOLTR 64-99; US Naval Ordnance Laboratory: White Oak, MD, USA, 1964.
33. Young, F. *Experimental Investigation of the Effects of Surface Roughness on Compressible Turbulent Boundary Layer Skin Friction and Heat Transfer*; Technical Report DLR-532, CR-21; Defense Research Laboratory, University of Texas: Austin, TX, USA, 1965; Also, AD621085.
34. Samuels, R.; Peterson, J.; Adcock, J. *Experimental Investigation of the Turbulent Boundary Layer at a Mach Number of 6 with Heat Transfer at High Reynolds Numbers*; NASA TN D-3858; NASA: Washington, DC, USA, 1967.
35. Horstman, C.; Owen, F. Turbulent Properties of a Compressible Boundary Layer. *AIAA J.* **1972**, *10*, 1418–1429. [CrossRef]
36. Owen, F.; Horstman, C. On the Structure of Hypersonic Turbulent Boundary Layers. *J. Fluid Mech.* **1972**, *53*, 611–636. [CrossRef]
37. Keener, E.; Hopkins, E. *Turbulent Boundary-Layer Velocity Profiles on a Nonadiabatic Flat Plate at Mach Number 6.5*; Technical Note TN D-6907; NASA: Washington, DC, USA, 1972.
38. Lund, T.; Wu, X.; Squires, K. Generation of Turbulent Inflow Data for Spatially-Developing Boundary Layer Simulations. *J. Comput. Phys.* **1998**, *140*, 233–258. [CrossRef]
39. Hopkins, E.; Inouye, M. An Evaluation of Theories for Predicting Skin Friction and Heat Transfer on Flat Plates at Supersonic and Hypersonic Mach Numbers. *AIAA J.* **1971**, *9*, 993–1003. [CrossRef]
40. Morkovin, M. Effects of Compressibility on Turbulent Flows. In *Mécanique de la Turbulence, Colloques Internationaux du Centre National de la Recherche Scientifique*; Favre, A., Ed.; Centre National de la Recherche Scientifique: Paris, France, 1962; pp. 367–380.
41. Huang, P.; Coleman, G.; Bradshaw, P. Compressible Turbulent Channel Flows: DNS Results and Modelling. *J. Fluid Mech.* **1995**, *305*, 185–218. [CrossRef]
42. Duan, L.; Beekman, I.; Martín, M.P. Direct Numerical Simulation of Hypersonic Turbulent Boundary Layers. Part 2. Effect of Wall Temperature. *J. Fluid Mech.* **2010**, *655*, 419–445. [CrossRef]
43. Zhang, C.; Duan, L.; Choudhari, M. Direct Numerical Simulation Database for Supersonic and Hypersonic Turbulent Boundary Layers. *AIAA J.* **2018**, *56*, 4297–4311. [CrossRef]
44. Klebanoff, P. *Characteristics of Turbulence in a Boundary Layer with Zero Pressure Gradient*; Technical Note 3178; NACA: Washington, DC, USA, 1954.
45. Keener, E.; Polek, T. Measurements of Reynolds Analogy for a Hypersonic Turbulent Boundary Layer on a Nonadiabatic Flat Plate. *AIAA J.* **1972**, *10*, 845–846. [CrossRef]

fluids MDPI

Article

Implementation of Flux Limiters in Simulation of External Aerodynamic Problems on Unstructured Meshes

A. V. Struchkov [1], A. S. Kozelkov [1,2,3], R. N. Zhuchkov [1,2], K. N. Volkov [4] and D. Yu. Strelets [3,*]

1 Federal State-Funded Educational Institution of Higher Education, Nizhniy Novgorod State Technical University, 603999 Nizhniy Novgorod, Russia
2 Russian Federal Nuclear Center, All-Russia Research Institute of Experimental Physics, Federal State Unitary Enterprise, 607188 Sarov, Russia
3 Moscow Aviation Institute, 125993 Moscow, Russia
4 Federal State-Funded Higher Education Institution, Baltic State Technological University, 199406 St. Petersburg, Russia
* Correspondence: maksmai33@gmail.com

Abstract: The study is dedicated to the peculiarities of implementing the flux limiter of the flow quantity gradient when solving 3D aerodynamic problems using the system of Navier–Stokes equations on unstructured meshes. The paper describes discretisation of the system of Navier–Stokes equations on a finite-volume method and a mathematical model including Spalart–Allmaras turbulence model and the Advection Upstream Splitting Method (AUSM+) computational scheme for convective fluxes that use a second-order approximation scheme for reconstruction of the solution on a facet. A solution of problems with shock wave structures is considered, where, to prevent oscillations at discontinuous solutions, the order of accuracy is reduced due to the implementation of the limiter function of the gradient. In particular, the Venkatakrishnan limiter was chosen. The study analyses this limiter as it impacts the accuracy of the results and monotonicity of the solution. It is shown that, when the limiter is used in a classical formulation, when the operation threshold is based on the characteristic size of the cell of the mesh, it facilitates suppression of non-physical oscillations in the solution and the upgrade of its monotonicity. However, when computing on unstructured meshes, the Venkatakrishnan limiter in this setup can result in the occurrence of the areas of its accidental activation, and that influences the accuracy of the produced result. The Venkatakrishnan limiter is proposed for unstructured meshes, where the formulation of the operation threshold is proposed based on the gas dynamics parameters of the flow. The proposed option of the function is characterized by the absence of parasite regions of accidental activation and ensures its operation only in the region of high gradients. Monotonicity properties, as compared to the classical formulation, are preserved. Constants of operation thresholds are compared for both options using the example of numerical solution of the problem with shock wave processes on different meshes. Recommendations regarding optimum values of these quantities are provided. Problems with a supersonic flow in a channel with a wedge and transonic flow over NACA0012 airfoil were selected for the examination of the limiter functions applicability. The computation was carried out using unstructured meshes consisting of tetrahedrons, truncated hexahedrons, and polyhedrons. The region of accidental activation of the Venkatakrishnan limiter in a classical formulation, and the absence of such regions in case a modified option of the limiter function, is implemented. The analysis of the flow field around a NACA0012 indicates that the proposed improved implementation of the Venkatakrishnan limiter enables an increase in the accuracy of the solution.

Keywords: numerical simulation; Navier–Stokes equations; flux limiter; shock waves; gradient; unstructured mesh

Citation: Struchkov, A.V.; Kozelkov, A.S.; Zhuchkov, R.N.; Volkov, K.N.; Strelets, D.Y. Implementation of Flux Limiters in Simulation of External Aerodynamic Problems on Unstructured Meshes. *Fluids* **2023**, *8*, 31. https://doi.org/10.3390/fluids8010031

Academic Editors: Olga A. Azarova and Mehrdad Massoudi

Received: 9 November 2022
Revised: 20 December 2022
Accepted: 6 January 2023
Published: 15 January 2023

1. Introduction

Unstructured meshes are most preferable currently in computations of aerodynamic flows for industrial applications [1,2]. Implementation of these meshes requires adaptation of the numerical methods being used, for example, in the part of numerical schemes construction, rated quantities approximation, and computation of flows and gradients.

One of specific features of the computations of aerodynamic flows is the possible availability of high-gradient areas in the flow, e.g., a shock wave. Robustness of the implemented algorithm depends on the stability of the discretisation scheme of convective fluxes when simulating such flows [3–6]. It is known that it is strictly recommended to implement schemes of an upgraded order of accuracy in the computations of the flows with shock waves and rarefaction waves, as the schemes of a lower order of accuracy tend to cause considerable "smearing" of the solution. However, they are more reliable in the way of fail-safe features than higher-order schemes.

It is possible to improve robustness of a high accuracy order scheme through the implementation of a scalar limiter of the gradient of the flow quantities. A gradient limiter is basically used in upwind schemes of the second order of accuracy. It allows preventing non-physical oscillations in the solution that are characteristic in computations of shock wave flows. A similar technique of limiter implementation is used in hydrodynamics [7]. Gradient limiters there prevent the value produced in gradient reconstruction of the quantity at the facet of the cell from violating the limits of its minimum and maximum in the cells neighboring the given one. The Venkatakrishnan limiter [8] is one of the most popular options in practice.

The experience in the implementation of limiters is long and initially appeared regarding structured meshes and the meshes with cells of a regular geometric shape [7,8]. It is evident, for example, from the summands as a part of the limiter expression, that they are related to the particular cell size of the computational mesh. In the case of unstructured meshes, it is difficult to find a characteristic size of a cell in the form of a random polyhedron. In this case, it appears logical and reasonable to set the operation threshold based solely on the flow quantities.

This work studies the implementation of the Venkatakrishnan limiter. Unstructured meshes consisting of tetrahedrons, truncated hexahedrons, and polyhedrons are used for the computation. A modified option of the Venkatakrishnan limiter is offered based on the results produced, and its advantages are shown in comparison to the initial option of the function when solving the problems with shock wave processes that occur at supersonic flow around the wedge and transonic flow around NACA0012 airfoil.

The paper is organized in three main sections. The section of basic equations describes the mathematical model used to simulate gas flow. The next section describes implementation of the flow limiter and solution of practical problems. The main conclusions are given in the final part of the paper.

2. Governing Equations

The physical and mathematical model to describe 3D flows is realized in the Russian software package called LOGOS. The LOGOS software package is designed for computational hydro- and aero-dynamics problems on parallel systems [9–12].

Non-stationary 3D turbulent flows of a viscous thermally conductive gas are described with Reynolds-averaged Navier–Stokes equations [13,14]. In a conservative form in Cartesian coordinates, the system of equations has the following form (averaging signs are skipped):

$$\begin{cases} \frac{\partial \rho}{\partial t} + \nabla\left(\rho \vec{u}\right) = 0, \\ \frac{\partial\left(\rho \vec{u}\right)}{\partial t} + \nabla\left(\rho \vec{u}\vec{u}\right) = -\nabla p + \nabla(\tau_\mu + \tau_t), \\ \frac{\partial(\rho E)}{\partial t} + \nabla\left(\rho \vec{u} h\right) = \nabla\left[\vec{u}(\tau_\mu + \tau_t) - \left(\vec{q}_\mu + \vec{q}_t\right)\right]. \end{cases} \tag{1}$$

Here, ρ is density; $\vec{u}$ is the vector of the flow velocity with components u, v, w; p is pressure; $E = C_v T + 0.5(u^2 + v^2 + w^2)$ is total energy; $h = C_p T + 0.5(u^2 + v^2 + w^2)$ is total enthalpy; τ_μ and τ_t are molecular and turbulent components of the tensor of tangential stresses, respectively; q_μ and q_t are molecular and turbulent heat flux, respectively; T is temperature; $C_v = (C_p T - R/m)$ is specific heat capacity at constant volume; C_p is specific heat capacity at constant pressure; R is a universal gas constant; and m is a molar mass of the gas.

The values of the molecular component of the tangential stress tensor of the Newtonian medium meet the rheological Newton's law, and the components of the density vector of the heat flow are connected with the local temperature gradient by Fourier's law [13,14]:

$$\tau_\mu = 2\mu(T)\left(S - \frac{1}{3}I\nabla\vec{u}\right), \tag{2}$$

$$S = \frac{1}{2}\left(\nabla\vec{u} + \left[\nabla\vec{u}\right]^t\right), \tag{3}$$

$$q_\mu = \lambda(T)\nabla T. \tag{4}$$

The dynamic viscosity, $\mu(T)$, and heat conductivity, $\lambda(T)$, are found from the Sutherland formula as a function of the flow temperature [14,15].

$$\mu = \mu_0\left(\frac{T}{T_0}\right)^{0.5}\frac{T_0 + T_s}{T + T_s}, \tag{5}$$

$$\lambda = \lambda_0\left(\frac{T}{T_0}\right)^{0.5}\frac{T_0 + T_s}{T + T_s}, \tag{6}$$

where μ_0 and λ_0 are dynamic viscosity and heat conductivity at temperature T_0; T_s is the Sutherland constant.

System of Equation (1) is open due to the unknown connection of some of the basic variables of this system with averaged parameters of the flow. This connection can be established with additional ratios that, in a general case, are called turbulent models. Linear differential models of turbulence use empirical ratios for the turbulent viscosity factor. Here, the Spalart–Allmaras model [16,17] proved to be effective.

In the Spalart–Allmaras model, a single transport equation is considered. The transport equation is written with respect to the modified kinematic turbulent viscosity $\widetilde{\nu}$.

$$\frac{\partial\rho\widetilde{\nu}}{\partial t} + \frac{\partial\rho u_j\widetilde{\nu}}{\partial x_j} = \frac{1}{\sigma}\left\{\frac{\partial}{\partial x_j}\left[(\mu + \rho\widetilde{\nu})\frac{\partial\widetilde{\nu}}{\partial x_j}\right] + c_{b2}\rho\left(\frac{\partial\widetilde{\nu}}{\partial x_j}\right)^2\right\} + P^v - D^v \tag{7}$$

The generation and dissipation terms in Equation (7) are the source terms and they are expressed

$$D^v = \left(c_{w1}f_w - \frac{c_{b1}}{\kappa^2}f_{t2}\right)\left(\frac{\widetilde{\nu}}{d}\right)^2 \tag{8}$$

$$P^v = c_{b1}\rho\widetilde{S}\widetilde{\nu} - c_{b1}\rho f_{t2}\widetilde{S}\widetilde{\nu} \tag{9}$$

where d is the closet distance to the rigid wall, k is the von Karman constant.

The other parameters in the transport equation of turbulent viscosity can be found as follows:

$$\widetilde{S} = \Omega + f_{v2}\frac{\widetilde{\nu}}{\kappa^2 d^2}. \tag{10}$$

Here, Ω is the rate of vorticity tensor

$$\Omega = \left(2\Omega_{ij}\Omega_{ij}\right)^{1/2} \tag{11}$$

$$\Omega_{ij} = \frac{1}{2}\left(\frac{\partial u_j}{\partial x_i} - \frac{\partial u_i}{\partial x_j}\right) \tag{12}$$

$$f_w = g\left(\frac{1 + C_{w3}^6}{g^6 + C_{w3}^6}\right)^{1/6} \tag{13}$$

$$g = r + C_{w2}\left(r^6 - r\right) \tag{14}$$

$$r = \frac{\widetilde{\nu}}{\widetilde{S}\kappa^2 d^2} \tag{15}$$

$$C_{w1} = \frac{c_{b1}}{\kappa^2} + \frac{(1 + c_{b1})}{\sigma} \tag{16}$$

where the f_{t2} function provides the suppression of the transition from the laminar flow calculations in the boundary layer to the turbulent flow calculations, and is expressed as

$$f_{t2} = C_{t3}\exp\left(-C_{t4}\chi^2\right). \tag{17}$$

Effective turbulent viscosity of the model is given next expression:

$$\mu_t = \rho\widetilde{\nu} f_{v1} \tag{18}$$

$$f_{v1} = \frac{\chi^3}{\chi^3 + C_{v1}^3}, \tag{19}$$

$$\chi = \frac{\widetilde{\nu}_t}{\nu}. \tag{20}$$

Empirical constants of the model are as follows: $\sigma = \frac{2}{3}, \kappa = 0.41, c_{b1} = 0.1355$, $c_{b2} = 0.622, C_{w2} = 0.3, C_{w3} = 2, C_{v1} = 7.1$, and $C_{t3} = 1.2, C_{t4} = 0.5$.

3. Numerical Method

The system above is approximated with the finite element method and it uses an integral formulation of the basic conservation laws. Discrete analogs of summands are written for the reference volume by summation over facets.

A finite volume method is based on integration of initial differential equations by the reference volume. Reference volumes (cells of the mesh) are arbitrary polyhedrons that cover the domain without gaps and overlaps. Each polyhedron is limited with a random number of facets. The vertices of the facets are the nodes of the mesh. A general view of the cell is given in Figure 1.

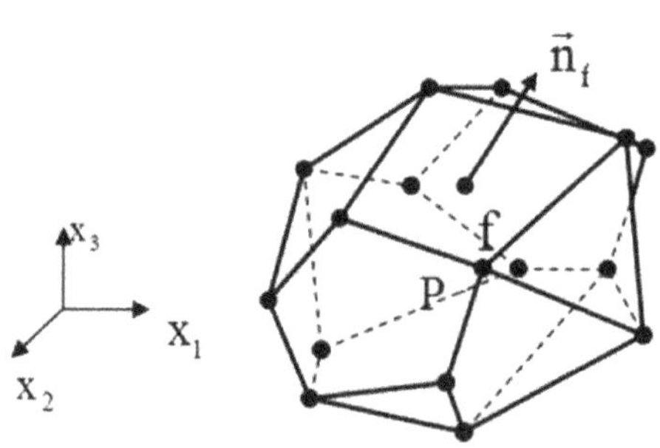

Figure 1. Control volume.

A system of the Navier–Stokes equation is written in a vector form for numerical solution using a finite volume method:

$$\frac{d}{dt}\int_{\Delta V} W dV + \oint_{\Delta \Sigma_P} (F - G) dS = \int_{\Delta V} H\, dV, \tag{21}$$

where W is a vector of conservative variables, F and G are vectors of convective and diffusion fluxes, and H is a source term

$$W = \begin{pmatrix} \rho \\ \rho u \\ \rho v \\ \rho w \\ \rho E \end{pmatrix}, F = \begin{pmatrix} \rho u_n \\ \rho u u_n + p n_x \\ \rho v u_n + p n_y \\ \rho w u_n + p n_z \\ \rho H u_n + p u_n \end{pmatrix}, G = \begin{pmatrix} 0 \\ \tau_{nx} \\ \tau_{ny} \\ \tau_{nz} \\ \tau u + q \end{pmatrix}, \tag{22}$$

where u_n is a normal component of the velocity, q is a heat flux, and τ_{ij} are components of the tensor of viscous stresses.

The full description of the way of approximation of the system of Equation (22) is given in [18,19].

4. Flux Limiters

When solving a problem, computation accuracy for the convective flows is of great importance. Schemes on the basis of the Riemann problem solution became quite popular in the case of aerodynamic problems. Such schemes comprise Advection Upstream Splitting Method (AUSM)–family schemes [20–24], and in particular AUSM+ scheme [21–24].

According to [25], a convective flow through the facet is computed in AUSM+ as follows:

$$F_f = c_f\left(\overline{M}_L^+ U_L + \overline{M}_R^- U_R\right) + \left(P_L^+\big|_{\alpha=\frac{3}{16}} P_L + P_R^-\big|_{\alpha=\frac{3}{16}} P_R\right) \tag{23}$$

where c_f is the sound velocity at the facet; U_L and U_R are vectors of primitive variables on the left and on the right facet f; P_L and P_R are vectors of pressure $P = P\{0, n_x, n_y, n_z, 0\}^T$ on the left and on the right facet f, $\overline{M}_L^+$, $\overline{M}_R^-$, $P_L^+\big|_{\alpha=\frac{3}{16}}$, and $P_R^-\big|_{\alpha=\frac{3}{16}}$ are parameters of the scheme.

- If $M_L^+ + M_R^- \geq 0$, where M_L and M_R are Mach numbers on the left and on the right of the facet, then
$$\overline{M}_L^+ = M_L^+ + M_R^-[(1-\omega)(1+f_R) + f_R - f_L]. \tag{24}$$
- If $M_L^+ + M_R^- < 0$, then
$$\overline{M}_R^- = M_R^- + M_L^+[(1-\omega)(1+f_L) + f_L - f_R]. \tag{25}$$

Parameter ω is set by the function that depends on cubes of relations of pressure, and takes a minimum value in the larger part of the domain, except for the areas with a high gradient of pressure, such as the areas of shock waves and discontinuities

$$\omega(p_L, p_R) = 1 - \min\left(\frac{p_L}{p_R}, \frac{p_R}{p_L}\right)^3. \tag{26}$$

Parameter $f_{L,R}$ also takes a minimum value, except for the areas with oscillations of the solution

$$f_{L,R} = \begin{cases} \left(\frac{p_{L,R}}{p_s} - 1\right) \\ 0 \end{cases} \min\left(1, \frac{\min(p_{1,L}, p_{1,R}, p_{2,L}, p_{2,R})}{\min(p_L, p_R)}\right)^2, \ P_L^+ p_L + P_R^- p_R \neq 0. \tag{27}$$

Second-order polynomials are used to find parameters on the facet:

$$M^{\pm} = \begin{cases} \pm\frac{1}{4}(M \pm 1)^2, & |M| \leq 1, \\ \frac{1}{2}(M \pm |M|), & |M| > 1, \end{cases}, P^{\pm}_{\alpha} = \begin{cases} \frac{1}{4}(M \pm 1)^2(2 \mp M) \pm \alpha M(M^2 - 1), & |M| \leq 1, \\ \frac{1}{2}(1 \pm \mathrm{sign}(M), & |M| > 1. \end{cases} \quad (28)$$

In order to compute convective fluxes, reconstruction of the solution is conducted. It lies in the definition of parameters on the left and on the right of facet f. When solving flow problems, reconstruction of the solution is performed with regard to primitive variables Q, conservative variables W, and with regard to acoustic invariants. For the first order of approximation, the values from the center of a respective cell (Figure 2) are taken as parameters on the right and on the left from the facet:

$$\phi_f^- = \phi_P, \quad \phi_f^+ = \phi_E. \quad (29)$$

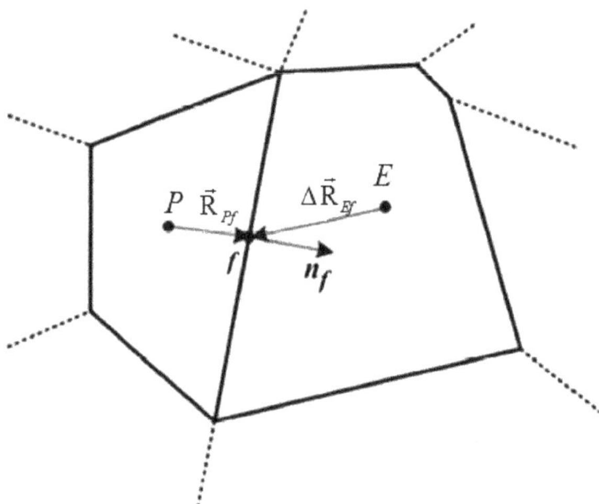

Figure 2. Reconstruction of values.

Reconstruction of the solution of the second-order approximation is usually taken to find the value on the facet [26,27].

$$\begin{aligned} \phi_f^- &= \phi_P + \alpha_f^-(\Delta\vec{R}_{Pf} \cdot \nabla\phi_P) \\ \phi_f^+ &= \phi_E + \alpha_f^+(\Delta\vec{R}_{Ef} \cdot \nabla\phi_E) \end{aligned} \quad (30)$$

where ϕ_f^- and ϕ_f^+ are values of the variable on the left and right facet, ϕ_P and ϕ_E are the values of the variable in the center of cells E and P (Figure 2), $\vec{R}_{Pf}$ and $\Delta\vec{R}_{Ef}$ are the distance from the center of cells E and P to the center of the facet, $\nabla\phi_E$ and $\nabla\phi_P$ are the values of the gradient in cells E and P, and α_f^- and α_f^+ are limiters designed to prevent oscillations at discontinuous solutions. Implementation of the limiter function allows for controlling the gradient value on unstructured grids (decreasing the gradient value multiplying it by value $\alpha_f \leq 1$); it is used to restore the value on the left and on the right from the facet [28,29].

5. Implementation of Flux Limiters

Implementation of so-called limiter functions is necessary to preserve monotonicity property of the solution in the areas with high gradients. In fact, introduction of the limiter of the scheme comes to "smoothening" of false maximum extremums of the values in the flow.

For example, in [28], as a result of limiter implementation, they managed to produce a smooth solution of trans-sonic flow without oscillations even on irregular meshes. The limiter function should be equal to zero in case of strong ruptures to produce the scheme of the first order that would guarantee monotonicity property, but in the areas of "monotonous" flow, the limiter function takes the value of a unit, and reconstruction of values at the facet is not limited. Transition from the limited value to the unlimited one should be smooth; only in this case would you expect upgraded convergence. Implementation of limiters is described in [29–35].

Correct behaviour of the limiter function is especially important when it is used in engineering codes to solve industrially specific problems on unstructured meshes.

More than a dozen different limiter functions were made available and published; they were reviewed in [36], for example. The most often used ones are:

- Barth–Jespersen limiter [37]: its general view is $\alpha_f = \min(r_f, 1)$;
- Van Albada–Leer [7]: its general view is $\alpha_f = \min(\frac{r_f^2 + r_f}{r_f^2 + 1}, 1)$;
- Venkatakrishnan [38]: its general view is $\alpha_f = \min(\frac{r_f^2 + 2r_f}{r_f^2 + r_f + 2}, 1)$.

Figure 3 shows a Sweby diagram [29] that demonstrates dependence of α_f values on factor r_f.

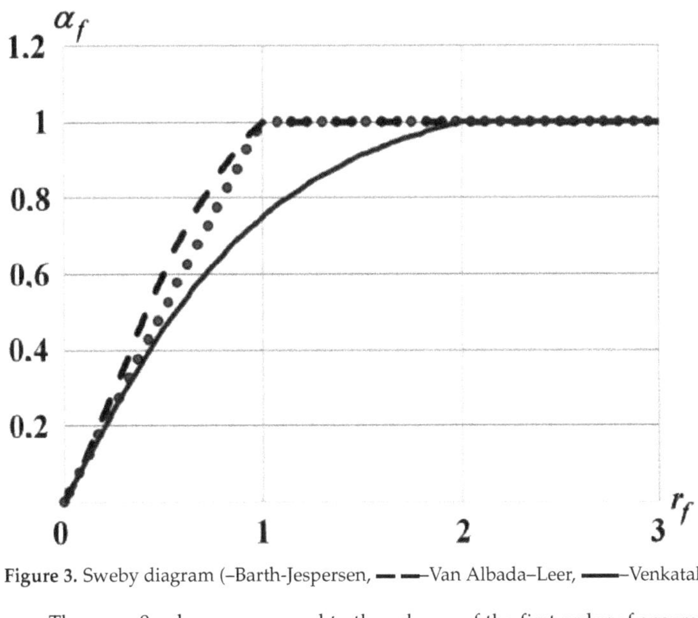

Figure 3. Sweby diagram (-Barth-Jespersen, — —-Van Albada–Leer, ——-Venkatakrishnan).

The $\alpha_f = 0$ values correspond to the scheme of the first order of accuracy, and $\alpha_f = 1$ to the scheme of the second order of accuracy. You see that that "strict" limiter function is the Venkatakrishnan limiter that makes transition to the scheme of the second order the latest of all (at $r_f = 2$), which will provide high stability and ensure good monotonous property [7,8,38].

The Venkatakrishnan limiter controls the value of gradient ϕ_E in cell E according to the expression:

$$\alpha_E = \begin{cases} \frac{1}{\Delta_2}\left[\frac{\left(\Delta_{l,\max}^2+\varepsilon^2\right)\Delta_2+2\Delta_2^2\Delta_{l,\max}}{\Delta_{l,\max}^2+2\Delta_2^2+\Delta_{l,\max}\Delta_2+\varepsilon^2}\right], & \Delta_2 > 0 \\ \frac{1}{\Delta_2}\left[\frac{\left(\Delta_{l,\min}^2+\varepsilon^2\right)\Delta_2+2\Delta_2^2\Delta_{l,\min}}{\Delta_{l,\min}^2+2\Delta_2^2+\Delta_{l,\min}\Delta_2+\varepsilon^2}\right], & \Delta_2 < 0 \\ 1, & \Delta_2 = 0 \end{cases} \tag{31}$$

$$\begin{aligned} \Delta_{l,\max} &= \phi_{\max} - \phi_E, \\ \Delta_{l,\min} &= \phi_{\min} - \phi_E, \\ \Delta_2 &= \tfrac{1}{2}\left(\nabla\phi_f \cdot \Delta\vec{R}_{Ef}\right). \end{aligned} \tag{32}$$

$$\varepsilon^2 = (K\Delta h)^3 \tag{33}$$

where $\phi_{\max}$ and $\phi_{\min}$ are maximum and minimum values in all neighboring cells, including the values in cell E itself, and $\Delta\vec{R}_{Ef}$ is the distance from the center of cell E (or P) to the center of the facet. Parameter ε^2 controls the value of the limiter, where K is a constant (a normalizing coefficient), Δh is a characteristic size of the cell.

In (31), ε^2 is a symbolic operation threshold of the limiting function. Oscillations lower than this threshold are allowed in the solution and are not considered by the limiter. A zero value of parameter ε^2 means that the limiter is active even in the near-constant regions, when a very high value of parameter ε^2 means practically no limit. Such modification of the limiting function makes it possible to protect from random operations and reach improvement of convergence and stable solution on unstructured meshes.

Let us mark

$$y = \frac{\Delta_{l,\max}}{\Delta_2} or \left(y = \frac{\Delta_{l,\min}}{\Delta_2}\right)$$

and write the function from expression (31) as follows [8,37]:

$$\alpha_E = \frac{y^2 + 2y + \varepsilon}{y^2 + y + 2 + \varepsilon} \tag{34}$$

where $\Delta_{l,\max}$ or $\Delta_{l,\min}$ from (32) is increment of the solution.

With quantity ε exceeding the increment value of the solution, or with the unlimited quantity ε, the limiter takes the value of a unity, i.e., the value of the gradient (in the expression to find the value per facet) is not limited to anything. Where the increment of the solution exceeds quantity ε (e.g., in the region with large gradients or at a small value of ε), the solution itself determines the value of the limiting function (summand $\frac{y^2+2y}{y^2+y+2}$ of expression (34)); in this way, it sets the degree of limitation.

6. Results and Discussion

Several CFD benchmark problems are considered to validate the robustness of the flux limiters and their parameters.

6.1. Flow in a Channel with Wedge

The operation of the limiter (at various values of parameter K) is considered using the example of the problem of supersonic flow around the wedge, where the parameters of the incident flow are as follows: Mach number is 2, the pressure is 101,325 Pa, and the temperature is 300 K [39]. Structured computational mesh is used for simulation with the number of cells at 95,000. General view of the computation domain and mesh are shown in Figure 4.

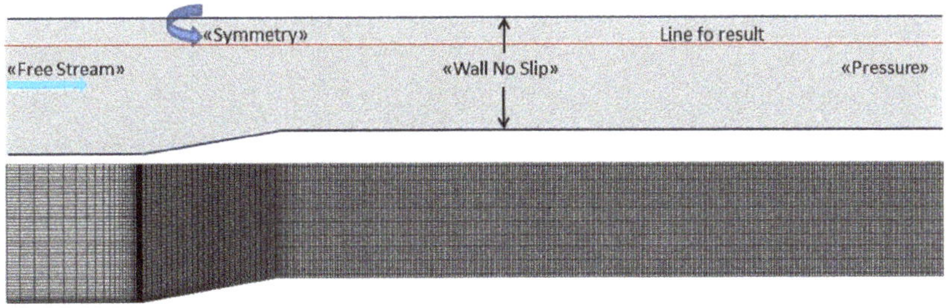

Figure 4. General view of the computation domain and mesh.

At supersonic flow in the channel with a wedge, an attached compression shock is formed, which results in the formation of the shock wave structure of the flow in the channel. Figure 5 shows the origination of the compression shock, its development and reflection from the walls of the channel, and its interaction with the system (finitary spread) of rarefaction waves [39].

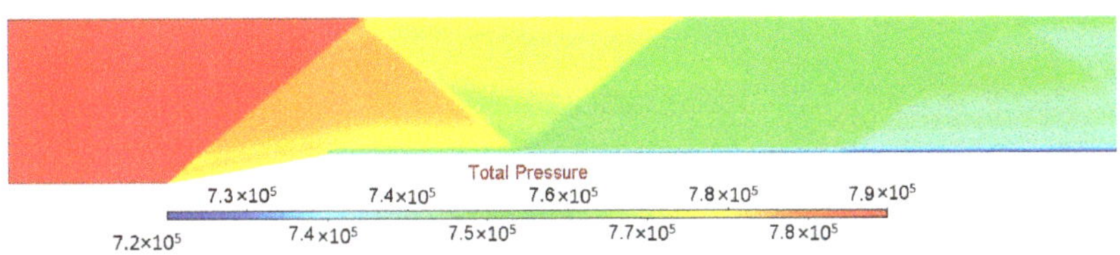

Figure 5. Contours of total pressure.

Each shock wave is characterized with its front (the surface where flow quantities have a jump in the development, whereas outside the front they change continuously). From the practical point of view, the implemented numerical scheme should provide stability and preserve monotonicity of the solution in all regions of the flow, including the shock wave fronts.

Let us study the section of the first series of compression shocks to explore the solution as a function of the limiter in Figure 5. Look at the plot in Figure 6, of the distribution of the full pressure along the line in Figure 4 in the specified region produced with and without the limiter (Venkatakrishnan limiter) for different values of parameter K in expression (33).

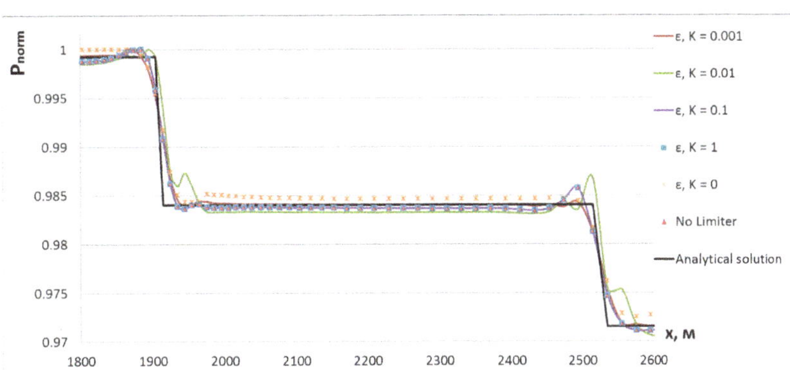

Figure 6. The distribution of full pressure.

According to the plot, at $K = 0.1$ and $K = 1$ the produced result complies with the computation result without a limiter, i.e., at such values of K, the solution is fully unlimited (complete computed value of the gradient is taken at the reconstruction step). The solution produced at $K = 0.01$ is characterized with local maximums that have a non-physical (fictitious) origination character that can cause a considerable error in the solution, as it acquires extremely non-monotonic behaviour.

The solution has the best accuracy (as compared to the analytical solution) and monotonicity properties at $K = 0$ and $K = 0.001$. However, $K = 0$ means probability of switching on the limiter in the entire domain, i.e., it has a random reaction character in Figure 7, and this is also intolerable.

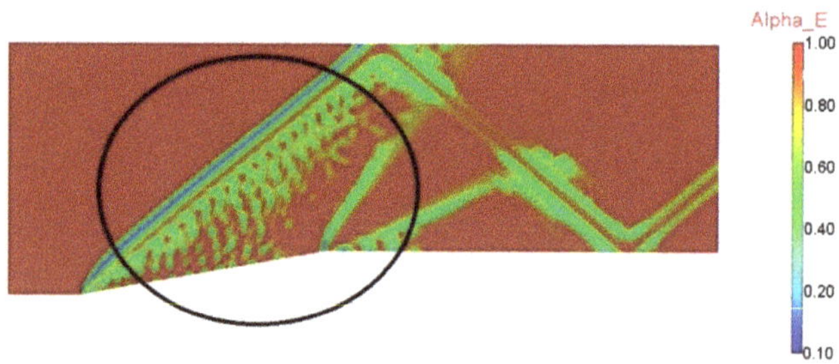

Figure 7. Random behaviour of the limiter (31) at $K = 0$.

The solution at $K = 0.001$ has a minimum amplitude of oscillations and, in general, is characterized with a good monotonicity property. However, due to a very small value of parameter K, the limiter is practically always activated (random cases of operation on other mesh models can be revealed), and the picture of the produced solution actually reflects the property of this limiter that depends on the size of the mesh cell.

Computation of parameter ε by formula (33) is related to the characteristic cell size only, and the function depends only on the geometric parameters of the mesh. In this case, the limiter operation depends only on the parameters of the computation model and increases the probability of the limiter's reaction in the regions of the local refinement mesh model. At the same time, in this formulation parameter, ε is not related in any way to the flow quantities. This also increases the probability of random operation of the limiter function for a particular flow quantity. The characteristic cell size is valid only in the case of cells of a regular geometric shape; whereas, in the case of a cell with a shape of a random polynomial, this value does not have a clear definition.

Modification of expression (33) to compute ε is necessary to use the limiter on cells of a random shape. The idea is to make parameter ε the function of the flow quantities, i.e., the function of that quantity, for which the limiter is implemented. Another option to define ε is

$$\widetilde{\varepsilon} = K\phi \tag{35}$$

where K is an operation threshold of function and ϕ is a flow variable. The value of the limiter is computed on the basis of the flow variable itself, for which the limiter is computed (when pressure is computed, ϕ shows pressure, and when density is computed, it shows density; same for velocity).

Introduction of a flow variable as a parameter when computing $\widetilde{\varepsilon}$ disconnects the limiter from geometric parameters of the mesh cells. In the current formulation parameter, $\widetilde{\varepsilon}$ has a physical sense. Changing constant K, the value of oscillations of computational flow quantities filtered with the limiter is found. For example, $K = 0.01$ means that the operation threshold of the limiter is equal to 1% from flow quantity ϕ, i.e., the limiter is switched on when the solution becomes oscillating and the increment of the solution in the cell is higher than 1% of the solution in the cell.

Let us analyse the implementation of the modified formula for the problem of a supersonic flow in the channel with a wedge in Figure 8, and look at the case when $K = 0.01$ in (35).

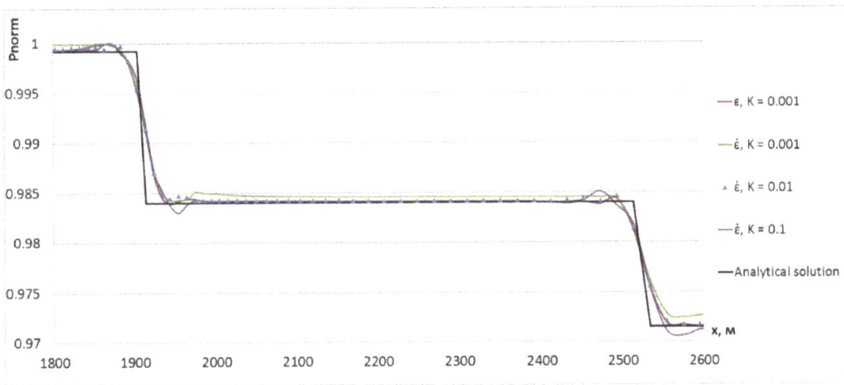

Figure 8. The distribution of total pressure.

Distribution of total pressure in case $K = 0.01$ from (35) has a comparable solution with the one produced in case $K = 0.001$ in (33), the property of monotonicity. However, if expression (35) is implemented, the limiter function has a clear physical interpretation. In this case, the region of activation of the limiter is characterized by the absence of random reactions in Figure 9.

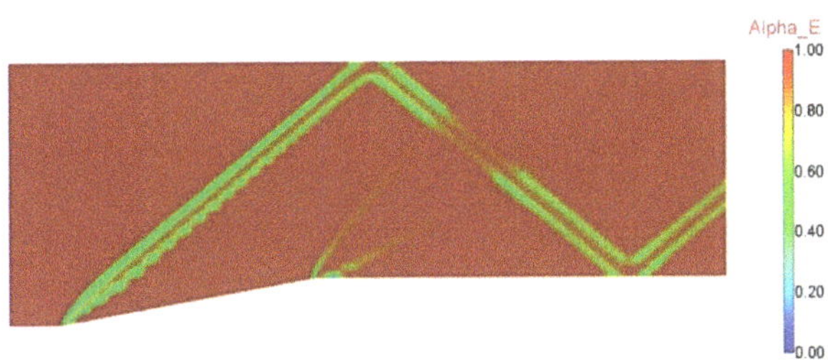

Figure 9. The region of activation of the limiter when $K = 0.01$ in (35).

To study the implementation on unstructured meshes, 3D meshes (one-cell thick) were generated for this geometry consisting of polynomials, tetrahedrons, and truncated hexahedrons. The behaviour of the limiters is shown in Figure 10. Note that, when $K = 0.001$ in (33), multiple areas of random reactions are observed in all options of mesh models.

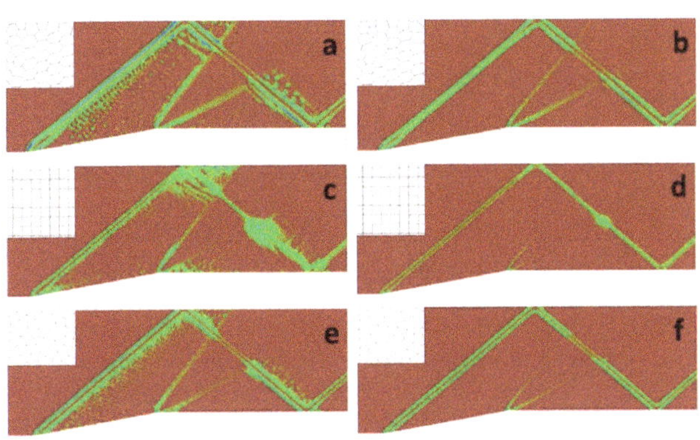

Figure 10. Activation area of the limiter. Fragments (**a**,**c**,**e**) correspond to the limiter (31) when $K = 0.001$ in (33). Fragments (**b**,**d**,**f**) correspond to the limiter (31) when $K = 0.01$ in (35).

So, having analysed quantity ε as it influences the behaviour of the solution, it can be concluded that implementation of the Venkatakrishnan limiter, where ε is computed by expression (33) at $K = 0.001$, contributes to the monotonicity of the convergence process of the solution and reduces the number of non-physical oscillations. However, ε depends on geometric sizes of the cell of the computation mesh. This could be the reason for the possible random character of the reaction of the function under consideration (revealed on unstructured meshes) and introduces a numerical error into the solution. To eliminate this problem, option (35) was suggested at $K = 0.01$, which has comparable monotonicity properties. However, it is characterized with no random character of the reaction of the limiter on random unstructured meshes. Form (35) provides dependence of quantity ε on the intensity of the flow, allowing for more precise definition of the activation threshold of the limiter.

6.2. Flow around Airfoil

Let us consider the limiter as it influences the accuracy of the produced solution on the example of transonic flow around NACA0012 airfoil [40,41]. The mesh is one cell thick in Figure 11. The total number of mesh elements is 731,000.

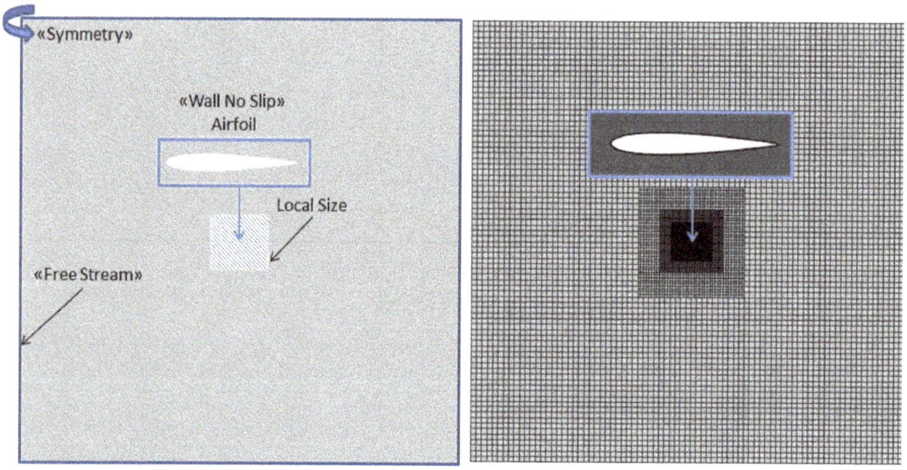

Figure 11. General view of the computation domain and mesh.

According to the research on the components of the drag of aerodynamic airfoils [13], the basic constituent of the drag for a medium-thick airfoil at small attack angles is friction resistance. The airfoil drag (pressure drag) due to the incomplete restoration of pressure in the tail part of the airfoil is 20–30% of the total profile drag (at small attack angles of 0–3° for symmetric and slightly bended airfoils with an average relative thickness of 11–15%).

As [13,39] show, the accuracy of the pressure drag computation depends directly on the level of so-called numerical viscosity (approximation errors that work as additional dissipation, resulting in the loss of complete pressure in the flow and growth of the resistance of the object under investigation). It can be caused by some peculiarities of the numerical method. For example, the selection of the limiter of the gradient increment could reduce the accuracy order greatly in the regions with large gradients of gas parameters. As a result of the incorrect operation of the limiter, not enough rarefaction can be observed in the region of minimum values of the pressure coefficient on the upper surface of the airfoil.

Let us consider this problem with the following boundary conditions. Parameters of the incident flow at the external boundary of the computational domain correspond to the values: the pressure is 46,066.2 Pa, the temperature is 248 K, the Mach number is 0.7, and the angle of attack is 1.489° and 3.046° [39–41]. The surface of the airfoil is considered to be a no-penetrated wall; a symmetric boundary condition is set at the sides.

At the set parameters of the airfoil overflow, the formation of the tear-off zone takes place on its leeside. A curvilinear shock wave is formed near the surface of the airfoil. It corresponds to the normal that transforms a supersonic flow into the subsonic flow. In case the angle of attack is 1.489°, a low-intensity compression shock is formed with smeared boundaries of the transfer of the supersonic flow into the subsonic flow in Figure 12.

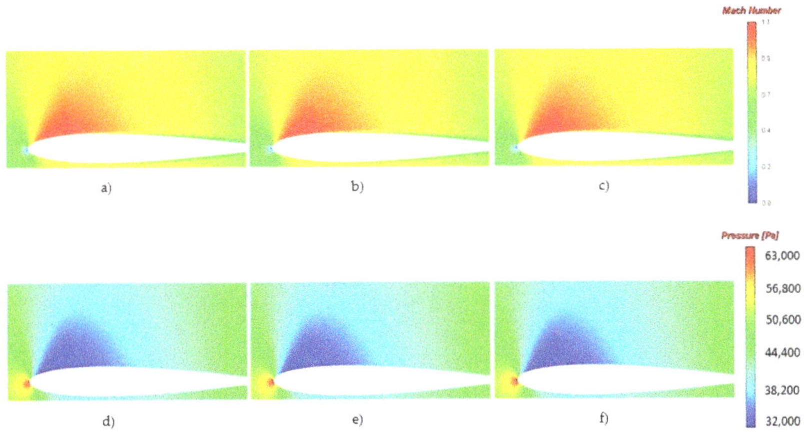

Figure 12. Contours of Mach number (**a**–**c**) and contours of pressure (**d**–**f**): computation without a limiter (**a**,**d**), a computation with a limiter function at K = 0.001 for (33) (**b**,**e**), and computation with a limiter at K = 0.01 for (35) (**c**,**f**).

When increasing the angle of attack to 3.046°, a more intensive compression shock is formed above the airfoil. It has a more expressed front and creates a large area of decreased pressure. Mach number and pressure contours are presented in Figure 13.

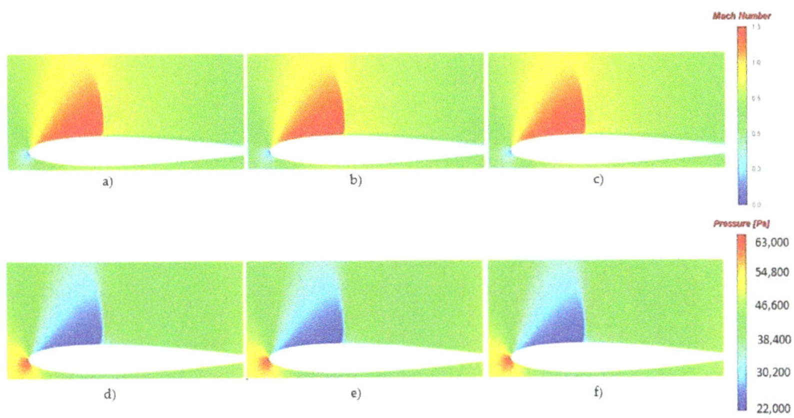

Figure 13. Contours of Mach number (**a**–**c**) and contours of pressure (**d**–**f**): computation without a limiter (**a**,**d**), a computation with a limiter function at K = 0.001 for (33) (**b**,**e**), and computation with a limiter at K = 0.01 for (35) (**c**,**f**).

No differences in the solution with different options of the limiter are observed in the general structure of the flow (in the shock wave generation above the airfoil). Cognominal fields correspond to each other. However, the differences in the solution are evident if integral characteristics are estimated, for example, the value of the drag force coefficient, C_{drag}, as compared to the experimental data in Table 1, [39–41].

Table 1. Drag coefficients.

No	Computation	$\alpha = 1.489°$		$\alpha = 3.046°$	
		C_{drag}	ΔC_{drag}, %	C_{drag}	ΔC_{drag}, %
0	Experiment	0.00819	–	0.01267	–
1	Without limiter	0.00848	3.6	0.01423	12.3
2	ε (33) at $K = 0.001$	0.00850	3.7	0.01425	12.5
3	$\dot{\varepsilon}$ (35) at $K = 0.01$	0.00838	2.3	0.01382	9.1

Maximum error in the solution for both angles of attack is produced in the case of computation No 2. The main contribution into the error in the resistance computation is due to the pressure force component that happens to be overestimated by more than 10%. One of the most probable reasons for these results could be the work of the gradients increment limiter, which is of a random reaction character in all the computation domains revealed in this mesh model in Figure 14.

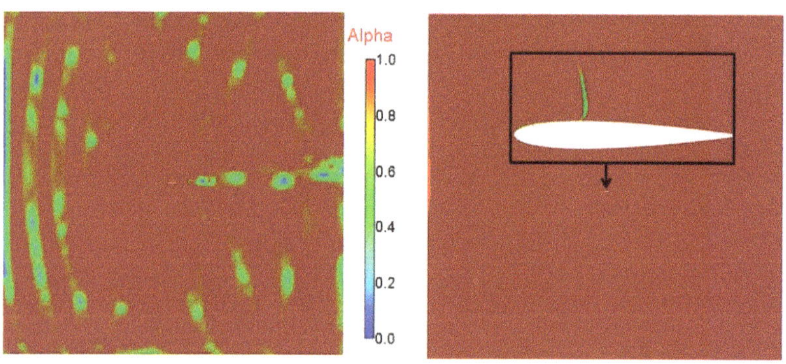

Figure 14. The region of the limiter activation (ε (33) at $K = 0.001$—on the left, $\dot{\varepsilon}$ (35) at $K = 0.01$—on the right).

As for computation No 3, the limiter in this case was active only in the region of the shock wave above the airfoil in Figure 14, which is completely true for form $\dot{\varepsilon}$ (35) dependent on the flow quantities.

As a result, a solution was produced that has a minimum deviation from the experimental data (2.3% and 9.1%). It is worth noting that the error in the computation grows with the increase in the angle of attack, and it is approaching the critical values. It is connected to the formation of a more intensive shock wave above the surface of the airfoil and a more complicated operation of the limiter in this region. In particular, it reduces the accuracy order of the scheme most precisely in the regions with large gradients of flow quantities and not to introduce additional numerical viscosity. So, evident advantages of implementation of the limiter in combination with parameter $\dot{\varepsilon}$ at $K = 0.01$ are shown using this problem.

6.3. *Flow around Bullet*

A modified option of the limiter can also be applied to compute supersonic flow around the "168 Grain Sierra International Bullet" [42]. An unstructured mesh of truncated hexahedrons with a general number of elements of 386 thousand was generated for the computation of this problem in Figure 15.

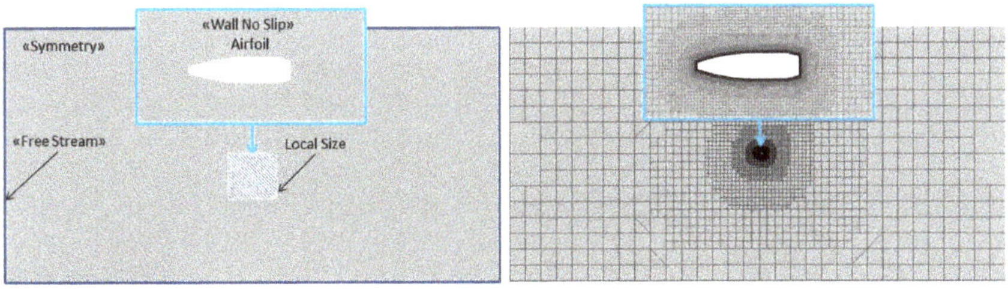

Figure 15. General view of the computation domain and mesh.

Regions of the flow, where the Mach number is close to 1, are characterized with complexity due to the formation of compression shocks and possible separation of the flow that influence aerodynamic properties of the object considerably. The results produced here can also be generalized for the solids geometrically similar to the object under consideration.

The geometry of the bullet, the properties of the flow, and experimental aerodynamic parameters are taken from the work of Author [42]. Let us take the problem with the following boundary conditions. Parameters of the striking flow at the external boundary of the domain correspond to the following quantities: the pressure is 101,325 Pa, the temperature is 288.15 K, the angle of attack is 0°, and the Mach number is 1.05 and 1.6 [42]. The surface of the bullet is taken as a solid wall, and a symmetric boundary condition is set on the side boundary.

In case Mach number is equal to 1.05 and 1.6, the flow is characterized by the presence of the head shock wave in Figure 16. Distribution of flow quantities near the surface of the object influences aerodynamic properties considerably. That is why it is necessary to provide correct operation of the limiter in this region to get a high-quality solution. There are no differences in the solutions with different variants of the limiter function in the general structure of the flow in Figures 16 and 17.

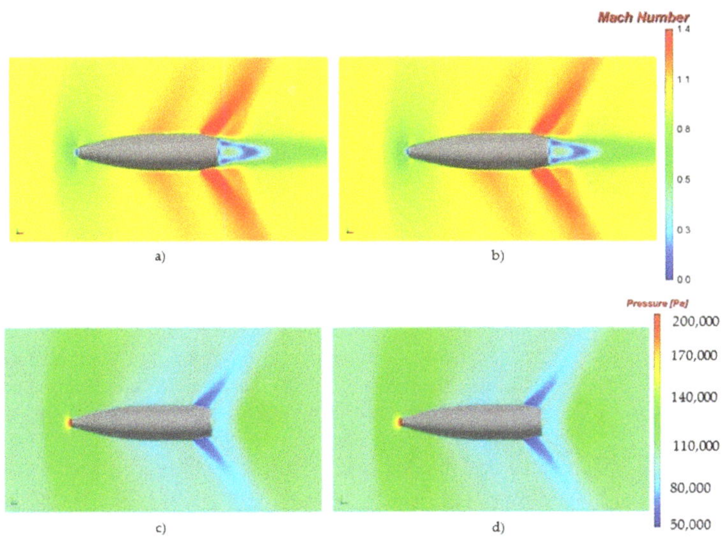

Figure 16. Contours of Mach number (**a**,**b**) and contours of pressure (**c**,**d**): computation with a limiter at K = 0.001 for (33) (**a**,**c**), and computation with a limiter at K = 0.01 for (35) (**b**,**d**).

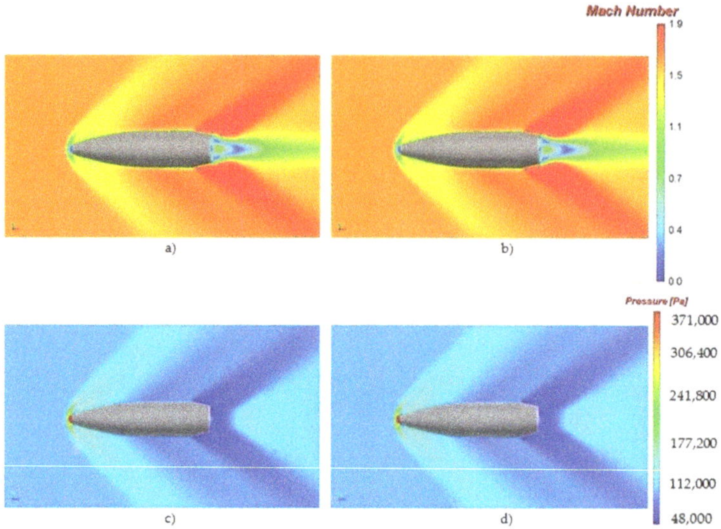

Figure 17. Contours of Mach number (**a**,**b**) and contours of pressure (**c**,**d**): computation with a limiter at K = 0.001 for (33) (**a**,**c**), and computation with a limiter at K = 0.01 for (35) (**b**,**d**).

Let us estimate the value of the drag force coefficient, C_{drag}, in comparison with experimental data in Table 2, [42]. The largest deviation of the results from the experimental data are observed in the computation with limiter ε (33) at $K = 0.001$ for both modes. So, the modified form of the limiter allows for producing a more accurate solution in case the flow with shock wave processes is considered on an unstructured mesh.

Table 2. Drag coefficient.

No	Computation	$M = 1.05$		$M = 1.6$	
		C_{drag}	ΔC_{drag}, %	C_{drag}	ΔC_{drag}, %
1	Experiment	0.449	–	0.385	–
2	ε (33) at $K = 0.001$	0.4589	2.2	0.4219	9.6
3	$\dot{\varepsilon}$ (35) at $K = 0.01$	0.4524	0.7	0.4121	7.1

7. Conclusions

The study analyses implementation of the system of Navier–Stokes equations to simulate the problems of aerodynamics. A mathematical model is supplemented with equations of the Spalart–Allmaras turbulence model and AUSM+ scheme to compute convective fluxes that use a second-order approximation scheme for reconstruction of the solution on the facet. The Venkatakrishnan limiter is implemented to prevent the generation of false oscillations of the solution when computing shock wave processes.

The work describes the research on the influence of this limiter on the behaviour of the numerical solution of aerodynamic problems. It shows that implementation of the gradient limiter helps to improve monotonicity of the convergence when simulating the problems with shock waves and local compression shocks. Venkatakrishnan limiter implementation was studied on structured and unstructured meshes. It was found out that the initial option of the Venkatakrishnan limiter on unstructured meshes can produce the regions of its random activation. To eliminate this phenomenon, it was suggested to modify the summand that controls the limiter operation threshold. The suggested option is based on the flow quantities, and it showed correct behaviour on unstructured meshes that helps to improve the accuracy of the produced solution.

The proposed option of the limiter allows for producing a more accurate solution for the problems of trans- and supersonic flows, and that was demonstrated on the example of overflowing an airfoil and a bullet.

Computations on hypersonic flows are planned, as well the study of their peculiarities.

Author Contributions: Conceptualization, A.S.K. and D.Y.S.; data curation, A.V.S. and R.N.Z.; formal analysis, A.S.K., R.N.Z. and K.N.V.; investigation, A.S.K., A.V.S., D.Y.S. and R.N.Z.; methodology, A.S.K. and R.N.Z.; software, A.V.S. and R.N.Z.; supervision, A.S.K.; validation, A.V.S.; visualization, A.V.S. and K.N.V.; writing—original draft, A.S.K. and D.Y.S. All authors have read and agreed to the published version of the manuscript.

Funding: The work was supported by the program for the creation and development of the World-Class Research Center "Supersonic" for 2020–2025 funded by the Ministry of Science and Higher Education of the Russian Federation (Grant agreement of 20 April 2022 № 075-15-2022-309).

Acknowledgments: The results have been obtained with financial support from the Science & Universities National Project under the Young Scientists Lab Program of the RF Ministry of Education and Science # FSWE-2021-0009 (Research Topic: Development of CFD methods, models and algorithms to simulate liquids and gases in natural and industrial environments under normal and critical conditions on petascale supercomputers) and the Council of the Grants of the President of the Russian Federation for state support of Leading Scientific Schools of the Russian Federation (Grant No. NSH-70.2022.1.5).

Conflicts of Interest: The authors declare no conflict of interest. All authors have read and agreed to the published version of the manuscript.

References

1. Kozelkov, A.S.; Krutyakova, O.L.; Kurulin, V.V.; Lashkin, S.V.; Tyatyushkina, E.S. Application of numerical schemes with singling out the boundary layer for the computation of turbulent flows using eddy-resolving approaches on unstructured grids. *Comput. Math. Math. Phys.* **2017**, *57*, 1036–1047. [CrossRef]
2. Kozelkov, A.S.; Lashkin, S.V.; Efremov, V.R.; Volkov, K.N.; Tsibereva, Y.A.; Tarasova, N.V. An implicit algorithm of solving Navier–Stokes equations to simulate flows in anisotropic porous media. *Comput. Fluids* **2018**, *160*, 164–174. [CrossRef]
3. Kumar, V.; Sharma, A.; Singh, R.K. Central upwind scheme based immersed boundary method for compressible flows around complex geometries. *Comput. Fluids* **2020**, *196*, 104349. [CrossRef]
4. Pirozzoli, S. Numerical methods for high-speed flows. *Annu. Rev. Fluid Mech.* **2011**, *43*, 163–194. [CrossRef]
5. Du, X.; Corre, C.; Lerat, A. A third-order finite-volume residual-based scheme for the 2D Euler equations on unstructured grids. *Comput. Phys.* **2011**, *230*, 4201–4215. [CrossRef]
6. Zhu, W.; Xiao, Z.; Fu, S. Studies of flight-velocity effects on near-field and intermittent properties of a subsonic jet. *Comput. Fluids* **2020**, *196*, 104351. [CrossRef]
7. Van Albada, G.D.; Van Leer, B.; Roberts, W.W., Jr. A comparative study of computational methods in cosmic gas dynamics. *Astron. Astrophys.* **1982**, *108*, 76–84.
8. Venkatakrishnan, V. Convergence to steady state solutions of the Euler equations on unstructured grids with limiters. *J. Comput. Phys.* **1995**, *118*, 120–130. [CrossRef]
9. Kozelkov, A.S.; Kurulin, V.V.; Lashkin, S.V.; Shagaliev, R.M.; Yalozo, A.V. Investigation of Supercomputer Capabilities for the Scalable Numerical Simulation of Computational Fluid Dynamics Problems in Industrial Applications. *Comput. Math. Math. Phys.* **2016**, *56*, 1506–1516. [CrossRef]
10. Kozelkov, A.S.; Pogosyan, M.A.; Strelets, D.Y.; Tarasova, N.V. Application of mathematical modeling to solve the emergency water landing task in the interests of passenger aircraft certification. *AS* **2021**, *4*, 75–89. [CrossRef]
11. Kozelkov, A.S. The Numerical Technique for the Landslide Tsunami Simulations Based on Navier-Stokes Equations. *J. Appl. Mech. Tech. Phys.* **2017**, *58*, 1192–1210. [CrossRef]
12. Tyatyushkina, E.S.; Kozelkov, A.S.; Kurkin, A.A.; Pelinovsky, E.N.; Kurulin, V.V.; Plygunova, K.S.; Utkin, D.A. Verification of the LOGOS Software Package for Tsunami Simulations. *Geosciences* **2020**, *10*, 385. [CrossRef]
13. Landau, L.D.; Lifshits, V.M. Theoretical physics. In *Hydrodynamics*; Nauka: Moscow, Russia, 1988; Volume IV, 736p.
14. Loytsyanskiy, L.G. *Mechanics of Fluids and Gases*; Nauka: Moscow, Russia, 1979; 904p.
15. Shlikhting, G. *Theory of a Boundary Layer*; Nauka: Moscow, Russia, 1974; 712p.
16. Spalart, P.R.; Allmaras, S.R. *A One-Equation Turbulence Model for Aerodynamic Flows*; No. 0439; AIAA Paper: Reno, NV, USA, 1992.
17. Shur, M.; Strelets, M.; Travin, A.; Spalart, P.R. Turbulence modeling in rotating and curved channels: Assessment of the Spalart—Shur correction term. *AIAA J.* **2000**, *38*, 784–792. [CrossRef]

18. Deryugin, Y.N.; Zhuchkov, R.N.; Zelenskiy, D.K.; Kozelkov, A.S.; Sarazov, A.V.; Kudimov, N.F.; Lipnickiy, Y.M.; Panasenko, A.V.; Safronov, A.V. Validation Results for the LOGOS Multifunction Software Package in Solving Problems of Aerodynamics and Gas Dynamics for the Lift-Off and Injection of Launch Vehicles. *Math. Model. Comput. Simul.* **2015**, *7*, 144–153. [CrossRef]
19. Kozelkov, A.S.; Struchkov, A.V.; Strelets, D.Y. Two Methods to Improve the Efficiency of Supersonic Flow Simulation on Unstructured Grids. *Fluids* **2022**, *7*, 136. [CrossRef]
20. Roe, P.L. Characteristic Based Schemes for the Euler Equations. *Annu. Rev. Fluid Mech.* **1986**, *18*, 337–365. [CrossRef]
21. Liou, M.-S. A Sequel to AUSM: AUSM+. *J. Comput. Phys.* **1996**, *129*, 364–382. [CrossRef]
22. Rodionov, A.V. Artificial viscosity to cure the carbuncle phenomenon: The three dimensional case. *J. Comput. Phys.* **2018**, *361*, 50–55. [CrossRef]
23. Kotov, D.V.; Surzhikov, S.T. Calculation of Viscous and Inviscid Gas Flows on Unstructured Grids Using the AUSM Scheme. *Fluid Dyn.* **2011**, *46*, 809–825. [CrossRef]
24. Shingo, M. Performance of all-speed AUSM-family schemes for DNS of low Mach number turbulent channel flow. *Comput. Fluids* **2014**, *91*, 130–143.
25. Jan, V.; Bart, M.; Erik, D. Blended AUSM+ Method for All Speeds and All Grid Aspect Ratios. *AIAA J.* **2001**, *39*, 2278–2282.
26. Kim, K.H.; Ch, K.; Rho, O.-H. Methods for the accurate computations of hypersonic flows. I AUSMPW+ scheme. *J. Comput. Phys.* **2001**, *174*, 38–80. [CrossRef]
27. Ferziger, J.H.; Peric, M. *Computational Method for Fluid Dynamics*; Springer: New York, NY, USA, 2002.
28. Jasak, H. Error Analysis and Estimation for the Finite Volume Method with Applications to Fluid Flow. Ph.D. Thesis, Department of Mechanical Engineering, Imperial College of Science, London, UK, 1996.
29. Sweby, P.K. High resolution using flux limiters for hyperbolic conservation laws. *SIAM J. Numer. Anal.* **1984**, *21*, 995–1011. [CrossRef]
30. Darwish, M.S.; Moukalled, F. TVD schemes for unstructured grids. *Int. J. Heat Mass Transf.* **2003**, *46*, 599–611. [CrossRef]
31. Balcázar, N.; Jofre, L.; Lehmkuhl, O.; Castro, J.; Rigola, J. A finite-volume/level-set method for simulating two-phase flows on unstructured grids. *Int. J. Multiph. Flow* **2014**, *64*, 55–72. [CrossRef]
32. Li, L.-X.; Liao, H.-S.; Qi, L.-J. An improved r-factor algorithm for TVD schemes. *Int. J. Heat Mass Transf.* **2008**, *51*, 610–617. [CrossRef]
33. Hou, J.; Simons, F.; Hinkelmann, R. Improved total variation diminishing schemes for advection simulation on arbitrary grids. *Int. J. Numer. Meth. Fluids* **2012**, *70*, 359–382. [CrossRef]
34. Denner, F.; van Wachem, B. TVD differencing on three-dimensional unstructured meshes with monotonicity-preserving correction of mesh skewness. *J. Comput. Phys.* **2015**, *298*, 466–479. [CrossRef]
35. Balcázar, N.; Antepara, O.; Rigola, J.; Oliva, A. A level-set model for mass transfer in bubbly flows. *Int. J. Heat Mass Transf.* **2019**, *138*, 335–356. [CrossRef]
36. Gradient Computation. Available online: http://www.cfd-online.com/Wiki/Gradient_computation (accessed on 25 August 2019).
37. Barth, T.J.; Jespersen, T.J. *The Design and Application of Upwind Schemes on Unstructured Grids*; No. 89-0366; AIAA Paper: Reno, NV, USA, 1989.
38. Venkatakrishnan, V. *On the Accuracy of Limiters and Convergence to Steady State Solution*; No. 93-0880; AIAA Paper: Reno, NV, USA, 1993.
39. Povkh, I.L. *Technical Hydromechanics*; Mashinostroenie: Leningrad, Russia, 1976; 504p.
40. Kolesnikov, G.A.; Markov, V.K.; Mikhaykyuk, A.A. *Aerodynamics of Flying Vehicles*; Mashinostroeniye: Moscow, Russia, 1993; 544p.
41. Pilipenko, A.A.; Polevoy, O.B.; Prihodko, A.A. Numerical simulation of Mach number and attack angle as they influence the modes of trans-sonic turbulent overflowing of aerodynamic airfoils. *Tech. Notes TsAGI* **2012**, *XLIII*, 1–110.
42. Chang, P. *Separated Flow*; Mir: Moscow, Russia, 1973; Volume 3, 335p.

fluids

Article

Numerical Simulation of Pressure Fluctuation near an Expansion Corner in a Supersonic Flow of M = 3.01

Lei Zhang † and Zi-Niu Wu *,†

Department of Engineering Mechanics, Tsinghua University, Beijing 100084, China; lei-zhan19@mails.tsinghua.edu.cn
* Correspondence: ziniuwu@tsinghua.edu.cn
† These authors contributed equally to this work.

Abstract: The influence of the expansion corner on pressure fluctuation is an important subject in supersonic flow around high-speed vehicles. Past studies have clarified how the expansion corner alters the root-mean-square of the fluctuating pressure coefficient (Cp_{rms}) and the power spectral density (PSD) without considering how these fluctuating properties are related to compressible waves. In this paper, we use characteristics to determine the direction of wave propagation and identified three zones—U-zone, M-zone and D-zone—within which both Cp_{rms} and PSD are likely to display different behaviors across the boundary layer. The U-zone is upstream of the characteristic line of the second family and passing through the corner. The D-zone is downstream of the characteristic line of the first family and passing through the corner. The middle zone lies between the U-zone and D-zone. The results of Cp_{rms} and PSD at different layers within the boundary layer are obtained using numerical computation through a Detached Eddy Simulation (DES). It is found that in the U-zone and D-zone, both Cp_{rms} and PSD are the same in different layers within the boundary layer. In the M-zone, however, both Cp_{rms} and PSD may vary in different layers and this variation occurs in the high-frequency band upstream of the corner and mid-frequency band downstream of the corner. A feedback mechanism is tentatively used to explain the difference of spatial distribution of fluctuation properties inside the M-zone.

Keywords: fluctuating pressure; expansion corner; supersonic flow; characteristics

Citation: Zhang, L.; Wu, Z.-N. Numerical Simulation of Pressure Fluctuation Near an Expansion Corner in a Supersonic Flow of M = 3.01. *Fluids* **2021**, *6*, 268. https://doi.org/10.3390/fluids6080268

Academic Editor: Olga A. Azarova

Received: 1 July 2021
Accepted: 23 July 2021
Published: 28 July 2021

1. Introduction

The fluctuating pressure acting on the surface of the vehicle induces structural vibration, which may cause damage to the aircraft structure and a strong noise environment that affects the normal operation of airborne instruments, including the reliability and safety of weapons and equipment [1]. The intensity of the pressure fluctuations is commonly characterized by the root-mean-square fluctuating pressure coefficient (Cp_{rms}) , Power Spectral Density (PSD) and correlation coefficient [2]. Pressure fluctuation beneath a supersonic turbulent boundary layer (c.f. S. Beresh and J. Henfling and R. Spillers and B. Pruett [2]) may be amplified by shock–boundary layer interaction (c.f. M. Holden [3], H. Babinsky and J. Harvey [4]) and altered by geometry, such as forward step (V.Bibko and B. Efimtsov and V. Kuznetsov [5]) and expansion corners (c.f. [6]). Here in this paper, we consider pressure fluctuations due to expansion corners, which are typical geometric configurations that change the local flow properties, leading to the formation of a complex fluctuating environment [6]. This topic has received a number of studies.

Fluctuating pressure for subsonic flow around expansion corners has been studied experimentally by J. Robertson [7], who measured the surface fluctuating pressure in the vicinity of the expansion corner and found that the peak of the fluctuating pressure exists at the reattachment point, and this point moves backwards when the Mach number increases. Moreover, the fluctuating pressure of the expansion corner has been studied analytically by K. Plotkin and J. Roberson [6] based on the experimental data on the wall, and they found the relationship between the Cp_{rms} and the local Mach number of the separated flow caused by several expansion corners (cone-cylinder, back step, etc.). They found that the Cp_{rms} is closely related to the local Mach number and that the fluctuating pressure is stronger at low Mach numbers and decreases at high Mach numbers. Later, X. Ligong and L. Zhenhuan [8] also used the same formula to predict the Cp_{rms} at the expansion corner, he took the characteristic length as the displacement thickness of the local boundary layer and obtained results that are more consistent with the previous experiment. Furthermore, the surface pressure fluctuations of subsonic turbulent flow downstream of small expansion corners were found to be normally distributed through the expansion process but were severely attenuated [9]. In addition, the fluctuating pressure on an expansion corner has been investigated numerically using the Detached Eddy Simulation (DES) method by S. Deck and P. Thorigny [10]. They focused on the surface fluctuating pressure statistical properties and found that the spectral analysis of the pressure fluctuations has shown different frequency contributions depending on the location considered in the recirculation bubble. Close to the expansion corner, the spectrum displays a peak near a normalized frequency, which has been shown to represent the footprint of the vertical motion of the bubble.

Fluctuating pressure for transonic flow around expansion corners has been studied experimentally by D. Depres and P. Reijasse and J. Dussauge [11]. The experiments have been carried out to investigate the unsteady fluctuating surface-pressure field on afterbodies of revolution at transonic speeds with a freestream Mach number 0.85, and it was found that the spectra of fluctuating pressure in the whole region exhibit a well-defined periodicity that corresponds to the formation of large-scale structures in the wake, and when the shear layer reattaches near the end of the protruding wall, the rms pressure fluctuations reach their maximum value. Moreover, Z. Rui, R. Jili and R. Fang [12] used the large eddy simulation (LES) method to simulate the structure of a typical expansion corner numerically, and they recorded data on how pressure changes over time on the surface. Additionally, they found that the fluctuating pressure in the turbulent boundary layer, separation zone and other flow structures changes significantly with space; the fluctuating pressure in the separation zone is generally higher than the turbulent boundary layer; and the separation reaction on the shock wave will cause a stronger fluctuating pressure environment and shock waves to self-oscillate [13]. Then, based on previous experimental data, they provided an empirical formula for the PSD of the fluctuating pressure of the expansion-reflection separation flow. Combining the calculation results of LES, they introduced spatial-related information based on the empirical formula published by K. Plotkin and J. Roberson [6], which improved the prediction accuracy of engineering algorithms for fluctuating pressure [14].

Fluctuating pressure for supersonic flow around expansion corners has been studied experimentally by C. Kungming and F. Lu [9]. In this work, surface pressure fluctuations of Mach 8 turbulent flow past a 2.5- and a 4.25-deg expansion corner maintained a Gaussian distribution but were severely attenuated by the expansion process. The pressure fluctuations did not recover to those of an equilibrium turbulent flow even though the mean pressure reached downstream inviscid values in four to six boundary-layer thicknesses. The fluctuations were convected with a velocity comparable to that on a flat plate, and they maintained their identities longer for the stronger expansion. The damping of pressure fluctuations at hypersonic Mach numbers, even by small corner angles, may be exploited in fatigue design. Moreover, fluctuating pressure near expansion corners has been studied experimentally by J. Dawson and M. Samimy [15], wherein multipoint wall pressure measurements were used to investigate the response of a Mach 3.01, fully developed, compressible, turbulent boundary layer to centered and gradual expansions of both 7- and 14-deg deflection. Furthermore, they found that although rms fluctuation levels decrease across the expansions, the rms normalized by the local static pressure remains nominally constant. Just downstream of the expansions, normalized power spectra are more concentrated at low frequencies than upstream, suggesting small-scale turbulence is quenched. This spectra alteration is more prominent for centered expansions and larger deflections.

The spectra evolve very quickly after the centered expansions and very slowly after the gradual expansions. In the area near the expansion corner, there are separation points and reattachment points, a shock wave will appear at each location, and the fluctuating pressure environment will be more complicated [16].

K. Plotkin and J. Roberson [6] analyzed the typical subsonic, transonic and supersonic flow field and its fluctuating pressure of the reentry body with two shapes: cone-cylinder and cone-cylinder-skirt. They pointed out that at subsonic speeds, fluctuating pressure is caused by the appearance of the separation zone, and at transonic speeds, the airflow behind the cone-cylinder shoulder may reach supersonic speed, and shock waves will induce the separation of the boundary layer and cause strong fluctuating pressure. At supersonic speeds, there is a certain degree of separation after the expansion corner, and the fluctuating pressure environment will be more complicated.

Past studies have focused on fluctuating pressure on the surface. Here in this paper, we consider the spatial distribution of fluctuation properties in the vicinity of the expansion corner when the flow is supersonic. The problem we will use is the experimental model of J. Dawson and M. Samimy [15]. In Section 2, we will give the details of this model and present the method we use to obtain the pressure fluctuation properties. We will use characteristic lines to split the region near the corner into three zones—the U-zone, M-zone and D-zone—and study how Cprms and PSD change across the boundary layer in different zones. The description of characteristics and definition of the U-zone, M-zone and D-zone will be given in Section 3. The computed spatial distribution of Cprms and PSD in three zones around the expansion corner is presented in Section 4. The behavior of spatial distribution of fluctuating properties in the three zones will be summarized in Section 5, where we tentatively provide a possible feedback mechanism to explain the observed behavior. The conclusion will be summarized in Section 6.

2. The Expansion Corner Problem and Numerical Method for Simulation

In this paper, the experimental model of J. Dawson and M. Samimy [15], with an expansion of 14 deg, as shown in Figure 1, is used. The upstream flow is supersonic, with a freestream Mach number $Ma_\infty = 3.01$. The Reynolds number based on the momentum thickness ($\delta_2 = 0.37$ mm) is $Re_{\delta_2} = 24{,}700$. The incoming pressure is $p_\infty = 22$ KPa. J. Dawson and M. Samimy [15] measured the normalized rms pressure fluctuations and power spectra on the body surface.

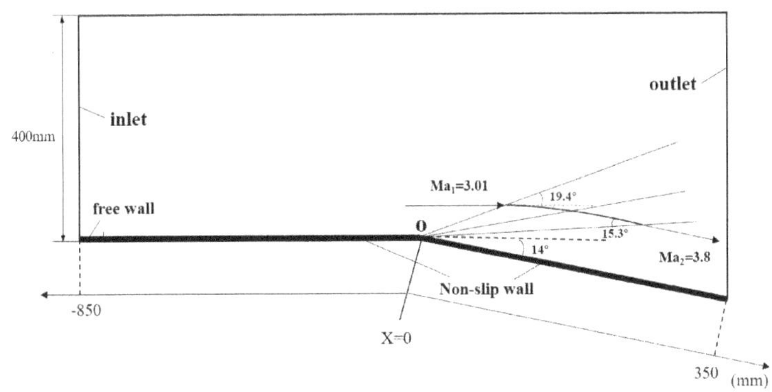

Figure 1. A model of the expansion probelm. The Mach angles $\mu_1 = arc\sin\frac{1}{Ma_1} = 19.4°$ and $\mu_2 = arc\sin\frac{1}{Ma_2} = 15.3°$ are marked in the figure.

Since we are interested in spatial distribution of the pressure fluctuation properties, we need numerical simulation to obtain these properties. Note that pressure fluctuation for flow around an expansion corner has been considered numerically using Reynolds Averaged Navier–Stokes (e.g., R. Soni and N. Arya and A. De [17]), Detached Eddy Simulation (DES) (e.g., S. Deck and P. Thorigny [10]), large eddy simulation (LES) (e.g., M. Grilli and S. Hickel and N. Adams [18]), and direct numerical simulation (DNS) (e.g., M. Kopera and R. Kerr and H. Blackburn and D. Barkley [19]).

To balance the accuracy and time requirement, we use the well-established DES method, which combines the features of the classical RANS formulations [20] with elements of LES method.

The domain and boundary conditions are shown in Figure 1. The left side is supersonic inlet, and the right boundary is a supersonic exit. The wall has a no-slip condition. A small segment of the free-sliding wall is set at the front edge of the non-slip wall. The calculation results are compared with J. Dawson and M. Samimy [15] in the paper to verify the accuracy and reliability of the CFD calculation method we chose.

The flow equation is solved by means of the second-order-accurate scheme of implicit Roe using finite difference simulation. The time step is fixed to $\Delta t = 5e^{-7}$ s, which corresponds to 10^4 time steps for one passage of the model at freestream velocity. The freestream turbulent in tensity is $Tu = 3\%$, and there are three grids containing 7.5, 10.36 and 23.6 million nodes for computation, respectively. Several grid spaces have been considered in successive refinement to study the influence of the mesh density on the fluctuating pressure of the expansion corner. Based on the results of analysis, a grid of 10.36 million nodes with refinement inside the boundary and near the expansion corner is finally used. The grid has a wall normal resolution $y^+ \sim 0.5$ and $x^+ \sim 0.5$ in the densified area (similarly as in the work of [21]), and the increasing rate is 1.1. A study on the time independence is performed and the results of fluctuating pressure at four different times are given in Figure 2. In the early stage of the calculation, the PSD keeps changing with time. Until 0.37 s, the PSD does not vary in time. Thus, we will extract the data of 0.45 s.

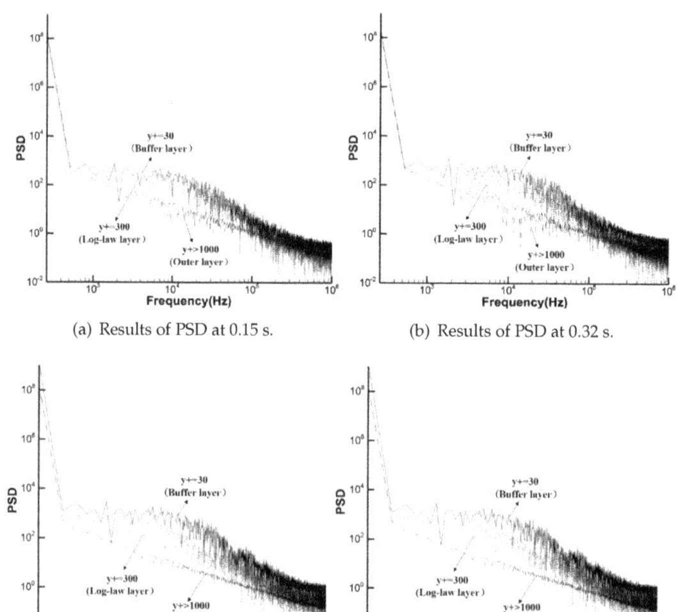

(a) Results of PSD at 0.15 s.

(b) Results of PSD at 0.32 s.

(c) Results of PSD at 0.37 s.

(d) Results of PSD at 0.45 s.

Figure 2. Time-independent check.

As displayed in Figure 3, the numerical results of static pressure are comparable with the experimental results of J. Dawson and M. Samimy [15].

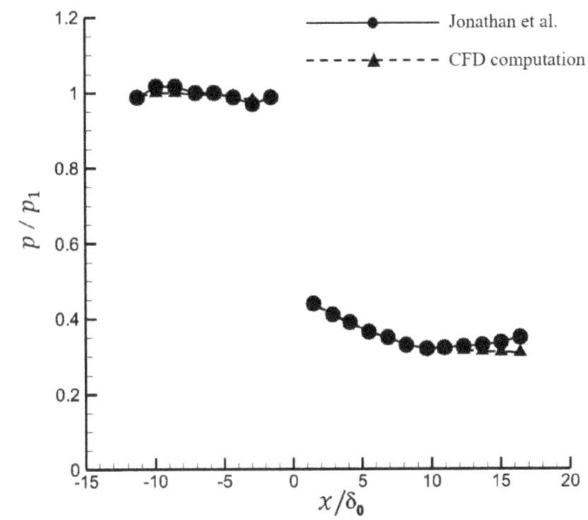

Figure 3. Normalized pressure p along the flow direction on the wall (p_1 is the pressure in the freestream).

As displayed in Figure 4, the numerical results of the normalized fluctuating pressure on the wall are comparable with the experimental results of J. Dawson and M. Samimy [15]. In Figure 4, x is the position of flow direction, $\sqrt{\overline{p}^2}$ is the root-mean-square fluctuating pressure, p is the static pressure, $x/\delta_0 = 0$ is the location of the expansion corner, and the thickness of the boundary layer δ_0 at the starting point of the expansion is 9.2 mm.

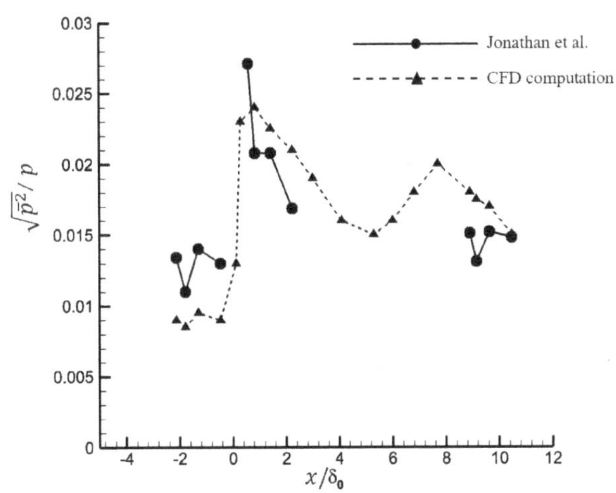

Figure 4. The distribution of normalized fluctuating pressure along the flow direction.

Here, we use a Detached Eddy Simulation (DES) to obtain the instantaneous flow field. The instantaneous pressure $p(t)$ at a given point in the flow field (not just on the body surface) is recorded, and Cp_{rms} is computed as $Cp_{rms} = \frac{\sqrt{\overline{p}^2}}{0.5\rho_\infty U_\infty^2}$, $\sqrt{\overline{p}^2} = \sqrt{\frac{1}{T}\int_0^T (p(t) - p_{avg})dt}$, where T is the time interval for sampling, p_{avg} is the average pressure during the entire sampling process, and $\sqrt{\overline{p}^2}$ is the root-mean-square (rms) of the pressure fluctuation, which represents the total intensity of pressure. The power spectral density (PSD) (named $\phi(\omega)$) is obtained by the Fourier transform of the pressure change in the time domain $\phi(\omega) = \frac{1}{T}\int_0^T p(t)e^{i\omega t}dt$, where ω is the frequency (Hz).

3. Characteristics and Definition of U-Zone, M-Zone and D-Zone

Characteristics are well defined for inviscid and isentropic flow. Outside the boundary layer, the flow may be regarded as inviscid and isentropic. For inviscid and isentropic flow, the characteristic line is defined by

$$\frac{dy}{dx} = \frac{1}{\lambda} \tag{1}$$

where λ is the eigenvalue. For the characteristic line of the first family, $\lambda = \lambda_1$, and for characteristic line of the second family, $\lambda = \lambda_2$, where

$$\begin{cases} \lambda_1 = \dfrac{uv - a^2\sqrt{Ma^2 - 1}}{v^2 - a^2} \\ \lambda_2 = \dfrac{uv + a^2\sqrt{Ma^2 - 1}}{v^2 - a^2} \end{cases} \tag{2}$$

Here, u and v are the local flow velocity components; Ma is the local Mach number; a is the local sound speed.

One way to obtain the characteristic lines is to use the local flow parameters obtained by inviscid CFD and then use Equations (1) and (2) to integrate the characteristic lines. Here, we use the local flow parameters from CFD results and apply Equations (1) and (2) to find the characteristic lines as if the flow is inviscid. Note that inside the boundary layer and very close to the wall, $Ma < 1$ so that we have no real values of λ_1 and λ_2. In such regions, we simply extend the characteristic lines to the wall using straight lines.

The characteristic lines and Mach contours in the vicinity of the expansion corner are displayed in Figure 5. Not surprisingly, the Mach number is constant along the characteristic line of the second family and connecting the expansion corner.

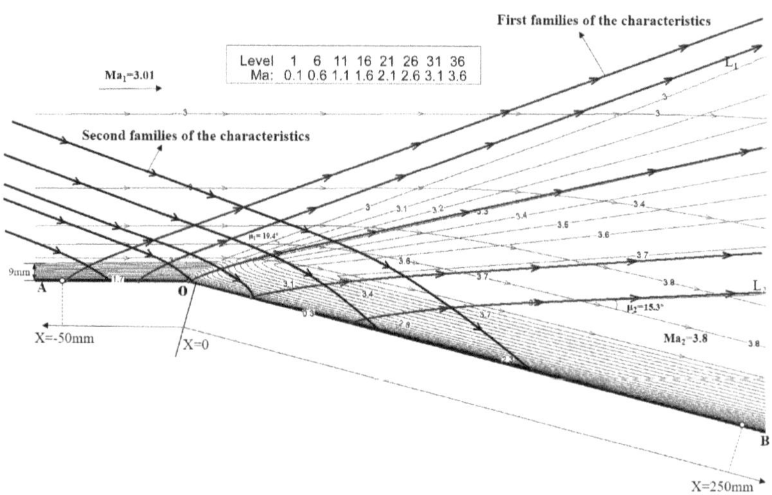

Figure 5. Characteristics in the Mach number contour map.

Now we define three zones using the two characteristic lines of the first and second families connecting the expansion corner O. These two characteristic lines, "C_U"and "C_D", are shown in Figure 6. The intersection of C_U and the edge of the boundary layer is denoted by "U"; the intersection of C_D and the edge of the boundary layer are denoted by "D".

The region near the expansion corner is divided into three zones:

(1) U-zone is the region upstream of the characteristic line C_U.
(2) M-zone is the region between the characteristic lines C_U and C_D.
(3) D-zone is the region downstream of the characteristic line C_D.

In the next section, we will display how Cp_{rms} and PSD vary across the boundary layer inside these three zones.

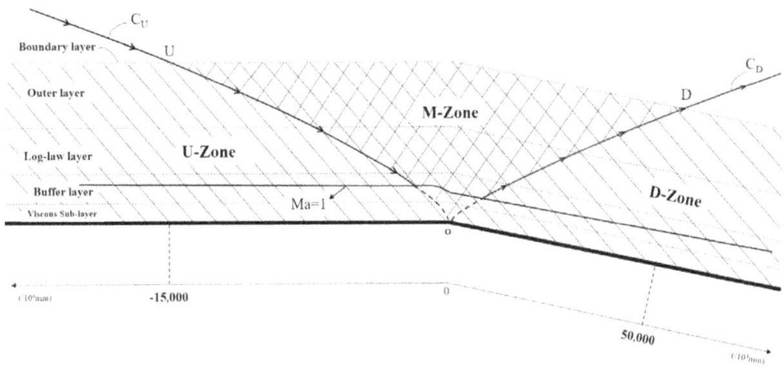

Figure 6. The U-zone, M-zone and D-zone.

The distribution of the measuring points is shown in Figure 7a,b. Measuring points are evenly distributed in U-zone, D-zone, M-zone, and along the vertical direction of the wall surface. The measuring points are arranged at different layers within the boundary layer in these three zones:

(a) wall surface, with $y^+ = 0$ and labelled point "1".
(b) buffer layer, with $y^+ \approx 30$ and labelled point "2".
(c) log-law layer, with $y^+ \approx 300$ and labelled point "3".
(d) outer layer, with $y^+ > 1000$ and labelled point "4".

For each layer, position "o" is at the expansion corner, positions "o−", "a", "b", "c", "d", "e", "f", "g", "h", "i", "j" and "k" are upstream of the expansion corner, and positions "o+", "l", "m", "n", "o", "p", "q", "r", "s", "t", "u", "v", "w", "x", "y" and "z" are downstream of the expansion corner. The subscripts 1, 2, 3 and 4 for these points correspond to the four layers.

For positions "j" and "k", all the four layer points (1, 2, 3, 4) are upstream of C_U and are inside the U-zone. For positions "y" and "z", the four layer points (1, 2, 3, 4) are downstream of C_D and are inside the D-zone.

Usually, the pressure fluctuation properties on the wall surfaces were considered. Here, we consider these properties at the four different layers. They have the same values at these four layers at a fixed horizontal location in the case of a pure boundary layer. We wonder if they may have different values in the vicinity of the expansion corner, and this will become clear in the next section.

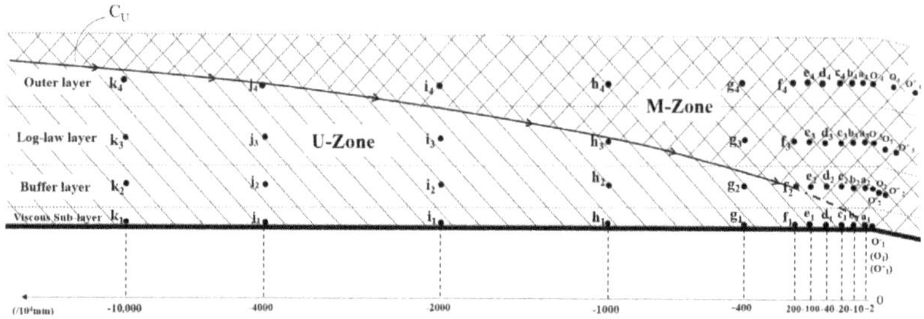

(a) The distribution of measuring points in the upstream area of the expansion corner.

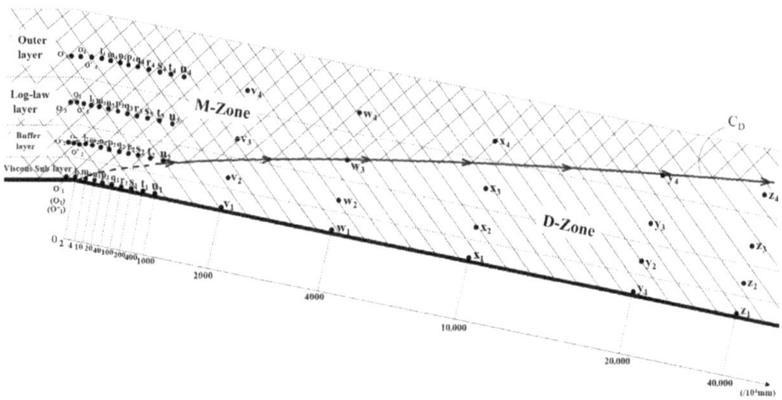

(b) The distribution of measuring points in the downstream area of the expansion corner.

Figure 7. The distribution of measuring points.

4. The Cp_{rms} and PSD in the U-Zone, M-Zone and D-Zone

Now we display the pressure fluctuation properties Cp_{rms} and PSD at the four different layers (wall, buffer layer, log-law layer and outer layer) in the U-zone, M-zone and D-zone.

4.1. The Distribution of Cp_{rms} in Various Zones

The distribution of Cp_{rms} at the four layers are displayed in Figure 8. The abscissa in Figure 8 corresponds to the different positions in Figure 7a,b. The ordinate corresponds to the value of Cp_{rms}. The four curves correspond to the four layers.

Looking at the positions of the points, as shown in Figure 7a,b, we observe that for points below the characteristic line C_U, i.e., inside the U-zone, the Cp_{rms} is almost constant in the vertical direction, i.e., they are the same at different layers of the boundary layers. For points below the characteristic line C_D, i.e., inside the D-zone, the Cp_{rms} is almost constant in the vertical direction, i.e., they are the same at different layers of the boundary layers.

In the M-zone, the Cp_{rms} changes in different layers of the boundary layer. Inside the M-zone, the Cp_{rms} is the highest in the buffer zone (point "2") and is the lowermost in the outer layer (point "4").

In approaching the expansion corner from the upstream along the flow direction, starting from position i for the measuring points that lie between characteristic line "C_U" and characteristic line "C_D", the values of Cp_{rms} began to differ across the vertical direction of the boundary layer, and the difference gradually increased along the flow direction. After the expansion corner, the difference of Cp_{rms} gradually decreases at different layers in the direction normal to the surface within boundary layer. Until position x, this difference almost shrinks to zero and continues to develop downstream along the flow direction. The values of Cp_{rms} across the vertical direction of the boundary layer do not change.

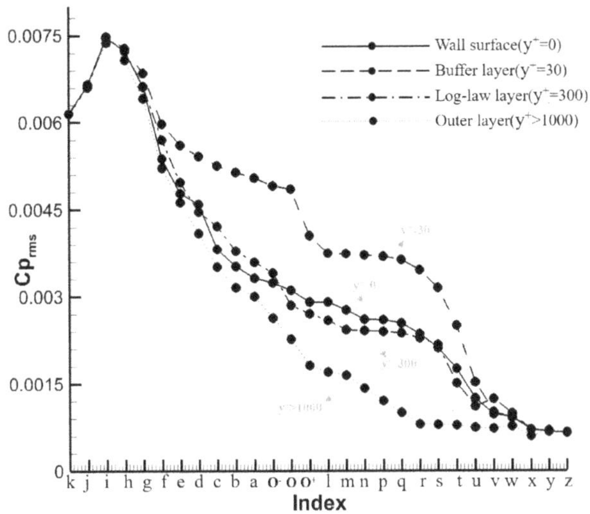

Figure 8. The distribution of Cp_{rms} in different layers within boundary layer near expansion corner.

4.2. The PSD in Various Zones

The distribution of the *PSD* at position "d~k" (sufficiently far upstream from the expansion corner) is displayed in Figure 9. For positions *k* and *j*, where all the four layer points are inside the U-zone, the curves of the *PSD* at $y^+ = 0$ (wall surface), $y^+ = 30$ (buffer layer), $y^+ = 300$ (log-law layer) and $y^+ > 1000$ (outer layer) almost overlap; that is, the frequency properties of fluctuating pressure do not change much in the vertical direction within the U-zone. For positions *i* and *h*, the values of *PSD* at $y^+ > 1000$ (outer layer) is lower than other layers in the high-frequency band, while for positions g~d, the values of *PSD* in the high-frequency band gradually decrease when the distance from the wall surface increases.

The distribution of *PSD* at positions "c~m" (in the vicinity of the expansion corner) is displayed in Figure 10. For positions c~o−, the *PSD* in the high-frequency band gradually decreases with the increasing distance from the wall surface. In approaching the expansion corner along the flow direction, the difference of PSD in the high-frequency band gradually increases, and at position o−, this difference reaches the maximum value. In return, along the streamwise direction starting from position o, the *PSD* becomes to show obvious differences in the mid-frequency band, and with increasing distance from the wall surface, the *PSD* in the mid-frequency band gradually decreases.

The distribution of *PSD* at positions "n~v" (sufficiently far downstream from the expansion corner) is displayed in Figure 11 and, at position "w~z", is displayed in Figure 12. For positions n~z, the *PSD* in the mid-frequency band gradually decreases with the increasing distance from the wall surface, and in the direction away from the expansion corner, the difference of PSD in the mid-frequency band gradually decreases. For positions *y* and *z*, where all the four layer points are inside the D-zone, the curves of *PSD* at $y^+ = 0$ (wall surface), $y^+ = 30$ (buffer layer), $y^+ = 300$ (log-law layer) and $y^+ > 1000$ (outer layer) almost overlap along the vertical direction of the wall surface within the boundary layer; that is, the frequency properties of fluctuating pressure do not change much in "D-zone".

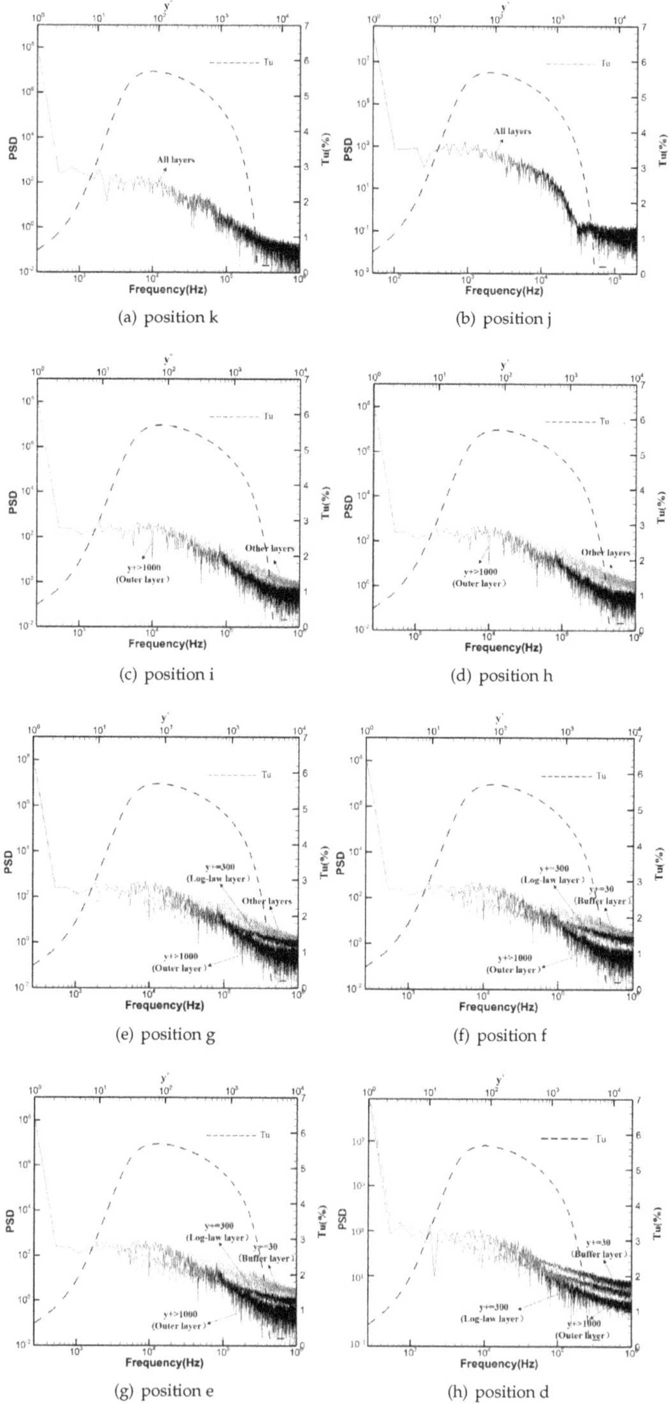

(a) position k (b) position j (c) position i (d) position h (e) position g (f) position f (g) position e (h) position d

Figure 9. PSD and turbulent intensity at positions "d~k".

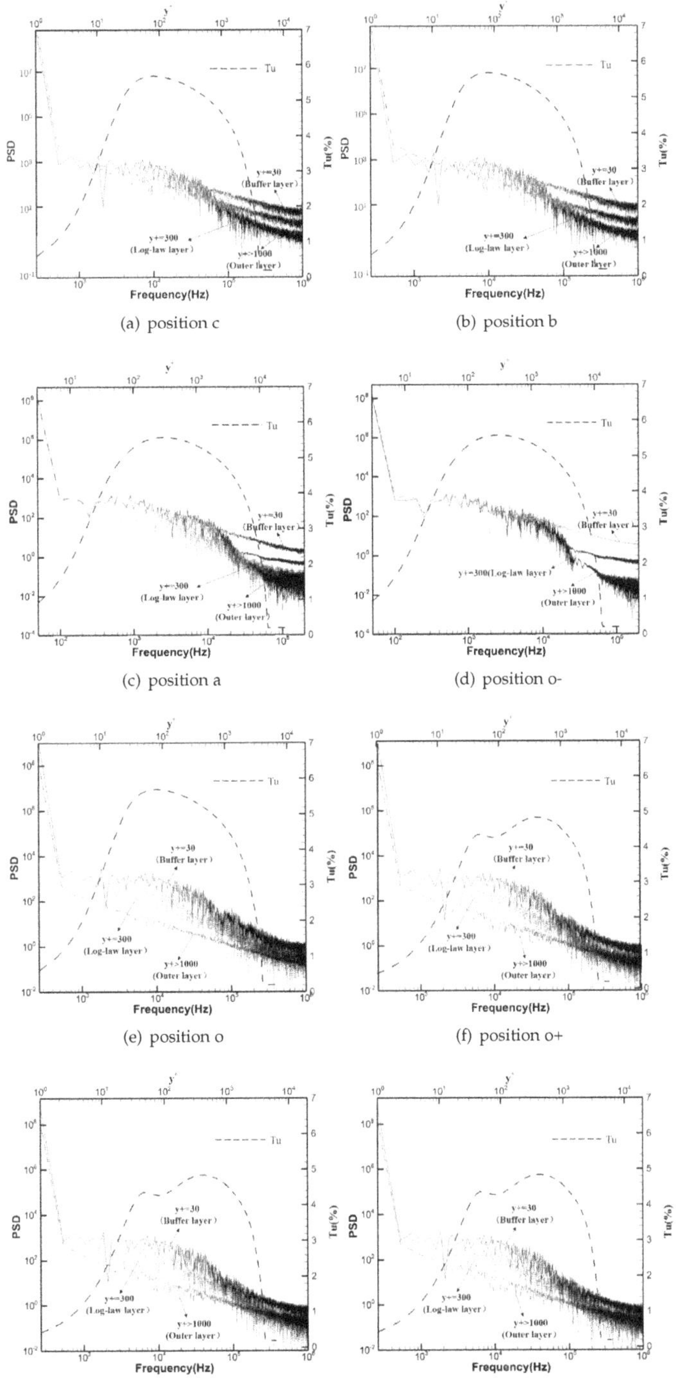

(a) position c
(b) position b
(c) position a
(d) position o-
(e) position o
(f) position o+
(g) position l
(h) position m

Figure 10. PSD and turbulent intensity at positions "c~m".

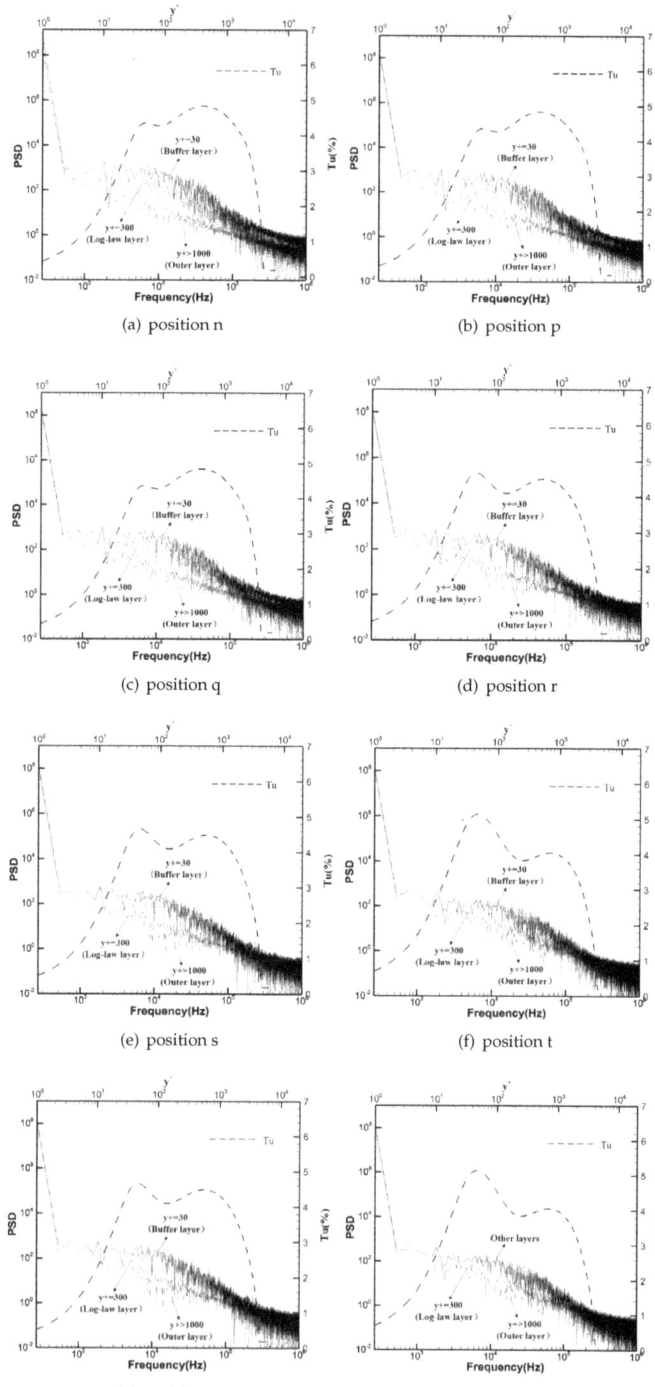

(a) position n

(b) position p

(c) position q

(d) position r

(e) position s

(f) position t

(g) position u

(h) position v

Figure 11. PSD and turbulent intensity at positions "n~v".

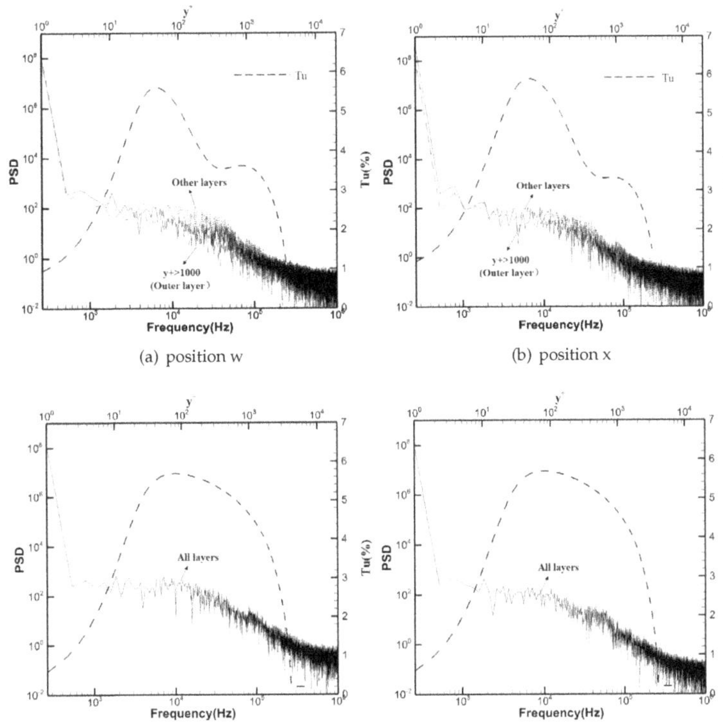

(a) position w (b) position x

(c) position y (d) position z

Figure 12. PSD and turbulent intensity at positions "w~z".

5. Summary of Fluctuating Properties and Feedback Mechanism

Based on the observation presented in Section 4, we use Figure 13 to summarize the distribution of the fluctuating properties in the vicinity of the expansion corner. There are three zones, U-zone, M-zone and D-zone, as defined in Section 3, using the characteristic lines C_U and C_D.

In the U-zone and D-zone, the two key parameters of fluctuating pressure—Cp_{rms} and PSD—are almost constant along the vertical direction of the wall surface within the boundary layer.

The M-zone can be divided into two subzones: M_1 zone is upstream of the expansion corner, and M_2 zone is downstream. The Cp_{rms} changes significantly along the vertical direction of the wall surface within boundary layer. In the M_1 zone, when approaching the expansion corner, the difference of the Cp_{rms} gradually increases along the vertical direction of the wall surface, and the PSD varies in the high-frequency band. In the M_2 zone, the difference of Cp_{rms} in each layer gradually decreases to zero when the distance from the expansion corner becomes large, and the variation of the PSD occurs in the mid-frequency band downstream of the corner. Moreover, in the upstream of the expansion corner, the difference of PSD in the high-frequency band gradually increases when approaching expansion corner and the difference of PSD in mid-frequency band gradually decreases to disappear along the flow direction.

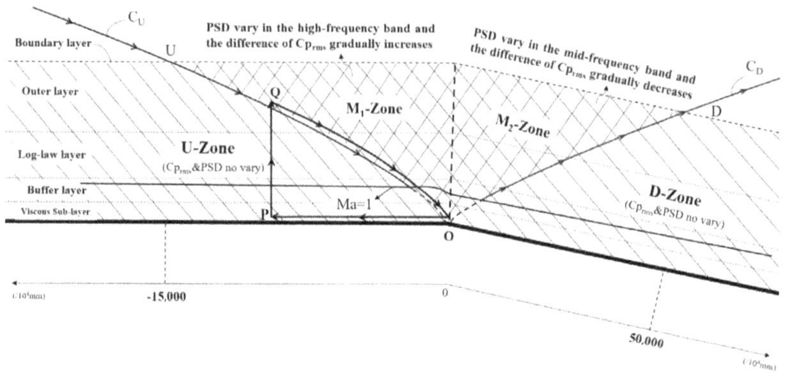

Figure 13. The pressure fluctuation in the U-zone, M-zone and D-zone .

It is desirable to understand why the fluctuation properties have different spatial distributions in the M-zone compared to the U-zone and D-zone. Here, we tentatively provide an explanation. For this purpose, we design a triangle (with one curved side) in Figure 13. Point O of this triangle is near the expansion corner. Point P is upstream so that the side OP lies inside the subsonic region within the boundary layer. The point Q is above and near the characteristic line C_U, such that PQ is perpendicular to the wall and the curved side QO is close and parallel to the characteristic line C_U.

The expansion corner is supposed to yield a perturbation of the pressure fluctuation, and this perturbation can propagate upstream along QP to the point P since OP is in the subsonic region within the boundary layer (information can propagate upstream in subsonic flow). This perturbation can then propagate to point Q along PQ. Since QO is along the propagation direction of the characteristic line C_U, the above-mentioned perturbation (originally from the expansion corner) can propagate back to O. This defines a feedback loop, which may explain why the pressure fluctuation properties change inside the M_1-zone compared to that inside the U-zone. The feedback mechanism can be similarly defined to explain the change of fluctuation properties in the M_2-zone compared to that inside the D-zone.

It is also interesting to see the correlation of instantaneous pressure at two different points once these points cross the characteristic line C_U or C_D.

The correlation coefficient r_{AB} of the instantaneous pressure at two different points A and B is

$$r_{AB} = \frac{cov(p_A(t), p_B(t))}{\sqrt{Var[p_A(t)] \cdot Var[p_B(t)]}} \tag{3}$$

where $p_A(t)$ is the instantaneous pressure at point A; $p_B(t)$ is the instantaneous pressure at point B; $cov(p_A(t), p_B(t))$ is the covariance of $p_A(t)$ and $p_B(t)$; $Var[p_A(t)]$ is the variance of $p_A(t)$; and $Var[p_B(t)]$ is the variance of $p_B(t)$.

The points A and B will be both inside zones U, M or D or in different zones among U, M and D. The correlation coefficients computed by Equation (3) are displayed in Figure 14, where three curves are displayed. For the wall surface and buffer layer curve, point A is in the wall layer and point B is in the buffer layer, for the buffer layer and log-law layer curve, point A is in the buffer layer, and point B is in the log-law layer. For the log-law layer and outer layer curve, point A is in the log-law layer, and point B is in the outer layer. The abscissa corresponds to different positions, and the ordinate corresponds to the correlation coefficient.

We observe that, inside the U-zone and D-zone, the correlation coefficient between any two layers is near 1.

If point A is inside the U-zone and point B is inside the M-zone, the correlation coefficient is less than 1. Consider, for instance, position c. At this position, the wall surface is in the "U-zone", and the buffer layer is in the "M-zone", so the correlation coefficient between wall surface and buffer layer is 0.96.

If both points are inside the M-zone, the correlation coefficient is much lower than 1. Consider, for instance, position c, the log-law layer and outer layer are in the "M-zone", so the correlation coefficient between buffer layer and log-law layer is 0.71, and the correlation coefficient between log-law layer and outer layer is 0.62.

If point A is inside the D-zone and point B is inside the M-zone, the correlation coefficient is less than 1. Consider, for instance, position n. At this position, the wall surface is in the "D-zone", and the buffer layer is in the "M-zone", so the correlation coefficient between wall surface and buffer layer is 0.94. Moreover, both the log-law layer and outer layer are in the "M-zone", so the correlation coefficient between buffer layer and log-law layer is 0.45, and the correlation coefficient between log-law layer and outer layer is 0.24.

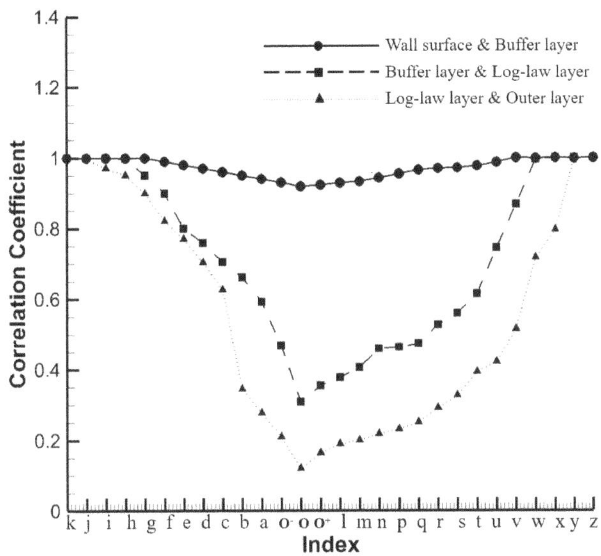

Figure 14. The distribution of the correlation coefficient between different layers near expansion corner.

6. Conclusions

In this work, the fluctuating pressure in the vicinity of the expansion corner in supersonic flow is studied using numerical results from DES simulation. We use characteristics (as if the flow is inviscid and isentropic) to determine the direction of wave propagation and identify three different zones — U-zone, M-zone and D-zone. These zones are bound by the characteristic lines of the first and second families and start from the expansion corner (in an approximative way).

In the U-zone, which is upstream of the upstream characteristic line connecting the expansion corner, both Cp_{rms} and PSD are almost constant along the vertical direction within the boundary layer, and the correlation coefficient between two vertical points within this zone is close to 1. In the D-zone, which is downstream of the downstream characteristic line connecting the expansion corner, both Cp_{rms} and PSD are almost constant along the vertical direction within the boundary layer, and the correlation coefficient between two vertical points within this zone is close to 1.

In the "M-zone", however, the values of Cp_{rms} differ across the vertical direction of the boundary layer, and the difference gradually increases along the flow direction in regions upstream of the expansion corner. Downstream of the expansion corner, the difference of

Cp_{rms} gradually decreases and almost shrinks to zero finally. Moreover, the variation of PSD occurs in the high-frequency band upstream of the corner and mid-frequency band downstream of the corner. The correlation of Cp_{rms} between two points with at least one point inside the M-zone is less than 1.

A feedback mechanism is tentatively used to explain the difference of spatial distribution of fluctuation properties inside the M-zone: the change of fluctuation properties by the expansion corner can propagate upstream from the subsonic region inside the boundary layer and then propagate vertically to the M-zone, where this perturbation can be brought back to the expansion corner.

Author Contributions: L.Z. and Z.-N.W. have equally contributed to the development of the present paper, including the conceptualization, methodology, data acquisition and analysis, interpretation of the results, and manuscript preparation. All authors have read and agreed to the published version of the manuscript.

Funding: This work was supported partly by the National Key Project (Grant No. GJXM92579) and the National Science and Technology Major Project 2017-II-003-0015.

Institutional Review Board Statement: Not applicable.

Informed Consent Statement: Not applicable.

Data Availability Statement: Not applicable.

Acknowledgments: We are grateful to all reviewers who provided valuable suggestions that improve the paper.

Conflicts of Interest: The authors declare no conflict of interest.

References

1. Shiwei, M.; Huabing, J. Discussion on prediction methods of fluctuating pressure environments of flow fields surrounding the aircraft. *Equip. Environ. Eng.* **2021**, *18*, 14–22.
2. Beresh, S.J.; Henfling, J.F.; Spillers, R.W.; Pruett, B.O.M. Fluctuating wall pressures measured beneath a supersonic turbulent boundary layer. *Phys. Fluids* **2011**, *23*, 75–110. [CrossRef]
3. Holden, M.S. Shock Wave-Turbulent Boundary Layer Interaction in Hypersonic Flow. *AIAA J.* **1972**, 1–30.
4. Babinsky, H.; Harvey, J.K. *Shock Wave-Boundary-Layer Interactions*; Cambridge University Press: Cambridge, UK, **2011**, 1–461.
5. Bibko, V.N.; Efimtsov, B.M.; Kuznetsov, V.B. Pressure fluctuations on a plate ahead of a step. *Uch. Zap. TsAGI* **1986**, *17*, 170–173
6. Plotkin, K.; Roberson, J. *Prediction of Space Shuttle Fluctuating Pressure Environments, Including Rocket Plume Effects*; NASA: San Jose, CA, USA, 1973; pp. 1–163
7. Robertson, J. *Wind Tunnel Investigation of the Effects of Reynolds Number and Model Size on the Steady and Fluctuating Pressures Experienced by Cone Cylinder Missile Configurations Attransonic Speeds*; AEDC: Ascension Parish, LA, USA, 1967; pp. 152–161
8. Ligong, X.; Zhenhuan, L. Analysis and prediction of fluctuating pressure environment of reentry aircraft. *Acta Aerodyn.* **1991**, *4*, 457–464.
9. Kungming, C.; Lu, F.K. Damping of surface pressure fluctuations in hypersonic turbulent flow past expansion corners. *AIAA J.* **1993**, *31*, 1229–1234.
10. Deck, S.; Thorigny, P. Unsteadiness of an axisymmetric separating-reattaching flow: Numerical investigation. *Phys. Fluids* **2007**, *19*, 65–103.
11. Depres, D.; Reijasse, P.; Dussauge, J.P. Analysis of Unsteadiness in Afterbody Transonic Flows. *AIAA J.* **2004**, *42*, 2541–2550. [CrossRef]
12. Rui, Z.; Jili, R.; Fang, R. An improved empirical prediction formula for fluctuating pressure on the wall of transonic rotating body. *J. Astron.* **2016**, *37*, 1179–1184.
13. Qiang, L.; Xiaolong, C.; Yi, L. Numerical Simulation Research on Transonic BuffetingLoad of Supercritical Airfoil. *Tactical Missile Technol.* **2018**, *192*, 50–57.
14. Shujie, R.; Shouyun, Z.; Guirong, Y. Forecast of rocket transonic pressure environmentbased on RANS/NLAS. *Solid Rocket Technol.* **2011**, *34*, 418–422.
15. Dawson, J.; Samimy, M. Effects of Expansions on a Supersonic Boundary Layer: Surface Pressure Measurements. *AIAA J.* **1994**, *32*, 2169–2177. [CrossRef]
16. Tsutsumi, S.; Takaki, R.; Takama, Y. Hybrid LES /RANS simulations of transonic flowfieldaround a rocket fairing. In Proceedings of the 30th AIAA Applied Aerodynamics Conference, New Orleans, LA, USA, 25–28 June 2012; pp. 25–41
17. Soni, R.K.; Arya, N.; De, A. Numerical simulation of supersonic separating-reattaching flow through RANS. *J. Phys. Conf. Ser.* **2017**, *822*, 12–37.

18. Grilli, M.; Hickel, S.; Adams, N.A. Large-eddy simulation of a supersonic turbulent boundary layer over a compression-expansion ramp. *Int. J. Heat Fluid Flow* **2013**, *42*, 79–93. [CrossRef]
19. Kopera, M.A.; Kerr, R.M.; Blackburn, H.M.; Barkley, D. Direct numerical simulation of turbulentflow over a backward-facing step. *J. Fluid Mech.* **2014**, 24–47. [CrossRef]
20. Spalart, P.R.; Jou, W.; Strelets, M.; Allmaras, S. Comments on the feasibility of LES for wings andon a hybrid RANS/LES approach. In Proceedings of the 1st AFOSR International Conferenceon DNS/LES, Ruston, LA, USA, 4–8 August 1997; pp. 181–195
21. Guilmineau, E.; Deng, G.; Wackers, J. Numerical simulation with a DES approach for automotive flows. *J. Fluids Struct.* **2011**, *27*, 807–816. [CrossRef]

fluids

MDPI

Article

Experimental and CFD Investigation of Directional Stability of a Box-Wing Aircraft Concept

Gueraiche Djahid [1,*], Karpovich Elena [1,*], Pikulev Maxim [1], Kuznetsov Alexander [1], Sergey Popov [1] and Manoranjan Sinha [2]

[1] Moscow Aviation Institute, 125993 Moscow, Russia
[2] Department of Aerospace Engineering, Indian Institute of Technology Kharagpur, Kharagpur 721302, West Bengal, India
* Correspondence: d.gueraiche@mai.ru (G.D.); karpovichea@mai.ru (K.E.)

Abstract: This study aimed to explore the directional stability issues of a previously studied light box-wing aircraft model with a pusher propeller engine in the fuselage aft section. Earlier configurations have included the use of fuselage together with a lifting system consisting of two wings joined together at their wingtips with vertical stabilizers. However, these side vertical surfaces failed to provide the aircraft with sufficient directional stability, thus prompting the quest in this study for novel solutions that would exclude the need for a fuselage extension and a typical fin. Solutions included the use of a ducted propeller and few configurations of small "fishtail" vertical fins, which formed part of the aft fuselage itself and coupled with vortex generators on the fuselage surface to improve their interference and heal flow separation at the fuselage aft cone. The results of wind tunnel testing were supported with CFD simulations to explain the flow behavior of each of the studied solutions. Tuft visualization and computed flow patterns allowed identification of the sources of the observed low efficiency in terms of directional stability of the fishtail against a simple idle duct without a propeller. A final configuration with a duct and a modified version of the fuselage fins was achieved that provides enough yaw stability margins for a safe flight.

Keywords: box-wing; tailless; flow visualization; directional stability; yaw moment; tufts

Citation: Djahid, G.; Elena, K.; Maxim, P.; Alexander, K.; Popov, S.; Sinha, M. Experimental and CFD Investigation of Directional Stability of a Box-Wing Aircraft Concept. *Fluids* **2022**, *7*, 340. https://doi.org/10.3390/fluids7110340

Academic Editors: Mehrdad Massoudi and Olga A. Azarova

Received: 16 September 2022
Accepted: 22 October 2022
Published: 27 October 2022

1. Introduction

Despite the long known aerodynamic and structural gains of joint wing configurations in allowing significant wing spans and higher lift-to-drag ratios, due in particular to a lower induced component of drag, only very few prototypes have reached serial production, with most concepts never leaving the drawing board. With the idea of using a second wing for both pitch control and increasing the total lifting surface, a particular operational issue of stability concerns in the yaw channel arises due to the absence of a vertical tail with sufficient lever arm and static moment. Wind tunnel investigations [1–5] have revealed that even vertical surfaces that are large enough at the wing tips and have a total area equivalent to a conventional tail will fail to provide enough moment to ensure directional stability. For this reason, most tandem-wing aircraft are equipped with a conventional vertical fin at the aft fuselage [6–8]. The fin and the unnecessary fuselage length extension lead to a significant wetted area and friction drag penalties, which hinder the induced drag benefits of the concept. Given Prandtl plane concepts are of particular interest for green transport aviation, most recent studies are dedicated to stability issues of large box-wing aircraft concepts [9–13], with extremely few articles about light box-wings. Stephen et al. [14,15] found that stability margins of heavy transport can be improved if the wings have the same area and their spacing is increased. In addition, in terms of stability and flight safety, box-wing concepts have been found to fit regional and smaller aircraft mission profiles rather than larger airliners. Part of the box-plane research project "PARSIFAL", a comprehensive study by Cipolla et al. [16–18], concluded that static longitudinal stability issues can be leveraged through simultaneous geometrical optimization of both wing sweep angle and spacing, i.e., dihedral as well as the area. However, directional stability details are missing in this concept due to it being equipped with a conventional twin tail with large vertical surfaces that ensures enough stability margin at the cost of a significant wetted area. As can be noted from most recent studies [9–18], longitudinal stability of box-planes has been studied very thoroughly, in particular for heavy transport mission profile at transonic speeds. There is, however, a significant knowledge gap concerning directional stability, especially for a light subsonic box-wing lacking a conventional fin. Taking into account the importance of this issue for flight safety, crosswind landings, spin tendency, and recovery, this topic is urgent. Therefore, this study aimed to fill this gap by providing both experimental and computational data.

In this research, novel concepts were studied in a wind tunnel experiment supplied with CFD flow pictures. These included modifying the shape of the aft fuselage to incorporate vertical "fishtail" fins, along with a ducted pushing propeller, which is already known to be safer and more efficient in generating thrust. The baseline test model was a tailless box-plane design with the wingtips joined with vertical fins. Previous conceptual design and wind tunnel experiments of this box-plane model have investigated the Prandtl plane layout for general aviation applications, confirming both its well-known advantages against an equivalent monoplane and providing solutions to some of its disadvantages [1–5]. This study represents a further development in this research intended to improve the aft fuselage local aerodynamics and solve directional stability issues. Previous experiments have focused on the general aerodynamics of lifting surfaces, fuselage, and the wing fuselage junction areas. This study examined secondary lifting surfaces consisting of a highly nonconventional tail assembly: a ducted fan, different shapes of upper and ventral fishtail fins, combined with few patterns of vortex generators. Even without a propeller, the large "idle" duct was found to generate enough stabilizing yaw moment for directional stability at small sideslips. This effect of the duct was compared to that of vertical stabilizing fishtail surfaces. The shape of these surfaces is designed to minimize the interference drag by integrating their geometry into that of the fuselage aft cone using bioinspired curvilinear shapes tangent to the fuselage line, hence the term "fishtail" (Figure 1). As previous tuft flow experiments have revealed a separation of the aft fuselage cone [4], vortex generators were added in an attempt to sustain an attached flow over the fuselage aft to both reduce its drag and maximize fin efficiency. It was found, however, that the idle duct still provided better directional stability than the combined effect of the vertical fins and vortex generators, probably due to the insufficient area of theses surfaces, which was limited by the ability of the material to withstand air pressure. Taking into account the fact that these are non-airfoiled flat plates with little structural depth and easy to bend under pressure, a few preliminary experiments with trial and error allowed estimation of the maximum feasible area.

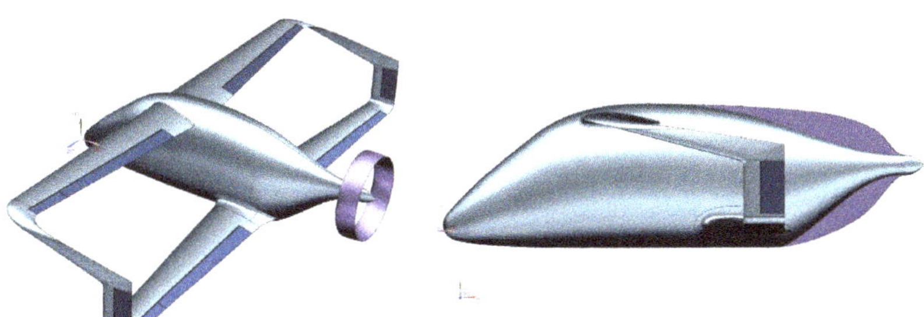

(a) (b)

Figure 1. Tested configurations of the box-wing model equipped with (**a**) an idle duct and (**b**) vertical fishtail surfaces.

Tuft flow visualization, together with computed pressure and velocity fields near the tail of the studied configurations, revealed insights into the reasons for the low efficiency of the fishtail surfaces, with these fins being strongly influenced by the aft fuselage local flow. The fuselage sidewash was found to significantly reduce the local angle of the attack of the upper fin, while the ventral portion at certain angles of sideslip had a quasi-complete flow separation of the aft cone. Attempts to use wing root fairings, large vortex generators on the fuselage top, and a few patterns of small- and medium-sized VG strakes on the fuselage sides resulted in better flow but insignificant improvements in directional stability. CFD streamlines revealed the size of 3D vortical structures and separation bubbles to be much larger than the vortices from the vortex generators; hence, the model remained unstable. Next, installing the duct allowed the model to gain neutral stability until sideslip angles of ~4°. The duct suffered less interference with the separated aft cone due to its outer section being in much cleaner air and its inner section being much further away from the fuselage surface. Despite that, streamlines, pressure, and velocity fields revealed that the fuselage sidewash caused the inner (shaded) section to experience a significant loss of angle of attack. Beyond sideslip of ~8°, the outer section of the duct separated and lost efficiency as well. Hence, the model remained unstable at sideslips larger than 4–6°. Finally, with the duct providing additional hard points, we investigated attaching to it a pair of large-sized flat plates similar to fishtail fins but with much larger area. With this final configuration of combining the duct and large fins, the model became stable well until large sideslips of ~10°.

It is worth noting that the experiments took place with an idle duct without a propeller. We believe that a rotating pusher propeller would provide enough sucking force to sustain an attached flow on the aft cone, leading to significant efficiency improvement of both the duct itself and the fishtail fins in ensuring more yaw stability. Hence, our current results could be used to validate a dead-stick landing and off-design condition of an idle duct. Future experiments with a rotating propeller might confirm these assumptions. The novelty of this research lies in using multiple nonconventional elements for directional stability of a tailless box-plane layout. These solutions exclude the necessity for a conventional tail fin with an extension of the aft fuselage, hence producing the least possible wetted area and friction drag penalties.

2. Materials and Methods

2.1. Wind Tunnel Experiment

2.1.1. Geometry

To improve the directional stability of the previously studied baseline model of a box-plane aircraft and to investigate innovative ways to boost the yaw channel performance of this aircraft concept, the baseline wind tunnel model was fitted with a duct fixed on 8 thin transversal spokes in the aft-most section of the fuselage, where a future piston engine with a pusher propeller would be installed. Thin spokes were dropped from the CAD model given their effect was neglected (Figure 1a). Another configuration was fitted with flat-plate "fishtail" surfaces tangent to the fuselage lines (Figure 1b), along with vortex generators. Both configurations included tufts for local flow visualization (Figure 2). As the model did not possess a classical tail empennage, yaw control was achieved by rotating rudders located on the vertical fins at the wing tips.

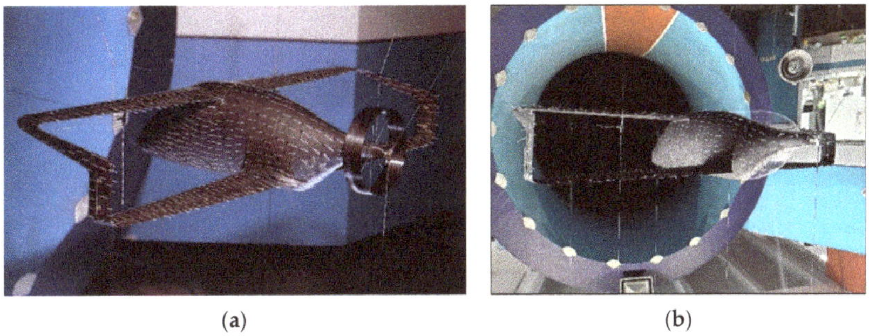

Figure 2. Models with tufts for flow visualization in T-1 wind tunnel 2.25 × 3 m test section (a) with an idle duct and (b) with vertical fishtail surfaces.

Details of the baseline model geometry of the wind tunnel model can be found in [3]. The following main parameters of the wing and fuselage are listed here for reference.

Dimensions (fuselage length × wing span × fuselage height)	0.709 × 1.1 × 0.207 m
Wing aspect ratio	12 (both wings)
Fuselage aspect ratio	3.42
Wing sweep angle at $\frac{1}{4}$ chord	1.6° (fore wing); 3.2° (aft wing)
Airfoil	NACA 3413 (fore wing); NACA 4415 (aft wing)
Airfoil relative thickness	15% (both wings)
Wing incidence angle	2.5° (fore wing); 2° (aft wing)
Elevator-to-wing area ratio	0.17 (both wings)
Flaperon-to-wing area ratio	0.03 (both wings)

Geometry of Fishtail Surfaces

Two configurations of vertical fins were tested. The first was a small version with minimum wetted area, henceforth designated as "S", and fins with a total area of 65.5 cm^2 (Figure 3a). The second had twice the total area, henceforth designated as "L", and fins with a total area 113 cm^2 (Figure 3b). Linear dimensions of both the upper and ventral "L" fins are shown in Figure 4 below.

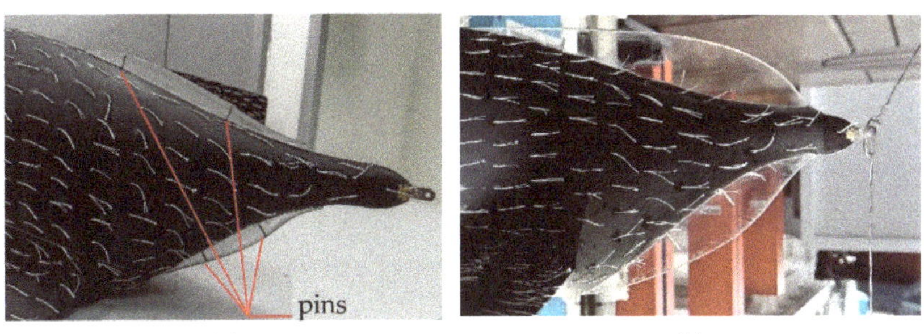

Figure 3. Wind tunnel model with tufts and different vertical fins installed to the aft cone with metallic pins: (a) small "S" fin configuration; (b) large "L" fin configuration.

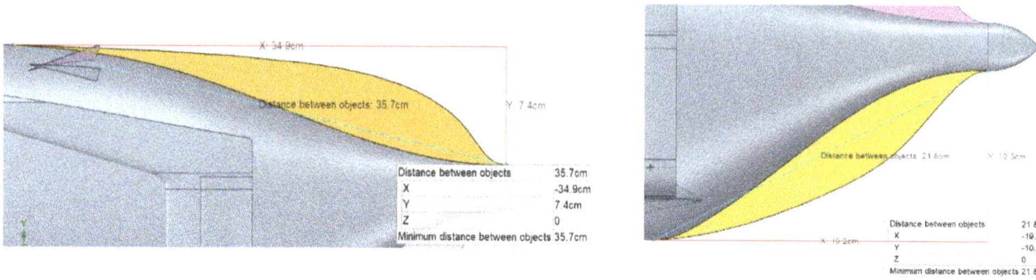

(**a**) (**b**)

Figure 4. Dimensions of the large configuration "L" of vertical fins: (**a**) upper portion of the "L" fin, (**b**) ventral part of the "L" fin.

Geometry of the Duct

Airfoil	NACA0012 (symmetric)
Airfoil thickness	0.5 cm
Duct diameter	17 cm
Hub diameter	3 cm
Spokes	7 × 0.3 cm (8 total)

The duct and its hub were 3D printed from high stiffness plastic; the hub was printed with embedded holes for metallic spoke installation (Figure 5b).

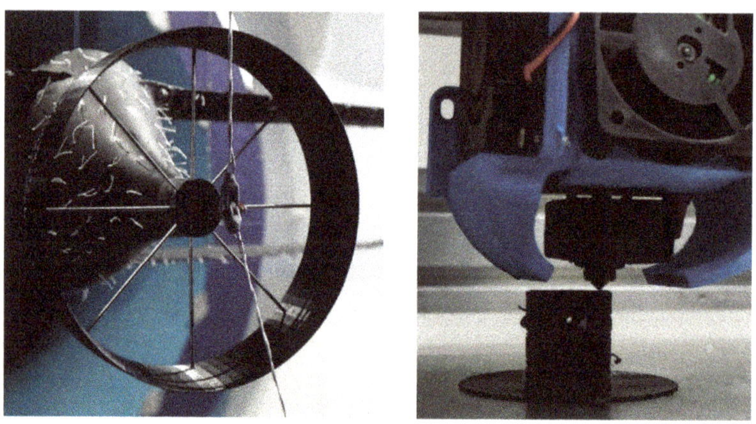

(**a**) (**b**)

Figure 5. (**a**) Aft duct fixed on spokes linked with a hub; (**b**) process of 3D printing of the duct hub.

Vortex Generators

The following problems could be solved using VGs:

- Lateral stability improvement;
- Increasing high-lift devices efficiency.

Suggested positioning:

- Bottom aft fuselage
- Outboard leading edges

Expected outcomes:

- Drag reduction;
- Attached flow on high-lift devices;
- Attached flow on the fuselage aft cone.

Hence, few patterns of small-sized (SVGs), medium-sized (MVGs), and large-sized (LVGs) vortex generators made from tin plate were tested in different locations, and their effect on the fuselage aft cone flow separation as well as on the directional stability were studied.

Table 1 presents the geometrical features of the tested configuration.

Table 1. Geometrical parameters of the tested configurations of vortex generators.

Configuration	Skew Angle $\delta°$	Height h cm	Length l cm	VGs Pitch in a Pattern L cm	Patterns Pitch λ	Min. Distance to Separation Line ΔX_{vg} cm
LVG_1	42	1.1	4.9	4.5	-	-
LVG_2	−42	1.1	4.9	4.5	-	-
MVG_1	35	0.75	1.5	1.5	-	2
MVG_2	35	0.75	1.5	1.5	-	0
MVG_3	−35	0.75	1.5	1.5	-	0
SVG_1	42	0.5	0.75	1	-	6
SVG_2	42	0.5	0.75	1	-	2
Illustration of the geometrical parameters of the vortex generator [19]	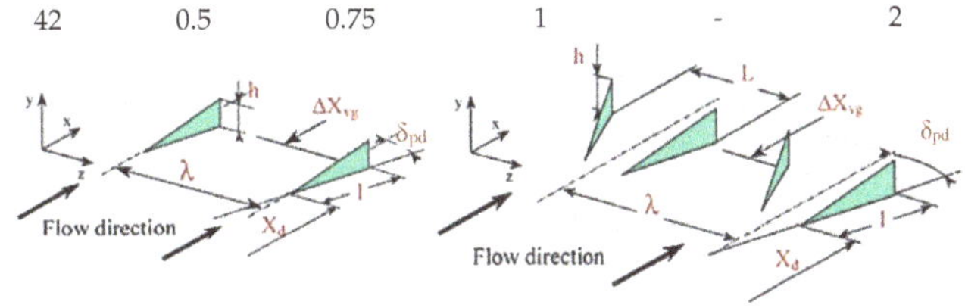					

Large vortex generators (LVGs) were installed on top of the fuselage center-aft section in order to delay separation at high angles of attack. Figure 6 shows the LVGs tested in different positions. In the figure, the converging position is denoted as "1" and the diverging as "2", and the geometrical dimensions of a single VG strake is also given. The LVGs on the wind tunnel test model is given in Figure 7.

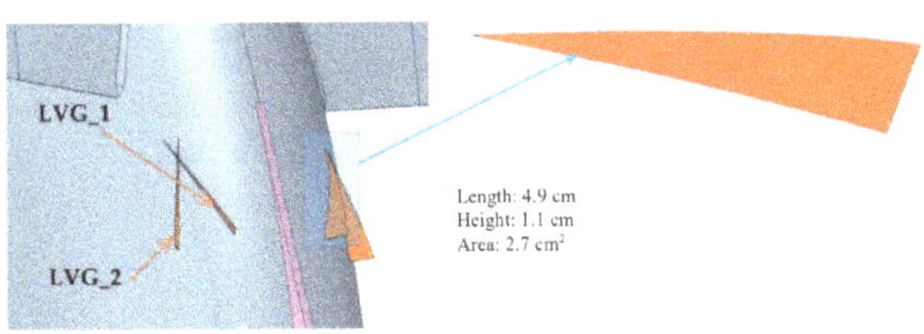

(a) (b)

Figure 6. (**a**) Large vortex generators (LVGs) in positions 1 (converging) and 2 (diverging); (**b**) single strake geometry and dimensions.

LVG_1 LVG_2

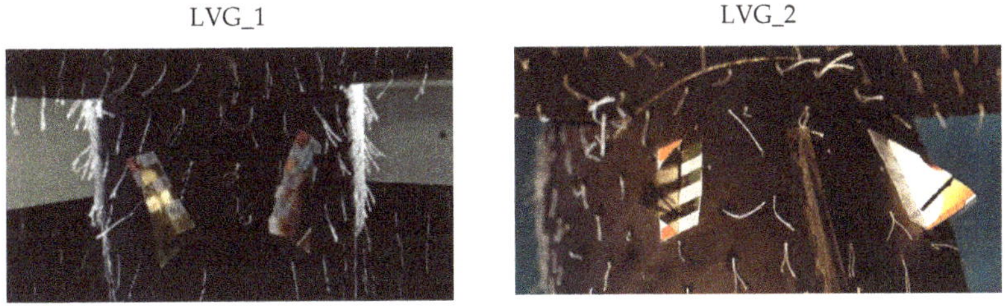

(a) (b)

Figure 7. Large vortex generators (LVGs) in positions: (**a**) "1" and (**b**) "2" on the wind tunnel model.

Medium-sized vortex generators (MVGs) had the following parameters:

Skew angle δ_{MVG}	~±35°
Pitch L_{MVG}	~1.5 cm
A single MVG strake height	~0.75 cm
A single MVG strake length	~1.5 cm

MVGs were also tested in a few configurations: along a vertical line with a positive β_{MVG}, denoted as MVG_1, and a pattern along an inclined line corresponding to the aft cone separation line with a positive β_{MVG} = +35°, denoted as MVG_2, and with a negative β_{MVG} = −35°, denoted as MVG_3. All configurations on the wind tunnel model are illustrated in Figure 8 below.

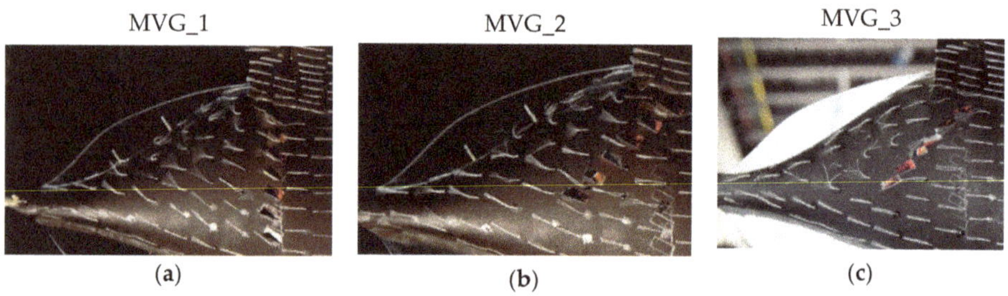

Figure 8. Medium-sized MVG in 3 positions: (**a**) vertical pattern upstream separation line, (**b**) inclined coincident with the separation line, and (**c**) inclined coincident with SL with a negative skew angle.

Small vortex generators (SVGs) had the following geometry:

Skew angle δ_{SVG}	~42°
Pitch L_{SVG}	~1 cm
A single SVG strake height	~0.5 cm
A single SVG strake length	~0.75 cm

Vertical patterns of SVGs were tested in 2 locations: SVG_1, which was close to the aft wing suction peak and well upstream the aft cone separation zone, and SVG_2, which was closer downstream to the separation line (similar to MVG_1), as shown in Figure 9 below.

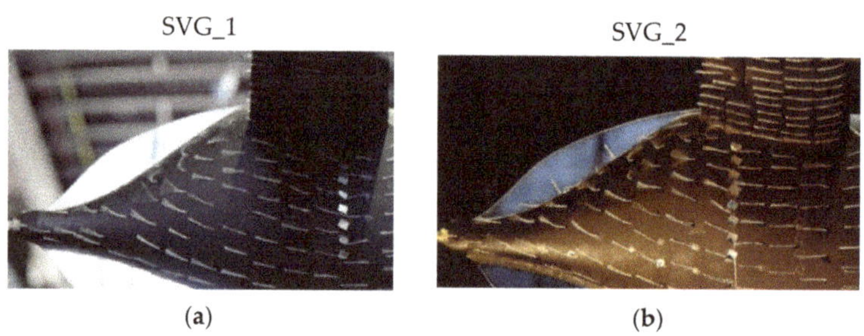

Figure 9. Small SVG in different locations: (**a**) near the aft wing suction peak, (**b**) near the separation line of the fuselage aft cone.

2.1.2. Wind Tunnel Test Conditions

The main geometry and structure of the test model are detailed in [3]. Fine thin silk tufts about 1 cm in length were glued to the surface for local flow observations. The initial tufts axes were coincident with the undisturbed velocity vector or the aircraft *X*-axis. Tests were carried out in wind tunnel T-1 of the Moscow Aviation Institute, which is a subsonic open return circuit type with an open test section. Measurement errors were in the range 3–5%. The main flow parameters at the test section were as follows.

Velocity V∞	38 m/s
Pressure p∞	100,500 Pa
Temperature T∞	293 K
Turbulence intensity ε	0.35%
Reynolds number Re	~10^6
Test section dimensions	Diameter 2.25 m * Length 3.5 m

2.2. CFD Model

2.2.1. Meshing

RANS CFD experiments using ANSYS Fluent were performed on a 30–50 million cells unstructured mesh generated in ANSYS Meshing (Figure 10). For boundary layer resolution, a prismatic layer of 15 layers was built around the wing, fuselage, and additional surfaces (i.e., the duct and fins), as shown in Figure 11. Given the significant impact of the aft fuselage local flow on tail aerodynamics, an extensive automatic refinement algorithm was set as face sizing with a minimum element size of ~10^{-4} m. Based on a similar mesh, the CFD model of the initial configuration was validated against wind tunnel data in [4]. In Figure 12, the wall Y+ function distribution is provided.

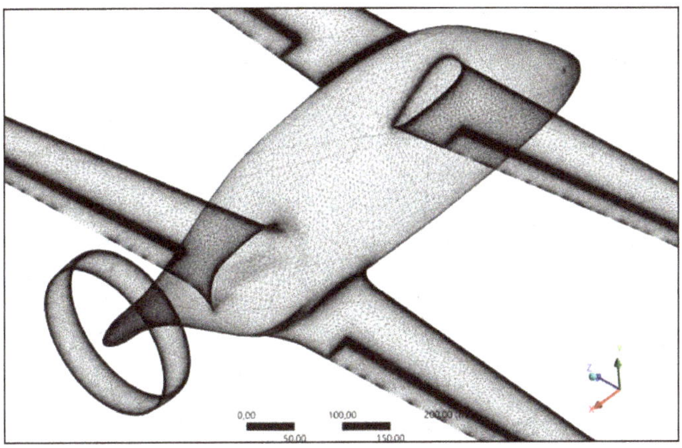

Figure 10. Fragment of the computational mesh of the CFD model with a tail duct.

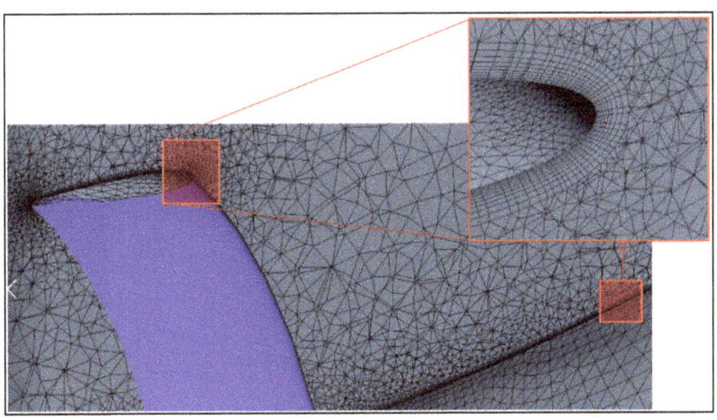

Figure 11. Prismatic boundary layer on both the fuselage and the duct surfaces.

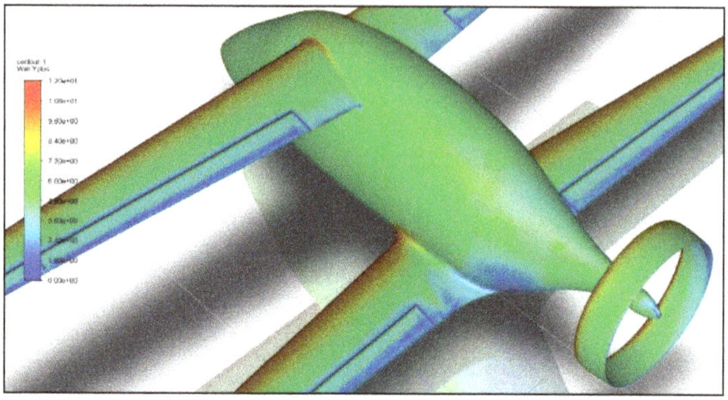

Figure 12. Distribution of wall Y+ values on the CFD model with a duct.

2.2.2. Governing Equations, Discretization Schemes, and Turbulence Modelling

Below are the governing equations solved by the RANS model for an incompressible, low subsonic flow:

Continuity equation:

$$\nabla.(\rho\vec{v}) = 0, \tag{1}$$

Momentum conservation:

$$\nabla.(\rho\vec{v}\vec{v}) = -\nabla p + \nabla.(\overline{\overline{\tau}}). \tag{2}$$

The low subsonic flow (M < 0.3) is incompressible: $\rho = const$; hence, the energy conservation equation was not included in the CFD setup. In addition, as the flow field was incompressible and the temperature was constant (T~293 K), viscosity was treated as a constant value and calculated by the solver based on the chosen fluid material "air": $\mu = const$.

Pressure–velocity coupling was performed through a "coupled" scheme in ANSYS Fluent. For spatial discretization, the Green–Gauss node-based gradient evaluation algorithm was applied; a second-order upwind convective scheme was used for the pressure, density, and momentum. For most of the tested geometry configurations, the solutions converged within 700–1500 iterations.

Turbulence Modelling

Based on the validated initial aircraft configuration [4], the shear stress transport (SST) k-ω turbulence model was applied in the current study as well, resulting in good agreement with the wind tunnel experiment of the visualized flow pictures, including the onset prediction and magnitude of the flow-separated areas at the fuselage aft cone (see Table 2 below). The standard k-ω model was based on the following transport equations for the turbulence kinetic energy, k:

$$\frac{\partial}{\partial t}(\rho k) + \frac{\partial}{\partial x_i}(\rho k u_i) = \frac{\partial}{\partial x_j}(\Gamma_k \frac{\partial k}{\partial x_j}) + G_k - Y_k \tag{3}$$

and the specific dissipation rate, ω:

$$\frac{\partial}{\partial t}(\rho \omega) + \frac{\partial}{\partial x_i}(\rho \omega u_i) = \frac{\partial}{\partial x_j}(\Gamma_\omega \frac{\partial \omega}{\partial x_j}) + G_\omega - Y_\omega \tag{4}$$

Table 2. Computed and tuft-visualized flow field near the aft fuselage and large vertical fins.

β°	4°	7°	9°
Tuft flow visualization			
Flow field of the inner (shaded) side of the fuselage		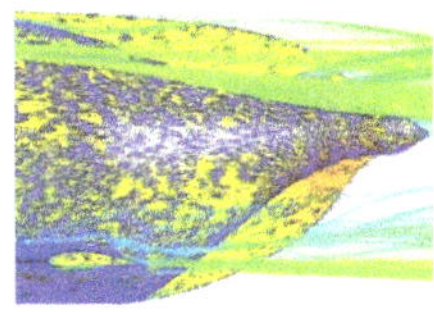	
Upper fin Ventral fin	Effective α~β° Weak separation	Effective α < β° partially separated	Effective α < 0 < β° Full separation

Apart from validation purposes, CFD-obtained values of moment coefficients were mostly dropped from the results as the wind tunnel coefficients were enough for the purpose of defining the values of total force and moment coefficients. Instead, CFD experiments were used to supply the wind tunnel data with better details of local and global three-dimensional flow fields, which are hard to achieve in the actual physical experiment.

For validation, the values of the yaw moment coefficient C_n at few points against the wind tunnel data are given in Figure 13. Overall, the model performed well at small angles of sideslip, although it slightly overestimated C_n at negative β. Increasing β led to a similar pattern, leading to greater coefficient misprediction at high α [4], with the CFD model experiencing premature stalling and discrepancy with the experiment at sideslips β~10°. The impact of the fuselage sidewash on directional stability was further revealed in the results and indicated the importance of proper viscosity treatment and Reynolds number. Viscosity was set to a constant based on the wind tunnel Reynolds number and air conditions. As a validation case for the applied SST k-ω turbulence model, we looked at its ability to accurately predict the onset and scale of separation bubbles. Table 2 presents a comparison of the computed flow field and tuft flow visualization.

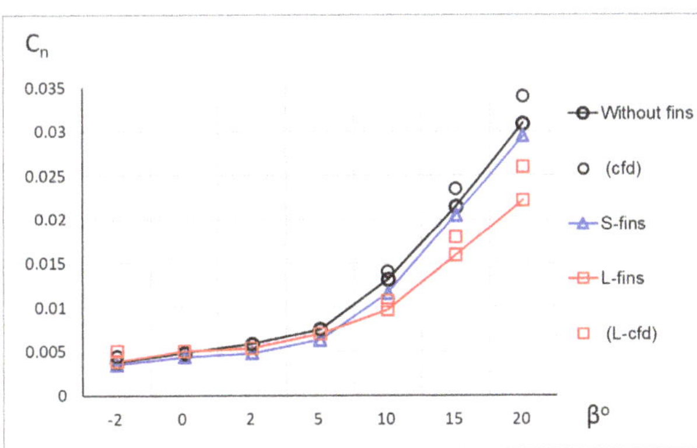

Figure 13. Yaw moment coefficient of the initial model versus the model equipped with small (S) and large (L) vertical fins. Results are provided from both wind tunnel experiment and CFD for validation.

3. Results

3.1. Vertical Fins

As can be seen from the plot in Figure 13, equipping the initial model with vertical fins did not change the overall picture of $C_n(\beta)$ dependency and even with large (L) fins, the model remained unstable in the yaw channel. At small angles of sideslip, both large and small configurations had very little effect on the yaw moment coefficient C_n values. With increasing β, only the large fins gradually started to reduce the yaw derivative. CFD data, which duplicates the wind tunnel values, is given in this plot for validation purpose only. In further results, they were dropped to avoid cluttering the graphs.

An insight into the inefficiency of vertical fins can be gained from the tuft flow visualization and the CFD flow field pictures near the tail at different angles of sideslip (Table 2 and Figure 14). A strong separation of the bottom aft cone at $\beta \sim 9°$ could be noticed, leading to a 3D vortex upstream the ventral fin and resulting in its virtually complete inefficiency. For the upper fin, the tufts and streamlines showed that it remained attached even at larger $\beta > 9°$ which is a good sign. However, the computed streamlines showed that its aft-most section received a side-washed flow from the fuselage, reducing its effective angle of attack at moderate β and eventually driving it negative at larger β. Hence, its overall efficiency was strongly affected by interference with the fuselage (see top view of streamlines in Figure 14).

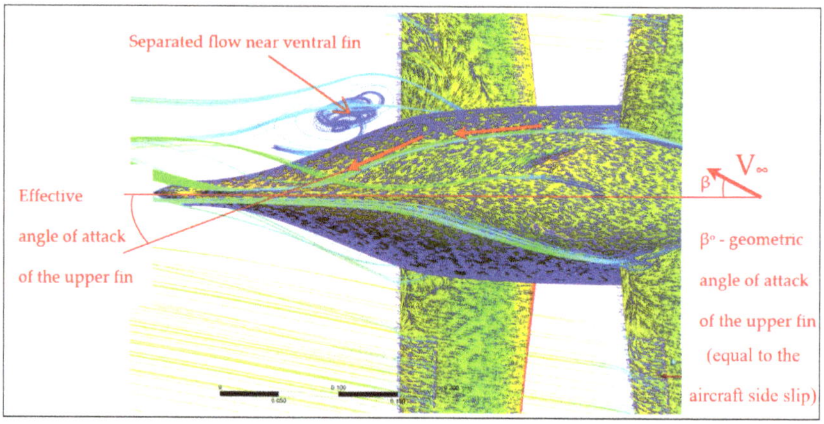

Figure 14. Top view of velocity streamlines at $\beta = +9°$ revealing a negative effective angle of attack of the upper fin and a 3D vortex near the separated ventral fin.

3.2. Vortex Generators

As has been deduced from the above flow pictures, the main reason for the vertical fin inefficiency can be traced to the boundary flow of the fuselage and its strong influence on the tail fins. Hence, an obvious solution is to use vortex generators to turbulate the flow in problematic areas. First, large LVGs were added to leverage the local effective angle of attack of the upper fin. As mentioned in the Materials and Methods section, we tested both convergent and divergent setting angles of LVG. As can be seen from the plots in Figure 15, the LVG_1 convergent configuration worked slightly better than LVG_2. The geometrical angle of attack of the outer generator, where the upper fin suffered a side-wash, increased as the model sideslip increased, which was probably the reason for the higher vortex intensity. Simultaneously, the inner generator lost the angle of attack with sideslip (see Table 3 below), eventually becoming useless at zero alpha. For LVG_2, this dependency was obviously inverted.

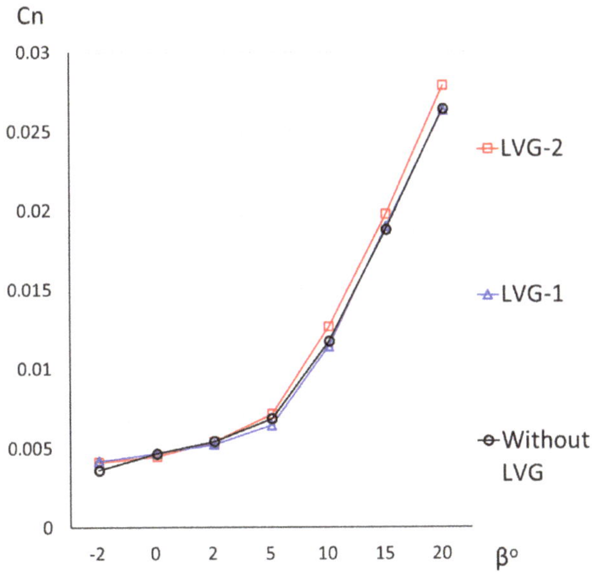

Figure 15. Yaw moment coefficient of the model without vortex generators and with large LVG in positions −1 and −2.

Table 3. Computed flow field near the vortex generators: configuration LVG_1.

β°	3°	7°	15°
Cp scale	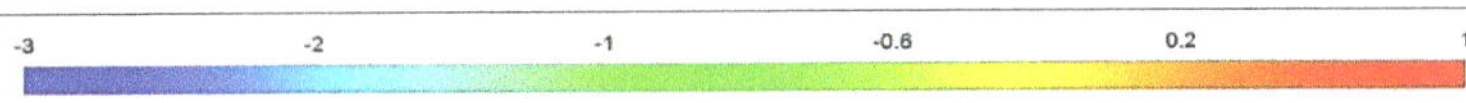		
Top View LVG_1	Outer strake  Inner strake		
Inner strake Outer strake	Moderate—small α Moderate α	Small α High α	α ~ 0 Post-stall α
Outer strake suction side			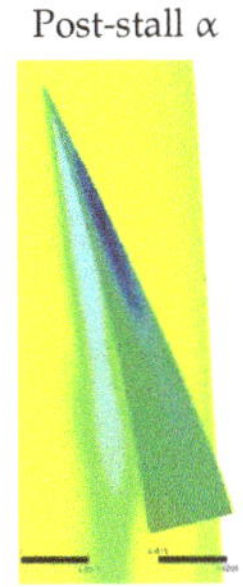
Outer strake remark	Small angle of attack—low vortex intensity	Low pressure covering the entire surface—highest intensity of the vortex.	Low pressure only near the leading edge indicating a stall—overall less vortex intensity

From Table 3 above, we can notice that only the outer VG worked well for the investigated (positive) range of sideslip angles, while the inner one failed to generate a vortex as its (geometric) angle of attack was reduced with sideslip. The geometric α of this inner VG was equal to the VG skew angle (δ, see Table 1) minus the aircraft angle of sideslip. Meanwhile, for the outer VG, sideslip β added to its α, and hence its efficiency increased and reached an optimal at $\beta = 7°$. At larger sideslips, a stall led to reduction of the overall vortex intensity. From Figure 16 below, we can notice that the fuselage side wash and its effect on the upper fin remained strong but was delayed to a further position downstream, and hence a smaller portion of the upper fin experienced a negative sidewash. Further moving the LVGs downstream could result in an even less sidewash and better tail efficiency.

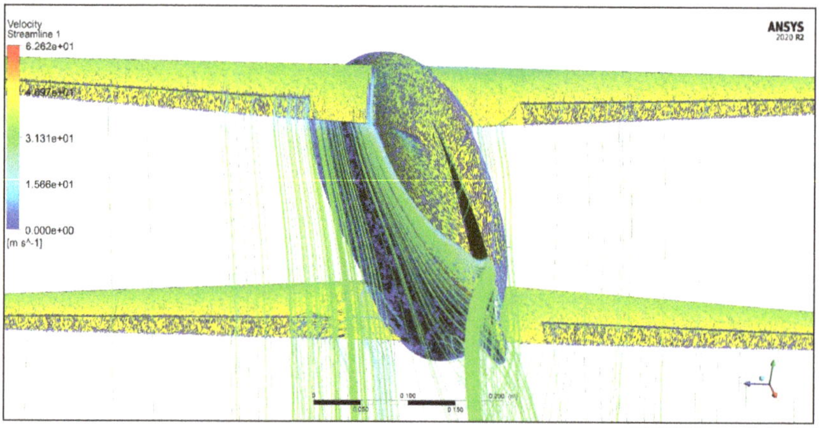

Figure 16. Velocity streamlines on model with LVG_1 and upper fin at $\beta = 15°$ showing the fuselage sidewash effect on the upper fin.

Next, in an attempt to improve the ventral fin efficiency by reducing separation of the bottom aft cone, small vortex generators (SVGs) were installed along the side perimeter of the fuselage near the aft wing. Given that separation happens on both the inner and outer sides of the aft cone (Figure 17), VGs were installed on both sides.

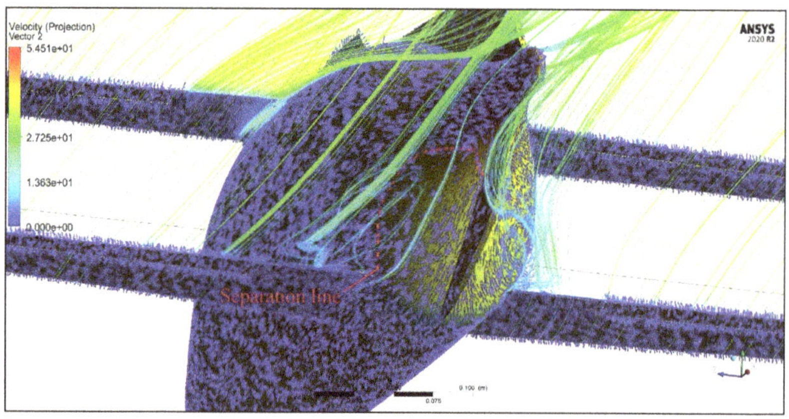

Figure 17. Flow field near the fuselage aft cone.

However, the tuft flow visualization provided in Table 4 revealed that even placing the SVG pattern immediately close upstream to the separation line (version SVG_2) resulted in little improvement of the separated aft cone area, probably due to the 3D vortex structure taking place (Figure 14) being much larger than vortices generated by the SVGs, hence dominating their local flow. With regard to the investigated velocities of ~37–40 m/s, SVGs were next replaced by medium-sized vortex generators (MVGs).

Table 4. Tuft visualization of the aft fuselage with small VG installed in configurations 1 and 2.

Configuration	L-Fins + LVG	L-Fins + LVG + SVG_1	L-Fins + LVG + SVG_2
Tuft visualization at $\alpha = 2^{\circ}$			
Remarks	Both SVG configurations did not lead to significant reduction of the aft cone separated area. Placing the VG pattern closer downstream led to a slight shift of separation below.		

MVG provided slightly better results in both reducing the separated area size and shifting it further downstream (Table 5). The best effect was achieved by placing an inclined pattern (MVG_2), coincident with the separation line, which is visible in Figure 17. Inverting the strakes (MVG_3) did not give better results, indicating that at the original position, MVG_2 strakes were at a much better angle of attack and generated higher vorticities. However, analysis of the impact of different configurations of MVGs on directional stability, as illustrated by the plots in Figure 18, showed that a good stability increment was achieved by equipping the model with a vertical pattern MVG_1. This could be explained by the fact that MVG_1 deflected the flow towards the ventral fin (see red arrows in Table 5), while MVG_2 deflected it slightly downward towards the separated area itself, giving more attached flow but having less direct effect on the ventral fin. MVG_3 gave the worst results both in terms of separation and stability as the flow was deflected upwards instead. A solution to flow deflection issues could be alternating inverted and positive strakes; however, in view of the increased drag coefficient caused by medium-sized VGs (plot in Figure 19) and their overall insignificant effect on directional stability, we decided to remove them in further experiments. For reducing drag, wing root fairings were added, and their effect was investigated in the subsequent item.

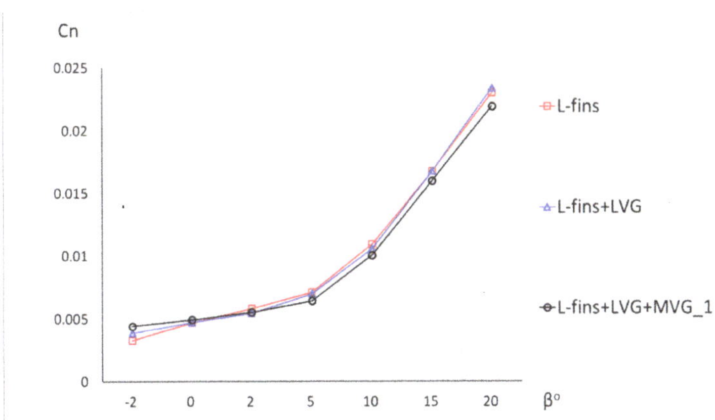

Figure 18. Yaw moment coefficient of the model equipped with combinations of L-fins, LVG, and MVG_1.

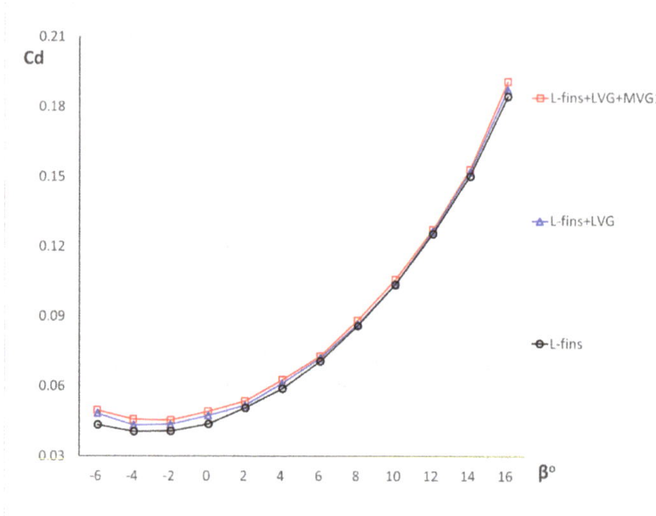

Figure 19. Drag coefficient (Cd) of the model equipped with combinations of L-fins, LVG, and MVG_1.

Table 5. Aft fuselage tuft flow visualization with medium VG installed in configurations 1, 2, and 3.

Configuration	L-Fins + LVG	L-Fins + LVG + MVG_1	L-Fins + LVG + MVG_2
Tuft visualization at $\alpha = 2°$		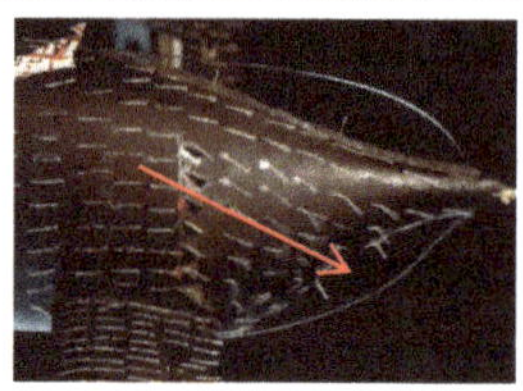	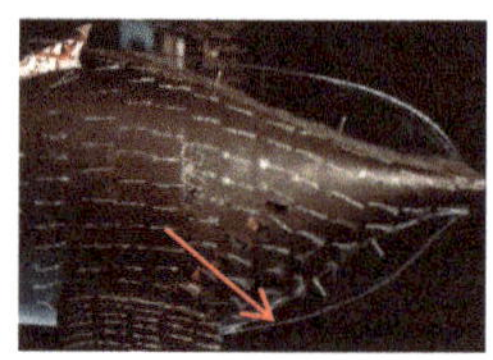
Remarks	• All medium-sized VGs provided a tangible effect on the aft fuselage local flow field. • MVG_2 had the best impact on the separated area as the flow was deflected downward. • MVG_1 provided better directional stability by deflecting the flow towards the ventral fin, thus improving its effect. • Inverting strakes at MVG_3 increased vorticity but led to deflecting the flow upward, exacerbating the aft cone separation and worsening the ventral fin efficiency.		**L-fins + LVG + MVG_3** 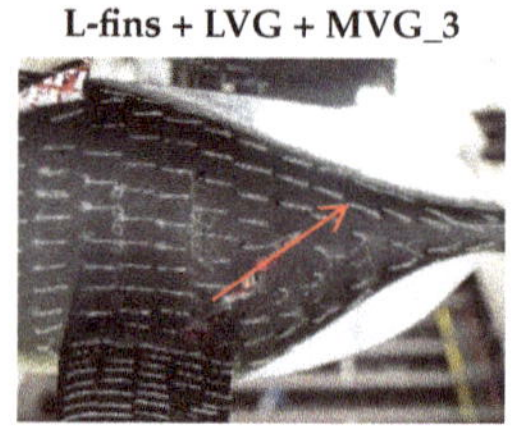

3.3. Wing Root Fairing

Fairings are a common solution to reducing interference drag in junction areas where a discontinuity in the geometry leads to a discontinuity in the flow field, which in turn induces strong pressure gradients, reverse flows, and separation bubbles. As investigated earlier [4], to a large extent, the aft cone separation bubble can be traced to the low pressure at the suction side of the aft wing propagating towards the aft cone and absorbing flow from the higher pressure upper section, causing more reverse flows. Although not directly linked to directional stability, 3D-printed wing root fairings were tested in the current research as a possible quick remedy to aft cone separation in conjunction with other directional stability solutions. As can be seen from the tuft flow visualization at large angles of attack (Table 6), the wing root faring worked as advertised in reducing the extent and intensity of the aft cone separation bubble. Hence, it was used in further experiments. The fairing impact on directional stability is illustrated in Figure 20. A slight improvement at small to moderate α in the lift/drag ratio of the model is visible in Figure 21. The little improvements in L/D can be explained by the increase in friction drag due to poor surface finishing of the fairing and overall increase in the wetted area.

Table 6. Aft fuselage tuft flow visualization with wing root fairing.

Configuration	Without Wing Root Fairing	Wing Root Fairing Installed
Tuft flow Visualization at $\alpha = 8°$		

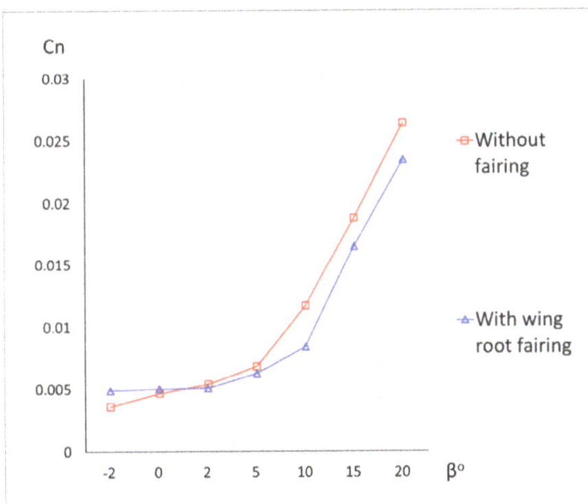

Figure 20. Yaw moment coefficient of the model with and without wing root fairing.

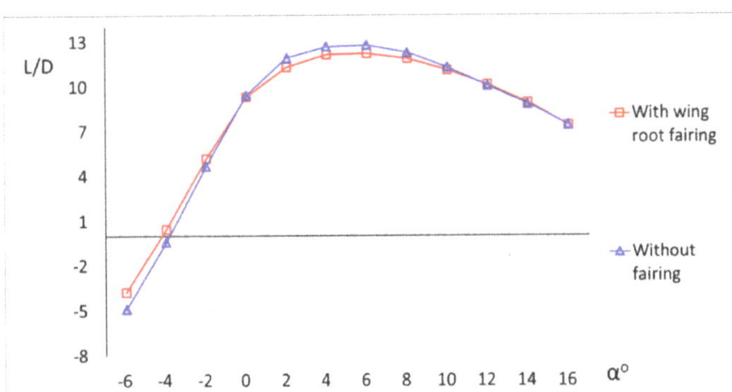

Figure 21. Lift/drag ratio of the model with and without wing root fairing.

3.4. Tail Duct

The above studied combinations of localized solutions did not provide enough directional stability nor completely healed the aft cone separation bubble, which is virtually impossible without globally changing the aft fuselage geometry. Therefore, large vertical surfaces that are far enough from the boundary flow are inevitable for achieving enough directional stability for a safe flight. Instead of a typical vertical fin with large wetted area, we investigated the effect of a ducted fan, which might simultaneously improve the propeller performance and generate enough yaw static moment, especially at significant sideslips. Experiments at the current stage were performed without a propeller. This helped assess directional stability margins of the model in an engine failure mode.

The results of yaw channel coefficient are given in Figure 22 below. The model was stable until angles of sideslip β~2° and then remained neutral until β~5°. This is a relatively much better result compared to vertical fins (Section 3.1). Still, a range of 2–5° of sideslip is too small for flight safety, and hence tuft and CFD flow visualizations were again used to analyze the local flow near the duct for any potential improvements. Streamlines over the model showed that the duct inner side was partially under the influence of the fuselage sidewash and less severe compared to the fin given the duct was at a fairly good distance away from the fuselage, but the local effective angle of attack of the inner shaded side of the duct was significantly reduced (see Figure 23 and Table 7 below).

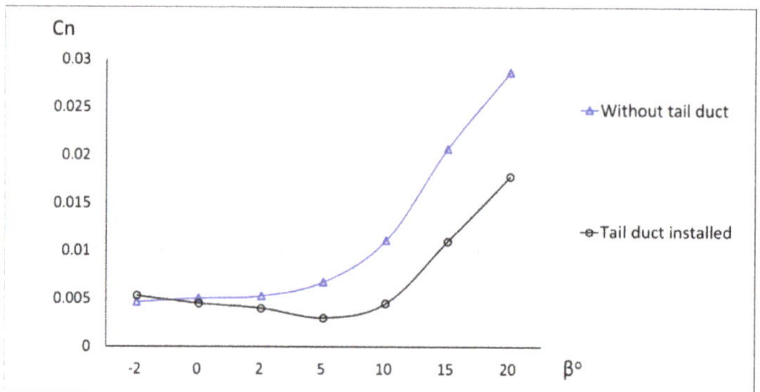

Figure 22. Yaw moment coefficient of the model equipped with a tail duct against the model without a duct.

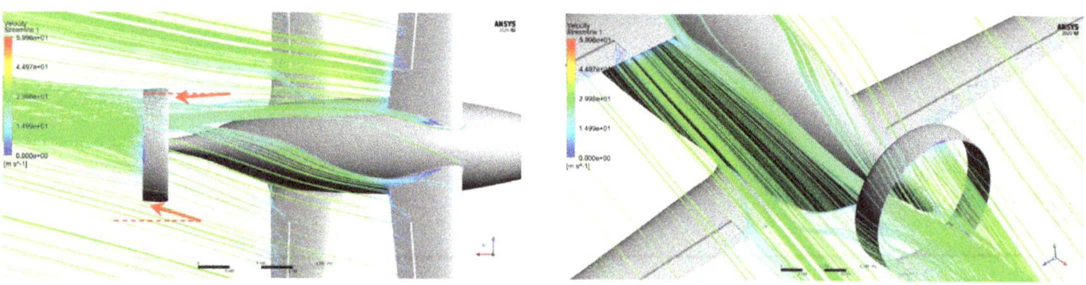

Figure 23. Velocity streamlines at a high sideslip β~15° revealing the impact of the fuselage sidewash on the tail duct inner (shaded) side local effective angle of attack.

Table 7. Pressure field of the tail duct in a horizontal section plane.

β°	3°	7°	15°
Cp scale			
Pressure coefficient visualization in a horizontal mid-section plane	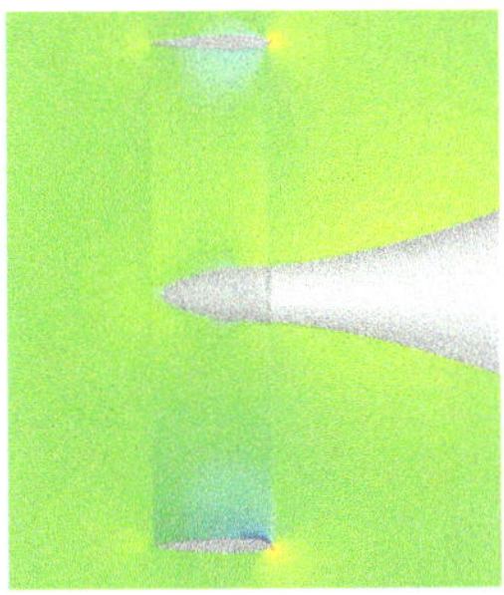	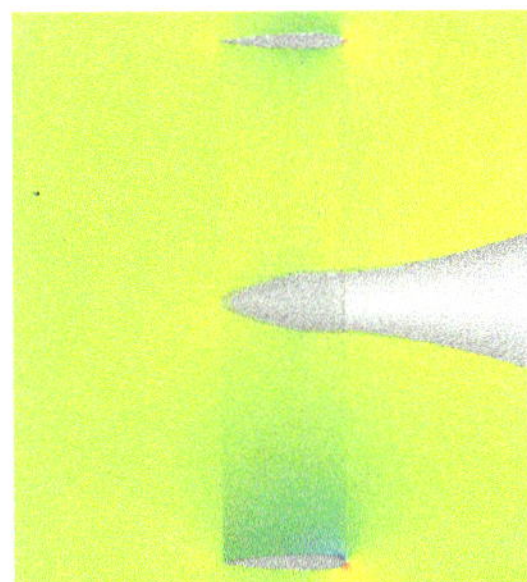	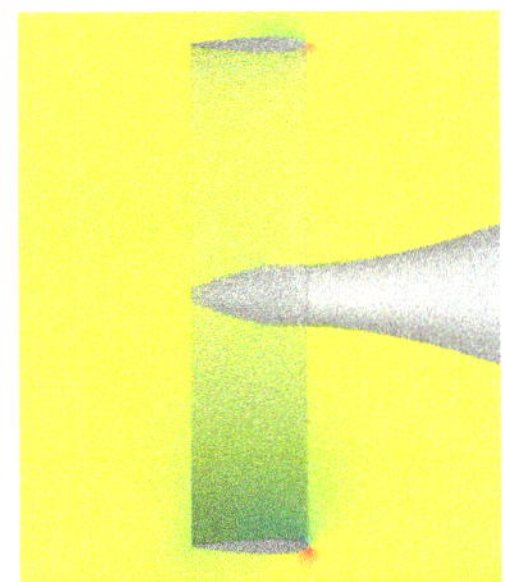
Inner (shaded) section	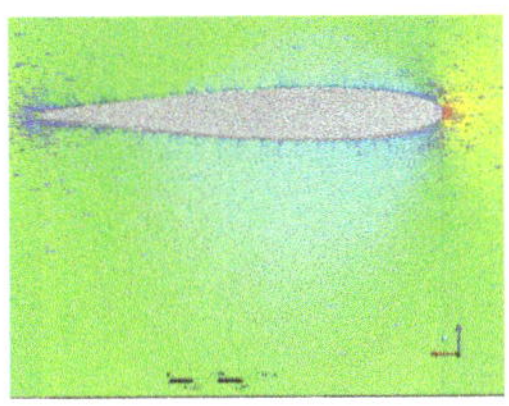	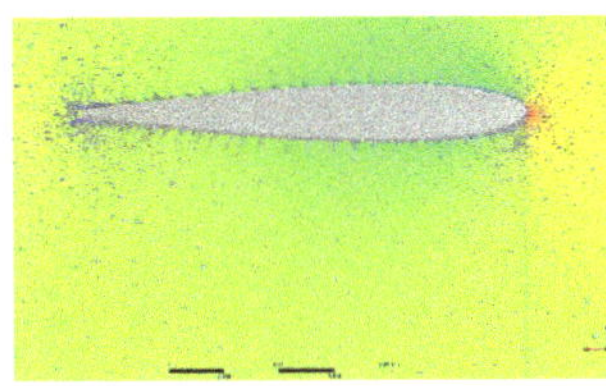	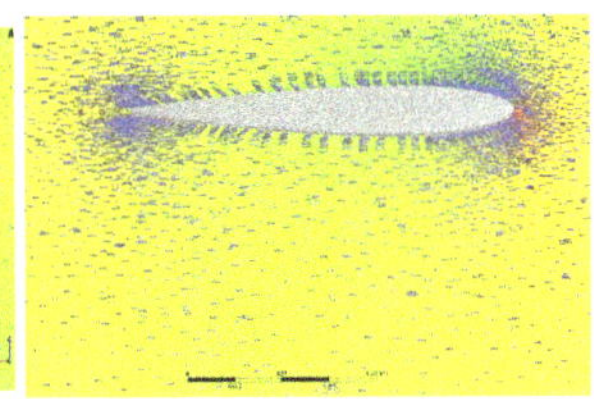
Flow Local effective α	Attached α ~ 0 (symmetric pressure field)	Attached α ~ 0 (symmetric pressure field)	Attached moderate α
Outer section	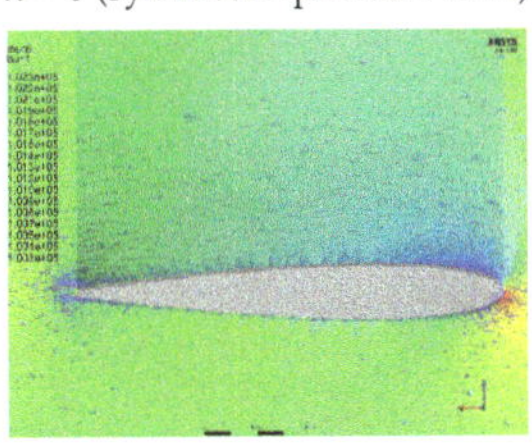		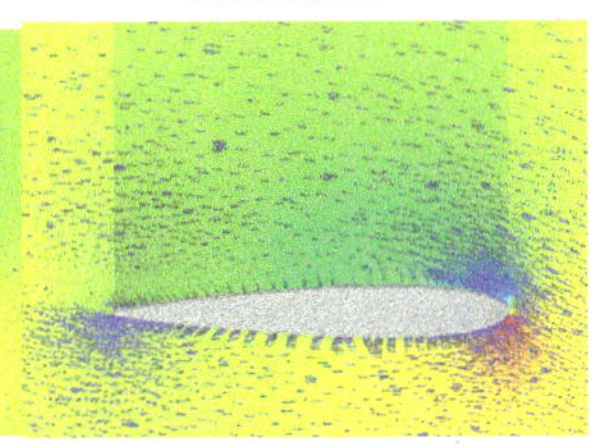
Flow Local effective α	Attached Moderate α ~ β ~ 3°	Partially separated High α ~ β ~ 7°	Fully separated Extremely high local effective angle of attack α ~ β ~ 15°
Conclusion	The outer section, which was outside the fuselage influence, had an effective α equal to the aircraft (undisturbed) sideslip angle β. The inner section, which was shaded by the fuselage sidewash, experienced a significant loss of effective α, which consequently did not increase with increasing β and hence did not contribute to generating a yaw moment.		

3.5. Final Configuration: Tail Duct Supplied with Large Fins

In view of the good results demonstrated by the tail duct regarding directional stability, at least at small sideslip angles, we decided to keep it for further experiments and to boost its performance with an enlarged version of the earlier studied (Section 3.1) vertical fishtail fins. Unlike the fins in the previous case, which had a small area limited by the ability of the flat plate material to withstand air pressure, this time, the presence of the duct gave us an additional hard point for attaching much larger fins. Supported with three points (Figure 24 below), even a flat zinc plate half a millimeter in thickness was able to withstand significant air pressure at high angles of sideslip.

(a) (b)

Figure 24. (**a**) Wind tunnel model of the final configuration with a tail duct and vertical fishtail fins with a large surface attached to it. (**b**) Attachment was through a 1.7 mm pin inserted in the fuselage and through metal wires tied to the vertical duct metallic spokes.

The results of the model directional stability are plotted below in Figure 25. As can be seen from the plot, the model in this final configuration became stable in the yaw channel well into sideslips of 6–8° and remained neutral beyond 10°. Future wind tunnel and computational experiments with a rotating propeller might reveal the extent of interference and mutual influence between the fins and the propeller, potentially leading to further geometrical optimization of the fins. On the actual aircraft, thicker airfoiled vertical fins might serve to attach the duct to the fuselage, along with similar horizontal surfaces, thus boosting the longitudinal stability performance.

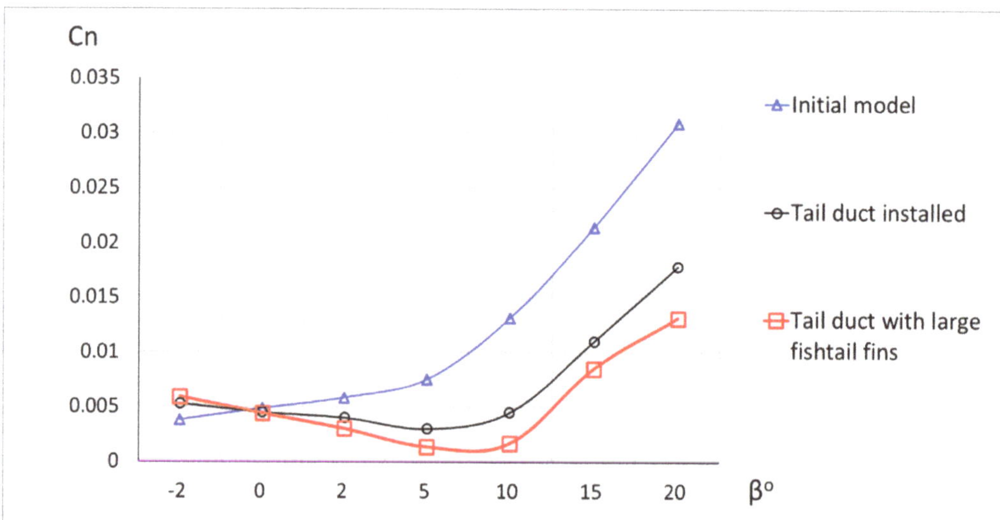

Figure 25. Yaw moment coefficient of the initial model without any additional elements against the model equipped with a tail duct and tail duct supplied with large vertical fins.

4. Discussion and Conclusions

This study experimentally and computationally investigated a tailless nonconventional box-wing aircraft concept with potential applications in general aviation, such as personal recreational planes, aerotaxis, UAV delivery, etc. An important issue that lies in the way of certification of such flying vehicles is their limited margin of both longitudinal and directional stability. This is a natural consequence of the core idea of the concept to get rid of conventional tails. Multiple studies in the past decades have focused on ways to tackle longitudinal stability, which is easy to achieve by carefully locating the aircraft center of mass and/or ensuring enough spacing of wings. Directional stability, however, is much more challenging in view of an extremely short fuselage. Given the studied concept has a pusher propeller in the tail, it is virtually impossible to locate a vertical fin that is far enough from the center of mass. In addition, extending the fuselage aft section to accommodate a conventional tail would result in significant wetted area and shift the fuselage mass further backwards, which would in effect reduce the lever arm by shifting the aircraft center of mass Hence, in this study, we investigated novel ways to solve this issue while keeping the fuselage geometry intact. A ducted fan and vertical fishtail fins were applied, which were tailored to fit inside the space between the duct and the aft fuselage cone.

The novelty of this research lies in using multiple nonconventional elements for directional stability of a box-plane layout. This excludes the necessity for a conventional tail fin with an extension of the aft fuselage, hence providing the least possible wetted area or "price" to pay for stability, both in terms of friction drag and structural mass. Among the investigated solutions were bioinspired novel surfaces presenting continuity of the aft fuselage geometry itself, combined with vortex generators to improve their efficiency. In addition, equipping the pushing propeller with an airfoiled duct was found to provide very satisfactory results and solved the issue of directional stability of this aircraft concept. The results of the physical experiments were supported with computed local flow fields of pressure and velocity streamlines. This approach revealed deep insights into local flow directions and effective angles of attack as well as the extent of interference zones between different elements, pointing towards sources of efficiency or inefficiency of different solutions and the potential solution.

Author Contributions: Conceptualization, G.D. and K.E.; Data curation, K.E.; Formal analysis, G.D. and K.E.; Funding acquisition, M.S.; Investigation, G.D., K.E., P.M. and K.A.; Methodology, G.D.; Supervision, K.E. and S.P.; Validation, G.D. and K.E.; Visualization, G.D. and K.E.; Writing—original draft, G.D.; Writing—review & editing, G.D., K.E. and S.P. All authors have read and agreed to the published version of the manuscript.

Funding: This work was financially supported by the Russian Science Foundation under Agreement No. 22-49-02047 from 09.03.2022.

Institutional Review Board Statement: Not applicable.

Informed Consent Statement: Not applicable.

Data Availability Statement: Not applicable.

Conflicts of Interest: The authors declare no conflict of interest.

References

1. Karpovich, E.A.; Liseitsev, N.K. Analytical Evaluation of Box Wing Aircraft Aerodynamic Characteristics at Early Design Stages. *Russ. Aeronaut.* **2019**, *62*, 417–422. [CrossRef]
2. Karpovich, E.; Liseitsev, N. Revisiting the Longitudinal Stability and Balance of a Boxwing Aircraft. In *Aerospace Engineering Bulletin*; Perm National Research Polytechnic University: Perm, Russia, 2019; pp. 29–44. [CrossRef]
3. Karpovich, E.A.; Kochurova, N.I.; Kuznetsov, A.V. Experimental Study of Aerodynamic Characteristics of a Boxplane Wind-Tunnel Model. *Russ. Aeronaut.* **2020**, *63*, 659–668. [CrossRef]
4. Karpovich, E.; Gueraiche, D.; Sergeeva, N.; Kuznetsov, A. Investigation of a Light Boxplane Model Using Tuft Flow Visualization and CFD. *Fluids* **2021**, *6*, 451. [CrossRef]
5. Karpovich, E.; Gueraiche, D.; Korotkov, D.A. A method for conceptual design of a light boxplane. *Aerosp. Syst.* **2022**, *5*, 349–356. [CrossRef]
6. Frediani, A.; Cipolla, V.; Oliviero, F. Design of a prototype of light amphibious PrandtlPlane. In Proceedings of the Florida 56th AIAA/ASCE/AHS/ASC Structures, Structural Dynamics, and Materials Conference, Kissimmee, FL, USA, 5–9 January 2015. [CrossRef]

7. IDINTOS: Light Amphibious PrandtlPlane. Available online: http://www.idintos.eu/eng (accessed on 26 October 2022).
8. RUTAN QUICKIE. Available online: https://web.archive.org/web/20150610213002/http://www.aeroresourcesinc.com/store_/images/classifieds/179-1.pdf (accessed on 26 October 2022).
9. Gagnon, H.; Zingg, D.W. Aerodynamic Optimization Trade Study of a Box-Wing Aircraft Configuration. *J. Aircr.* **2016**, *53*, 1–11. [CrossRef]
10. Andrews, S.; Perez, R. Parametric Study of Box-Wing Aerodynamics for Minimum Drag under Stability and Maneuverability Constraints. In Proceedings of the 33rd AIAA Applied Aerodynamics Conference, Dallas, TX, USA, 22–26 June 2015. [CrossRef]
11. Chau, T.; Zingg, D.W. Aerodynamic Shape Optimization of a Box-Wing Regional Aircraft Based on the Reynolds-Averaged Navier-Stokes Equations. In Proceedings of the 35th AIAA Applied Aerodynamics Conference, Denver, CO, USA, 5–9 June 2017. [CrossRef]
12. Cipolla, V.; Frediani, A.; Abu Salem, K.; Binante, V.; Rizzo, E.; Maganzi, M. Conceptual design of Prandtl Plane transport aircraft and preliminary CFD investigation of transonic flight. In Proceedings of the American Institute of Aeronautics and Astronautics 2018 Aviation Technology, Integration, and Operations Conference, Atlanta, Georgia Aviation Technology, Integration, and Operations Conference, Atlanta, GA, USA, 25–29 June 2018. [CrossRef]
13. Andrews, S.A.; Perez, R.E. Analytic Study of the Conditions Required for Longitudinal Stability of Dual-Wing Aircraft. *Proc. Inst. Mech. Eng. J. Aerosp. Eng.* **2017**, *232*, 958–972. [CrossRef]
14. Stephen, A.A.; Ruben, E.P. Stability and Control Effects on the Design Optimization of a Box-Wing Aircraft. In Proceedings of the 14th AIAA Aviation Technology, Integration, and Operations Conference, Atlanta, GA, USA, 16–20 June 2014. [CrossRef]
15. Stephen, A.A.; Ruben, E.P. Multidisciplinary Analysis of a Box-Wing Aircraft Designed for a Regional-Jet Mission. In Proceedings of the 16th AIAA/ISSMO Multidisciplinary Analysis and Optimization Conference, Dallas, TX, USA, 22–26 June 2015. [CrossRef]
16. Cipolla, V.; Abu Salem, K.; Picchi, S.M.; Vincenzo, B. Preliminary design and performance analysis of a box-wing transport aircraft. In Proceedings of the AIAA Scitech 2020 Forum, Orlando, FL, USA, 6–10 January 2020. [CrossRef]
17. Abu Salem, K.; Binante, V.; Cipolla, V.; Maganzi, M. PARSIFAL Project: A Breakthrough Innovation in Air Transport. *Aerotec. Missili Spaz.* **2018**, *97*, 40–46. [CrossRef]
18. PARSIFAL Project. Available online: https://parsifalproject.eu/ (accessed on 26 October 2022).
19. Godard, G.; Stanislas, M. Control of a decelerating boundary layer. Part 1: Optimization of passive vortex generators. *Aerosp. Sci. Technol.* **2006**, *10*, 181–191. [CrossRef]

fluids MDPI

Article

Principles of Unsteady High-Speed Flow Control Using a Time-Limited Thermally Stratified Energy Source §

Olga A. Azarova * and Oleg V. Kravchenko

Federal Research Center "Computer Science and Control" of the Russian Academy of Sciences, 119333 Moscow, Russia
* Correspondence: olgazarov@gmail.com

§ A part of the results was presented at the International Conference "High-Speed Transport Development (HSTD)", Moscow Aviation Institute, Alushta, Russia, 29 August–2 September 2022.

Abstract: This study focused on the development of the unsteady impact of a thermally stratified energy source on a supersonic flow around an aerodynamic (AD) body in a viscous heat-conducting gas (air). Research was based on the Navier-Stokes equations. The freestream Mach number was 2. A new multi-vortex mechanism of the impact of a time-limited stratified energy source on the aerodynamic characteristics of a body was described. Almost complete destruction of the bow shock wave in the density field, due to the multiple generation of Richtmyer-Meshkov instabilities in the region of a stratified energy source, was obtained. The dependences of the dynamics of frontal drag and lift forces of a streamlined body on temperature in the source layers were studied. It was determined that, by changing the temperature in the layers of a stratified energy source, it was possible to obtain more intense vortices accompanying the Richtmyer-Meshkov instabilities, causing a temporary decrease in the drag force of an AD body and ensuring the emergence and unsteady change in the magnitude of the lift (pitch) forces. The main principles of unsteady flow control using a stratified energy source were established.

Keywords: supersonic flow; bow shock wave; time-limited stratified energy source; multiple Richtmyer-Meshkov instabilities; shock-wave structure; drag force reduction; lift force generation

Citation: Azarova, O.A.; Kravchenko, O.V. Principles of Unsteady High-Speed Flow Control Using a Time-Limited Thermally Stratified Energy Source. *Fluids* **2022**, 7, 326. https://doi.org/10.3390/fluids7100326

Academic Editors: Giuliano De Stefano and Mehrdad Massoudi

Received: 12 September 2022
Accepted: 7 October 2022
Published: 12 October 2022

Publisher's Note: MDPI stays neutral with regard to jurisdictional claims in published maps and institutional affiliations.

1. Introduction

The problem of high-speed flow control using non-mechanical approaches, and remote energy deposition in particular, currently occupies a leading place among the problems in flow/flight control [1]. Control of supersonic flows by means of electrical discharges, microwave and laser impulses is currently a well-developed area of aerospace engineering (see surveys in [2–4]). A historical review of the ideas pertaining to the control of supersonic flow by energy deposition to different points of the flow, which arose several decades ago, was presented in [5]. In a significant number of studies, the effectiveness of energy supply in the form of extended filaments ("hot spikes") was established for reducing aerodynamic drag [6–8]. In air, the effect of microwave discharge was determined by a decrease in stagnation pressure, along with the reduction of a drag force of a blunt cylinder. In the calculations, a vortex mechanism of these phenomena was established [9]. In [10], the curvature of the shock wave was observed during the passage through the region of a plasma zone created by longitudinal pulsed discharge.

The influence of inhomogeneous layered plasma on a reflected shock wave in a supersonic flow was studied in [11]. In these experiments, the blurring and suppression of the reflected shock wave were obtained due to the organization of a system of plasma filaments created by a high-frequency discharge [12,13]. In [14], an array of surface arc plasma actuators were used to control the interaction of the shock wave with the boundary layer in a flow with Mach number 2. As a result, the disappearance of a fragment of the separation shock wave was established. In [15], the authors used a set of heated thin wires for the creation of thermal and density inhomogeneities, which led to the generation of the Richtmyer-Meshkov instability and the formation of a line of vortices due to the Kelvin-Helmholtz instability.

The impact of the ionization strata obtained in the gas discharge plasma region on a plane shock wave was researched in [16]. These experiments attained the curvature, and in some cases, complete disappearance of the shock wave front. In the numerical simulation of the experiment, generation in many points of the Richtmyer-Meshkov instabilities was shown under the action of which the shock wave front (in density field) practically ceased to exist, which explained the results of the experiment [16].

In [17,18], the vortex structure was obtained under the action of a combined energy source, and a double-vortex mechanism of its action on the body was proposed, explaining the additional decrease in the front drag force. It was shown that the generation of the vortices is a result of the manifestation of the Richtmyer-Meshkov instability. A thermally stratified energy source was shown to initiate multiple generation of the Richtmyer-Meshkov instabilities during the interaction with the shock wave front, causing significant density and temperature fluctuations [19]. The redistribution of energy types in a curved shock wave under the conditions of the experiment [16] was estimated for M = 2 and 5 [20], and for hypersonic speeds up to M = 12 [21]. In addition, the influence of a thermally stratified energy source on the supersonic flow around an AD body was investigated, and a new multi-vortex mechanism of the action of the energy source on the body surface has been established [22].

This study focuses on the development of a thermally stratified energy source impacting the flow around supersonic AD bodies in a viscous heat-conducting gas (air). The paper focuses on an unsteady temporary action of a thermally stratified energy source. The research is based on the system of Navier-Stokes equations. The freestream Mach number is 2. Almost complete destruction of the bow shock wave in the density field, through the multiple generation of the Richtmyer-Meshkov instabilities in the region of heated layers of a stratified energy source, is obtained. The dependences of drag and lift (pitch) forces of a streamlined body on temperature in the layers of a stratified energy source are studied. The temperature values in the layers are analyzed and show the possibility of influencing the drag force and the ability to cause the formation and change of the lifting (pitch) force (at zero angle of attack). The main principles of flow control using a stratified energy source are established.

2. Methodology and Statement of the Problem

The impact of a thermally stratified energy source on a supersonic flow past a plate sharpened by a wedge is considered. The angle at the apex of the body is 90° (Figure 1).

The simulations are based on the Navier-Stokes equations for perfect viscous heat conductive gas (air); the ratio of specific heats γ = 1.4. The full Navier-Stokes system of equations in the divergent form for the dimensionless variables [23] is solved numerically:

$$\frac{\partial \mathbf{A}}{\partial t}+\frac{\partial(\mathbf{B}+\mathbf{V})}{\partial x}+\frac{\partial(\mathbf{C}+\mathbf{W})}{\partial y}=0, \tag{1}$$

$$\mathbf{A}=\begin{pmatrix}\rho\\ \rho u\\ \rho v\\ E\end{pmatrix},\mathbf{B}=\begin{pmatrix}\rho u\\ p+\rho u^2\\ \rho uv\\ u(E+p)\end{pmatrix},\mathbf{C}=\begin{pmatrix}\rho v\\ \rho uv\\ p+\rho v^2\\ v(E+p)\end{pmatrix},$$

$$\mathbf{V}=-\begin{pmatrix}0\\ \mu/\mathrm{Re}(4/3u_x-2/3v_y)\\ \mu/\mathrm{Re}(v_x+u_y)\\ \mu\pi_1/\mathrm{Re}+(1/\mathrm{N})kT_x\end{pmatrix},\mathbf{W}=-\begin{pmatrix}0\\ \mu/\mathrm{Re}(v_x+u_y)\\ \mu/\mathrm{Re}(4/3v_y-2/3u_x)\\ \mu\pi_2/\mathrm{Re}+(1/\mathrm{N})kT_y\end{pmatrix},$$

$$\pi_1=u(4/3u_x-2/3v_y)+v(v_x+u_y),\pi_2=v(4/3v_y-2/3u_x)+u(v_x+u_y),$$

$$E = \rho\left(\varepsilon + 0.5\left(u^2 + v^2\right)\right), \mathrm{N} = \mathrm{RePr}(\gamma - 1)/\gamma,\ \varepsilon = p/(\rho(\gamma - 1)).$$

Here, ε is the specific internal energy. The following normalizing values for the parameters are accepted:

$$\rho_n = \rho_\infty, p_n = p_\infty, l_n = k_l^{-1}D, T_n = T_\infty, u_n = (p_\infty/\rho_\infty)^{0.5}, t_n = l_n/u_n.$$

where k_l is the dimensionless value of D. The freestream Mach number is $M_\infty = 2$, the Reynolds number is Re = 9500, and the Prandtl number is Pr = 0.703.

Sutherland's law in nondimensional form is used for the dependence of dynamic viscosity on temperature:

$$\mu = T^{1.5}(1 + s_1)/(T + s_1),$$

$s_1 = 0.409556$ (120 K). The coefficient of heat conductivity k is supposed to depend on temperature in nondimensional form as follows:

$$k = T^{0.5}.$$

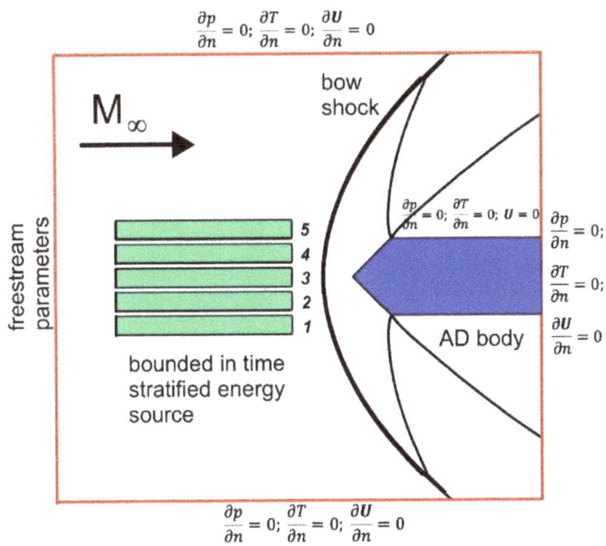

Figure 1. Statement of the problem (schematic).

Initial conditions for the problem are the fields of gas parameters in a steady supersonic flow past the body, $t = 0.6$. At this time, the pressure and density at the apex of the body differ from their theoretical values evaluated with the use of the Bernoulli's relation by 1.81% and 1.75%, respectively. Here, the converging criterion for the evaluation of the relative errors is used in a form:

$$\mathrm{abs}(f_t - f_{t\ \mathrm{theor}})/\ f_{t\ \mathrm{theor}} * 100(\%),$$

where f_t and $f_{t\ \mathrm{theor}}$ are the calculated value at the apex of the body and the theoretical one evaluated from the Bernoulli's relation.

The boundary conditions provide no-slip conditions for the adiabatic wall at the horizontal boundaries and at the wedge boundaries, and establish the absence of according normal flows on the boundaries of the body:

$$\frac{\partial p}{\partial n} = 0;\ \frac{\partial T}{\partial n} = 0;\ \boldsymbol{U} = 0.$$

At the exit boundaries of the computation domain, the absence of reflection in the normal directions is set as follows:

$$\frac{\partial p}{\partial n} = 0; \ \frac{\partial T}{\partial n} = 0; \ \frac{\partial \mathbf{U}}{\partial n} = 0.$$

The stratified energy source is modeled by a region of rarefied gas layers of the same width located ahead of the bow shock wave front in its immediate vicinity (see Figure 1). The distances between the layers were equal to half the width of the layer. Inside the layers (indicated by the index j), gas density was applied to be reduced,

$$\rho_j = \alpha_j \rho_\infty, \alpha_j < 1, \ j = 1 \div N,$$

N is a number of layers in the energy source. The pressure and velocity in the domain of the energy source are set equal to their values in the oncoming flow (indicated by the index ∞),

$$p_j = p_\infty, u_j = u_\infty, v_j = 0.$$

Therefore, the temperature inside the layers is increased compared to its value in the oncoming flow,

$$T_j = \alpha_j^{-1} T_\infty.$$

Thus, a stratified energy source is specified by a set of rarefaction parameters $\{\alpha_j\} = \alpha_1, \alpha_2, \ldots \ \alpha_N$ in its layers. The axis of symmetry of the stratified source is supposed to coincide with the axis of symmetry of the body. The energy source arises instantly in the steady flow at the time instant t_i, and it is assumed that it has a limited duration in time.

A domestic code based on the complex conservative difference schemes of the second order of approximation in space and in time is used in the simulations. Details of construction of the schemes in the computational domain and in the vicinity of the body's boundaries, are presented in [24]. For increasing the order of approximation in the development of the schemes, the differential consequences of system (1) for the spatial derivatives on x and y are used. The five-point stencil (the stencil of Lax's scheme) is applied for the scheme construction; therefore the grids used are staggered and uniform everywhere in the computational domain.

The boundaries of a body are introduced into the calculation area without breaking the conservation laws in it (in space and time). For this purpose, the boundaries of the body are included to a structural staggered orthogonal grid, and discrete conservation laws are written for each arising grid configuration. Thus, the schemes are written for 1/4 of a difference cell, 1/2 of a cell and 3/4 of a cell, and are used, as necessary, in accordance with the position of the body's boundary on the difference grid. In this case, the reflected versions of the schemes are possible, as well as schemes on half cells in the x- and y- directions. This allows the carrying out of calculations conservatively in the entire computational domain, including the regions adjacent to the boundaries of the body [24].

The position of the sharpened part of the body on the grid in an enlarged form is shown in Figure 2. In the calculations, the staggered numerical grids are used with the distance between the nodes at each time level equal to $2h_x$, $2h_y$ (h_x, h_y are the space steps in x- and y-directions). To select the time step, the Courant-Friedrichs-Levy criterion are used.

The numerous test examples for the used numerical methods and developed software are presented in [24]. Comparison with the experimental results was conducted in [25]. Nevertheless, we present some test case analysis below.

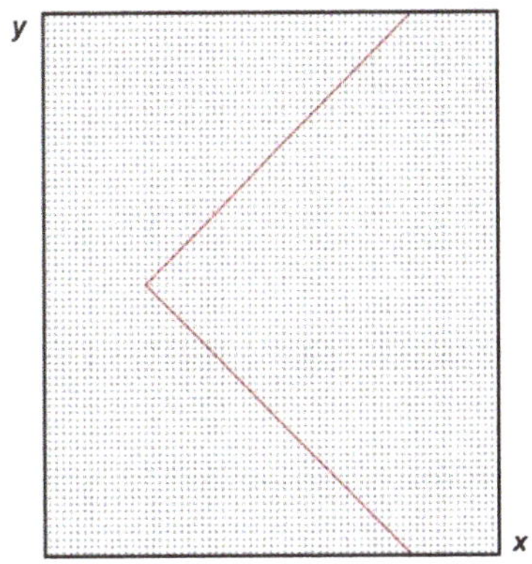

Figure 2. The position of the sharpened part of the body on a computational grid (enlarged).

3. Method Validation and Grid Convergence Analysis

Figure 3 demonstrates the analysis of the possibilities of the applied numerical code for the resolution of shock waves and contact surfaces (shear layers). The "quasi-one-dimensional" Riemann problem of the decay of an arbitrary discontinuity was chosen for validation. The statement of the problem is schematically shown in Figure 3a. At the initial moment of time $t = 0$, a heated gas region is set with the rarefaction parameter of $\alpha = 0.5$ at $0 < x < 1.45$, and a plane shock wave with the Mach number of 2 at the point with $x = 1.55$. At $t > 0$, the shock wave begins to move from right to left, and at time $t = 0.04226$ the shock wave begins to interact with the boundary of the heated region at the point with $x = 1.45$. The emerging flow in the inviscid case is described by the Riemann problem with the formation of a shock wave and a contact discontinuity moving from right to left (and a rarefaction wave propagating to the right) [26]. The numerical density profiles at the axis of symmetry are presented in Figure 3b. Comparison of the locations of shock fronts at different times with the analytical solution (dashed lines) shows a high computational accuracy of the used difference schemes.

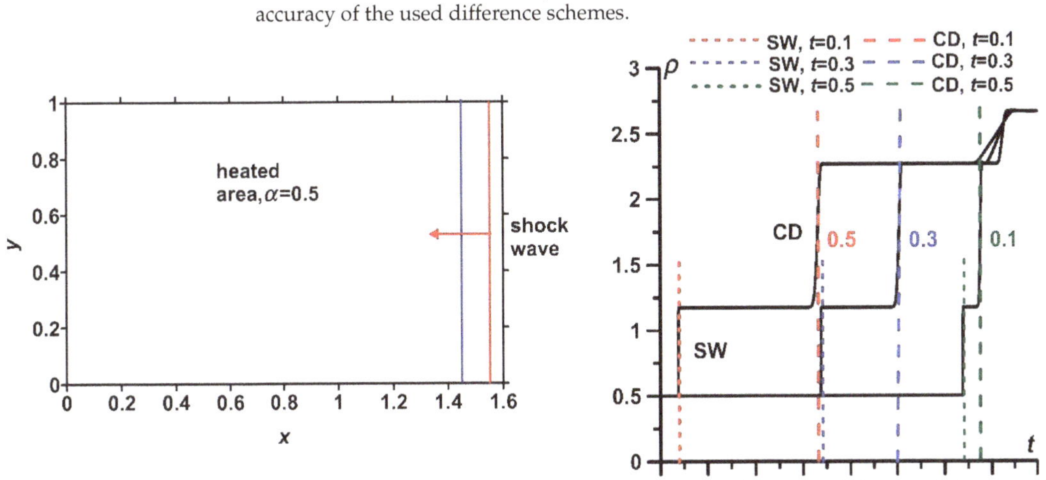

(a) (b)

Figure 3. Analysis of the Riemann problem: (**a**) problem statement scheme; (**b**) density profiles on the axis of symmetry for different moments of time.

For the analysis of grid convergence, the calculations of flow dynamics during the steady state establishment for three difference grids were conducted (Table 1, $t = 0.6$). The characteristics of these grids are presented, as well as the relative errors for the stagnation parameters at the apex of the body, in comparison with the theoretically obtained ones from the Bernoulli's relation.

Table 1. Characteristics of the grids and relative errors.

Grid	Steps $h_x = h_y$	Sizes	Relative Error *, p_t: $abs(p_t - p_{t\ theor})/p_{t\ theor} \times 100\%$	Relative Error *, ρ_t: $abs(\rho_t - \rho_{t\ theor})/\rho_{t\ theor} \times 100\%$
Grid1	0.0005	3400 × 2800	1.81%	1.75%
Grid2	0.001	1700 × 1400	1.17%	2.52%
Grid3	0.002	850 × 700	0.34%	3.61%

* $p_{t\ theor}$, $\rho_{t\ theor}$—the values calculated using the Bernoulli's relation.

Figure 4 demonstrates the flow fields in isochores (Figure 4a) and the dynamics of the parameters at the stagnation point (Figure 4b) obtained using these three difference grids. The number of nodes of Grid1 and Grid3 differ by 16 times, and the number of nodes of Grid1 and Grid2 by four times, but nevertheless, one can see that the values at the apex of the body differ from their theoretical values less than by 2% (for Grid1). Additionally, the positions and shapes of the bow shock waves almost coincide (see Figure 4a).

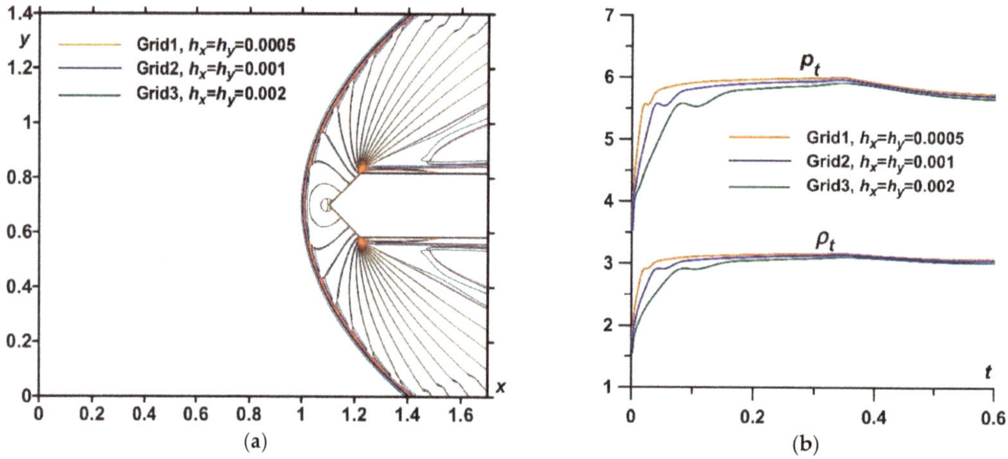

Figure 4. Analysis of the grid convergence on three different grids: (**a**) density fields (superposed); (**b**) dynamics of the pressure p_t and density ρ_t at the apex of the body.

Therefore, all these factors show that grid convergence has taken place. For the simulations, we use Grid1 ($h_x = h_y = 0.0005$) on two computation domains which contain 9.52×10^6 nodes (3400 × 2800, coordinate of the body's axis of symmetry $y_0 = 0.7$) and 12.96×10^6 nodes (3600 × 3600, $y_0 = 0.9$). Here, the dimensions of the grids are specified counting the middle node of the stencil.

4. Results

The defining flow parameters used in the simulations are presented in Table 2. The energy source specified, as described above, is supposed to instantly arise in the steady flow at the time $t_i = 0.601$ moves together with the oncoming flow and begins to interact with the bow shock wave. The dynamics of the density fields following this interaction for different symmetrical sets of the values of α_j in the layers of the energy source is presented in Figure 5. Here, the results of two series of the calculations are presented, with sets $\{\alpha_j\}$

with a heated central layer α_3, and sets $\{\alpha_j\}$ with the additional layers of reduced density α_2 and α_4.

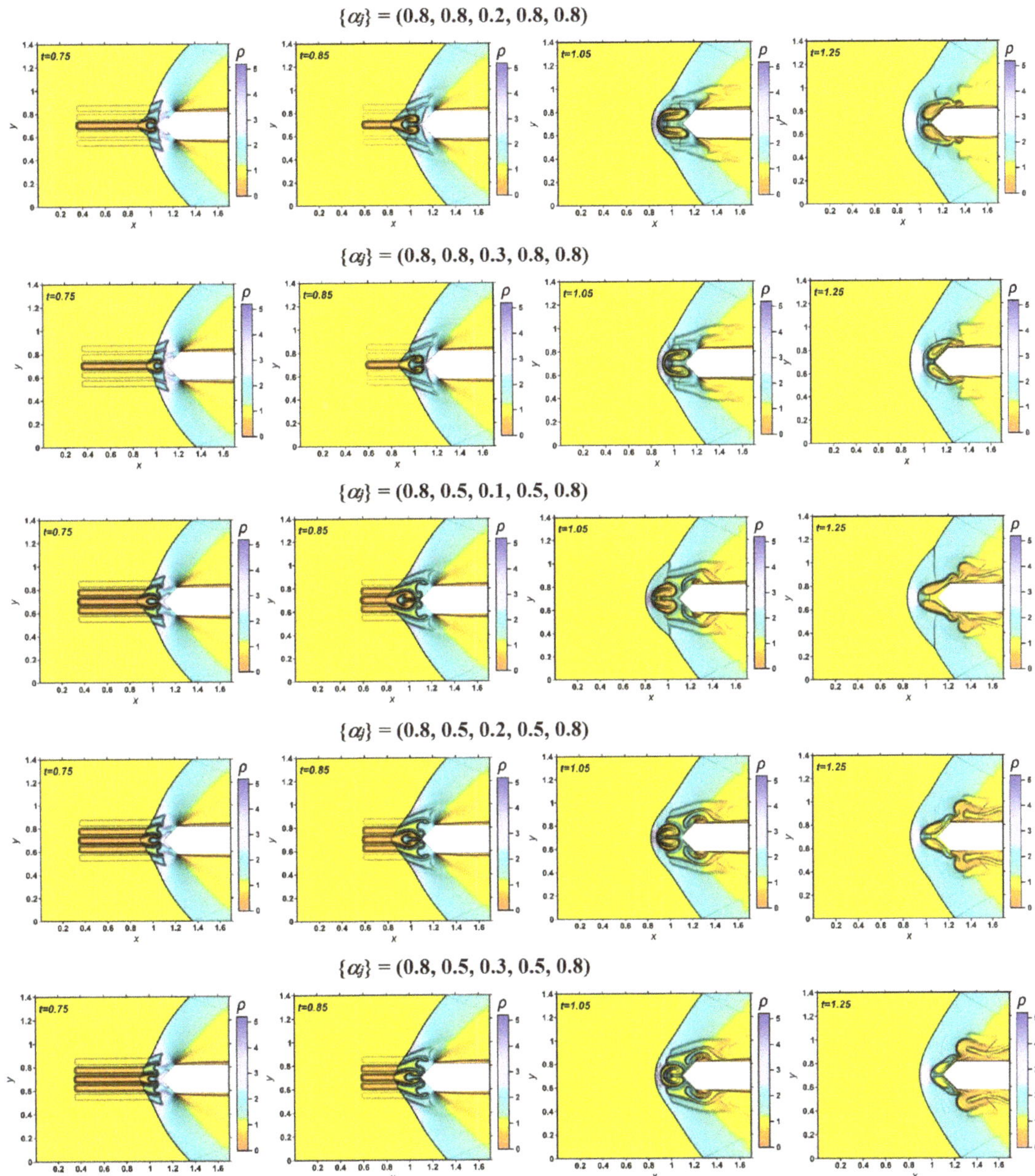

Figure 5. Dynamics of density fields for different symmetrical sets $\{\alpha_j\}$ in the thermally stratified energy source (planar view).

Table 2. Parameters of the oncoming flow, aerodynamic body and the energy source.

Description	Definition	Value
Freestream Mach number	M_∞	2
Ratio of specific heats	γ	1.4
Reynolds number	Re	9500
Prandtl number (20 °C)	Pr	0.703
The body's width	D	0.24
Coordinate of the body's axis of symmetry	y_0	0.7; 0.9
Number of layers in the energy source	N	5
The width of the layers in the energy source	h_j	0.05
Coordinate of the lower boundary of the energy source	y_{es}	0.525; 0.725
Rarefaction parameter in the layer j in the energy source	α_j	0.1–0.8

The first and second lines of images demonstrate the dynamic of the originated vortex structures, which are caused by the sets of layers $\{\alpha_j\}$ with a heated central layer. One can see the dynamic of a complicated shock structure with two vortices with the decreased density accompanying the Richtmyer-Meshkov instability. The front of the bow shock wave undergoes a significant transformation. As the stratified region of the energy source passes, the diffraction of the shock wave occurs, accompanied by the formation of two symmetrical triple configurations (t = 0.85). During the passage of the stratified pulse, one can see the curvature of the bow shock wave (t = 1.05), which is greater the smaller the value of α_3 (or the higher temperature) in the central layer. After the passage of the stratified pulse, the flow returns to the undisturbed steady mode. These results are consistent with the results of numerous studies on the effect of a heated longitudinal region on a bow shock wave.

Images on other lines in Figure 5 correspond to sets $\{\alpha_j\}$ with the additional layers of reduced density α_2 and α_4. In this case, the formation of additional vortices and more complex shock-wave structures is seen (t = 0.85, 1.05). The bending of the bow shock wave is almost rectilinear in the source zone. After passing the stratified region, the bending of the bow shock wave is also noticeable, which is greater the lower the values of α_j in the source layers (t = 1.25); at the end of the interaction process, the flow returns to the undisturbed steady flow mode.

The dynamics of the density fields in surface view is presented in Figure 6. One can see the blurring of the front of the bow shock wave under the action of thermal layers, and for the given several heated layers in the energy source, the front of the bow shock wave practically ceases to exist. This occurs under the influence of the development of multiple Richtmyer-Meshkov instabilities [17,22], which are characterized by the appearance of sharp fluctuations (peaks) of parameters and the formation of the accompanying "mushroom" structures.

The corresponding dynamics of the pressure fields is presented in Figure 7. Since, in the layers of the stratified source, pressure is equal to the freestream pressure, the source layers are not visible in the pressure field. It can be seen that at the beginning of the interaction, the front of the bow shock wave becomes wavy (Figure 7; the fourth and fifth lines of images, t = 0.75). Further, it is deformed, and after the passage of the stratified impulse, the flow comes to an unperturbed stationary state. It can also be seen that inside the formed vortices, the pressure, as well as the density, is reduced (Figure 7 1st line of images, t = 1.05 and t = 1.25).

The dynamics of the temperature fields are presented in Figure 8. One can see that the action of the stratified energy source causes complicated temperature structures inside the shock layer. Temperature fluctuations with high-temperature values (approximately four times higher than the temperature of the oncoming flow) are visible inside the resulting vortex structures (t = 0.85, 1.05). Temperature fluctuations are stronger for hotter layers, and for more number of heated layers in the energy source (third line of the images, t = 1.05); however, these can be expected.

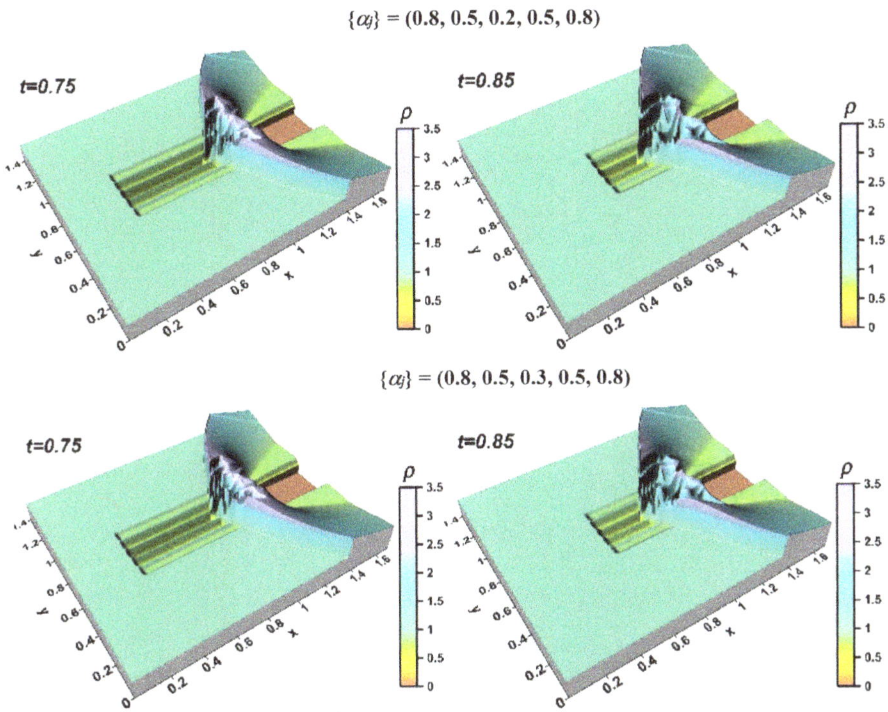

Figure 6. Dynamics of density fields for different symmetrical sets $\{\alpha_j\}$ (surface view).

The dynamics of drag force for different symmetrical sets $\{\alpha_j\}$ are presented in Figure 9. It is seen that the drag force reduction is greater for more rarefied layers in the energy source (with a greater temperature) (Figure 9a). The addition of the heated layers to the stratified source adds oscillation to the drag force dynamic, increasing the time interval on which the drag force is decreased (Figure 9b). Drag reduction is greater (and the longer action in time) the more layers are in the source with the reduced values of α_j (or higher temperatures) in the layers. Thus, it is possible to control the drag force of an AD body surface by changing the temperature values and the number of heated layers in a thermally stratified energy source. Note, that in this case, for symmetrical sets $\{\alpha_j\}$ in the source, lift forces do not arise, in contrast to asymmetric sets $\{\alpha_j\}$ in the source, as will be shown below.

The dynamics of the density fields for different asymmetric sets $\{\alpha_j\}$ in the stratified energy source are presented in Figure 10. One can see the initiation of the asymmetric vortex structures; the drop in densities in these vortices is greater for smaller values of α_j (t = 0.75, 0.85). On hotter layers, the bow shock wave diffracts with the formation of asymmetric triple shock configurations (t = 0.85). After the passage of the stratified pulse, the instabilities weaken, and the front of the bow shock wave is restored, retaining the asymmetric shape (t = 1.05). Furthermore, the shape of the bow shock wave approaches the body and its shape becomes close to symmetrical (t = 1.25). At the end of the process, when the impulse leaves the computational area, the flow returns to its original unperturbed state. Comparing the flow patterns for different sets of $\{\alpha_j\}$ presented in Figure 10, we can conclude that the drops in densities in the resulting vortices are larger for smaller values of α_j (see the third row of images). In addition, for smaller values of α_j, the action of resulting vortex structures on the lower body surface is stronger for smaller values of α_j, or larger the values of temperature in the layers (see the fourth row).

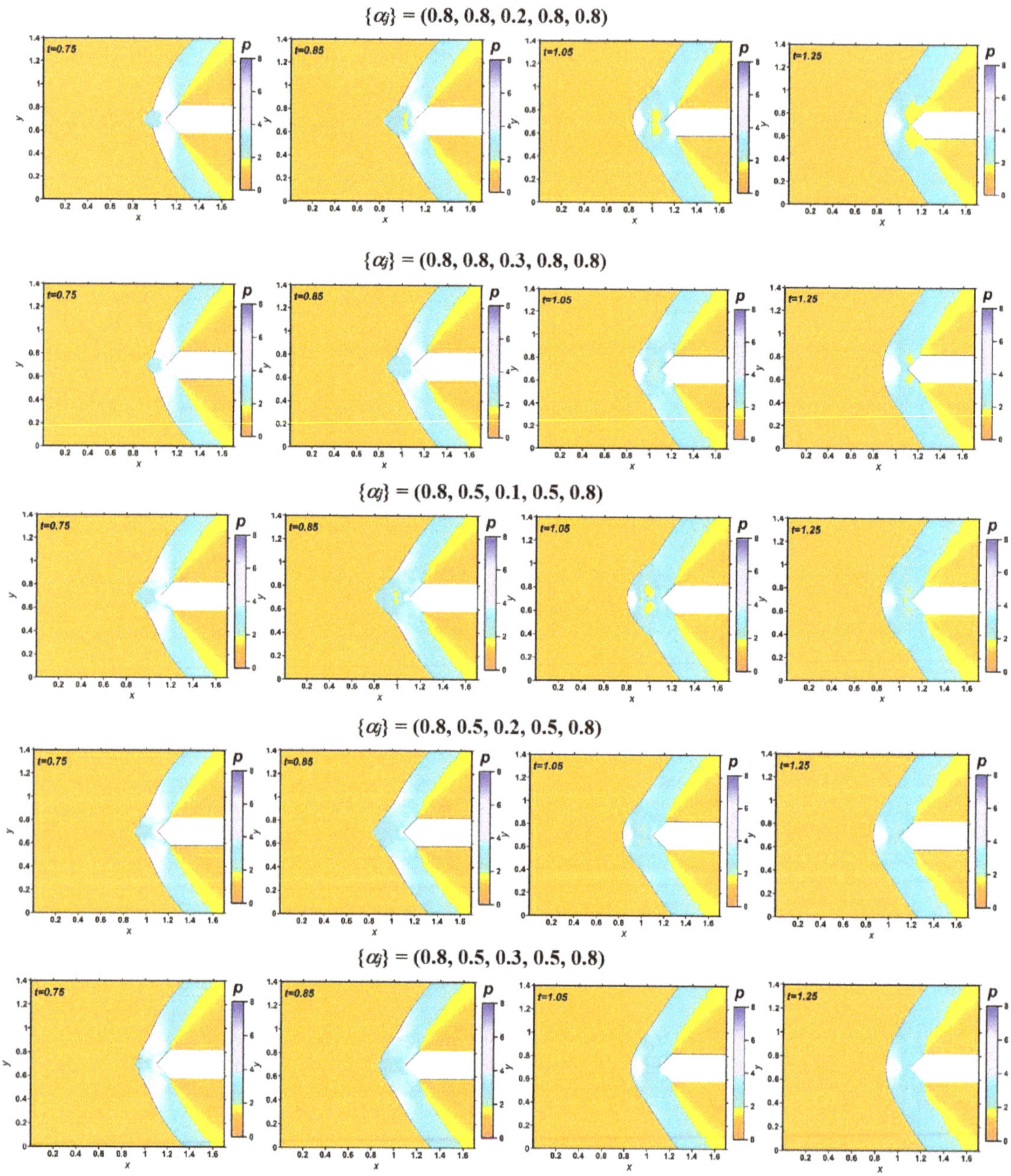

Figure 7. Dynamics of pressure fields for different symmetrical sets $\{\alpha_j\}$.

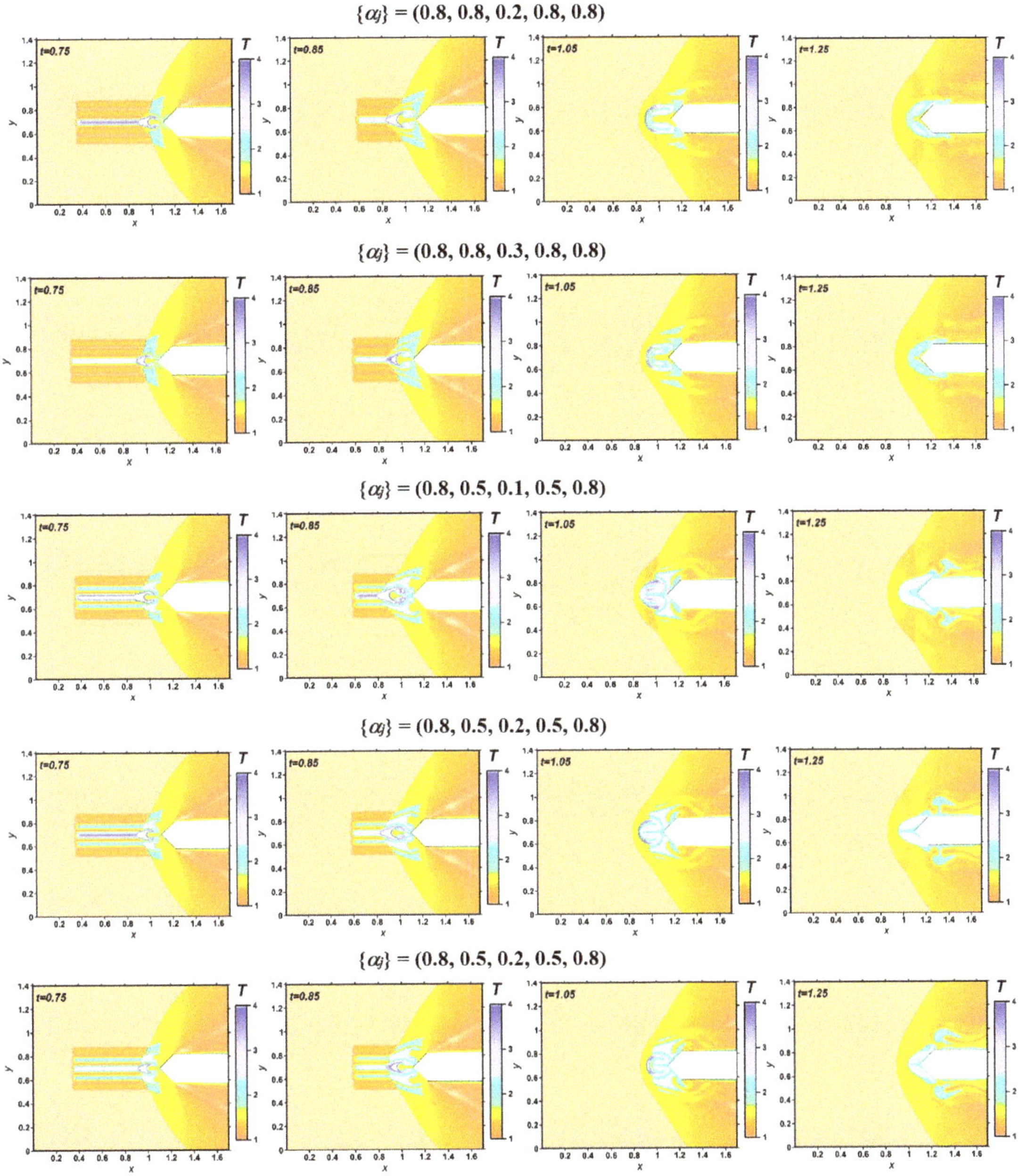

Figure 8. Dynamics of temperature fields for different symmetrical sets $\{\alpha_j\}$.

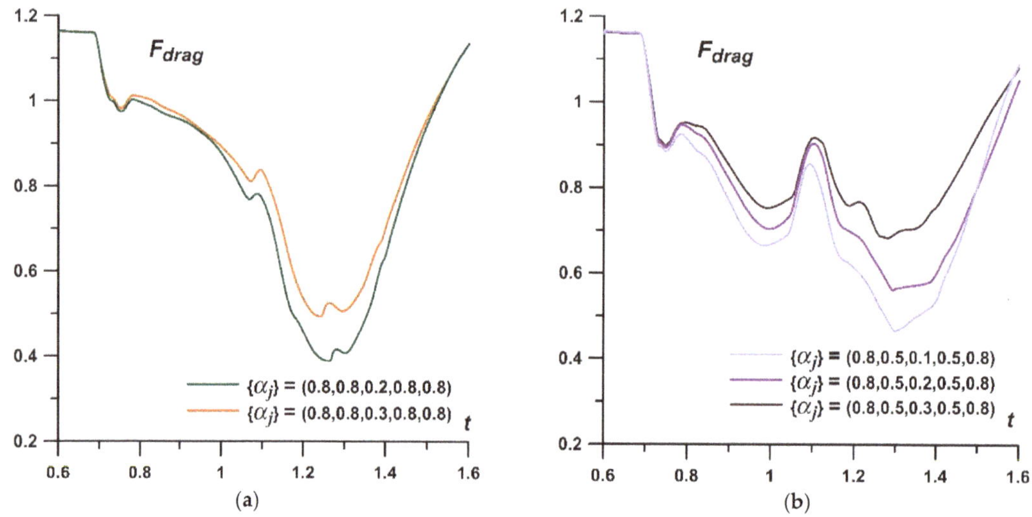

Figure 9. Dynamics of drag force for different symmetrical sets $\{\alpha_j\}$: (**a**) for the sets with one hotter layer; (**b**) for the sets with three hotter layers.

In Figure 11, the fields of density for asymmetric sets $\{\alpha_j\}$ are presented. Here, asymmetric (left row of images) and "reflected" asymmetric (right row of images) sets $\{\alpha_j\}$ are considered; where, in asymmetric sets $\{\alpha_j\} = (\alpha_1, \alpha_2, \alpha_3, \alpha_4, \alpha_5)$ and in "reflected" asymmetric sets $\{\alpha_j\} = (\alpha_5, \alpha_4, \alpha_3, \alpha_2, \alpha_1)$. In addition, the structure of the flow is demonstrated by the vector fields of the flow velocity $\boldsymbol{U} = (u,v)$. It is seen that inside the vortex structures the flow is circular. The corresponding flow patterns obtained are directly opposite, which is due to the coincidence of the symmetry axes of the energy source and the body, i.e., it is assumed that the source is rigidly installed symmetrically with respect to the AD body.

The dynamics of unsteady drag forces for different asymmetric sets $\{\alpha_j\}$ in the stratified energy source are presented in Figure 12. We consider the drag force formed by the wedge part of the body *Fdrag*, and the drag forces formed by the top and bottom surfaces of the wedge, $Fdrag_{top}$ and $Fdrag_{bottom}$. It can be seen that, as in the case of symmetrical sets $\{\alpha_j\}$, the drops in drag force *Fdrag* are greater for smaller α_j (compare the green, orange and blue curves); here, the rate of change in drag is almost the same. However, using the layers with different α_j, it is possible to set a different rate of change in the drag force, which is greater the greater the difference in α_j (in the temperature values) in the layers (see green, purple and olive curves). One can also see that for asymmetric and "reflected" asymmetric sets $\{\alpha_j\}$ the drag forces of top and bottom surfaces are of the opposite values, and the total drag forces are the same. This is connected with the fact that the axis of symmetry of the stratified source is supposed to coincide with the axis of symmetry of the body.

In Figures 13 and 14, the dynamics of unsteady lift (pitch) forces for different asymmetric sets $\{\alpha_j\}$ in the stratified energy source are presented. We consider the lift force formed by the wedge part of the body $Flift_{wedge}$, the lift force formed by the horizontal surfaces of the body $Flift_{horizontal}$, and the total lift force of the body $Flift_{total}$. The dynamics of lift forces $Flift_{wedge}$ and $Flift_{horizontal}$, for asymmetric and "reflected" asymmetric sets $\{\alpha_j\}$, are shown in Figure 13a,b, accordingly. It can be seen that the lift (pitch) forces are defined mostly by the wedge part of the body $Flift_{wedge}$, and the absolute value of these forces are greater for the sets $\{\alpha_j\}$ with the more rarefied layers (compare green, orange and blue curves).

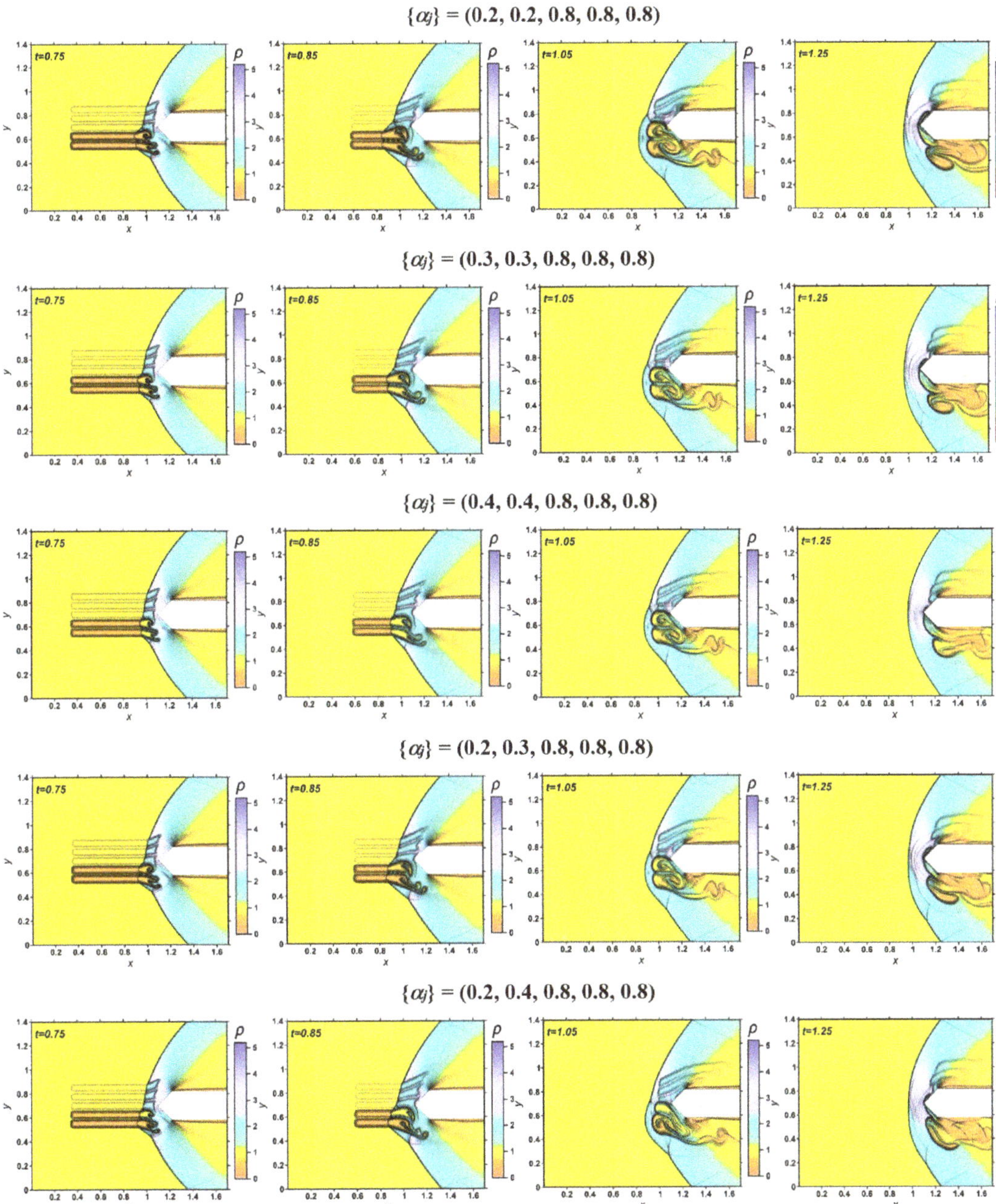

Figure 10. Dynamics of density fields for different asymmetric sets $\{\alpha_j\}$.

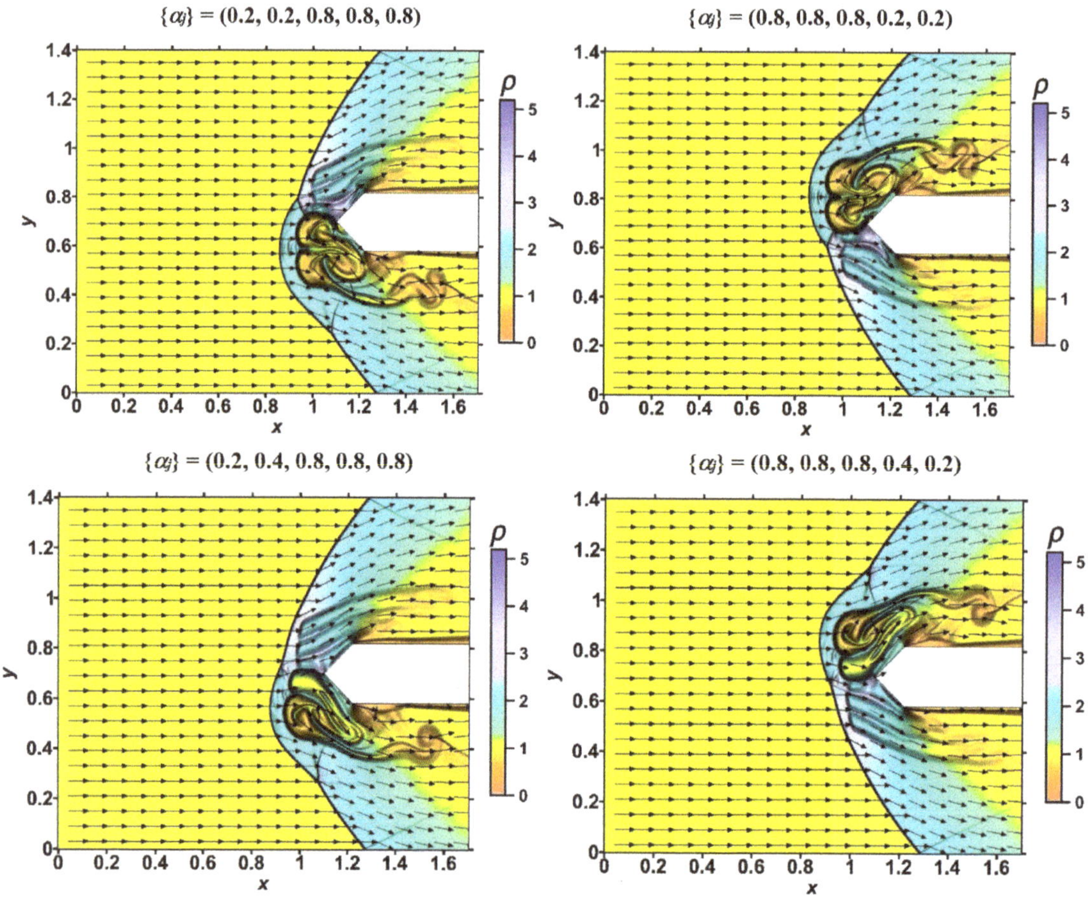

Figure 11. Fields of density and velocity for asymmetric sets $\{\alpha_j\}$ (left row) and for according "reflected" asymmetric sets $\{\alpha_j\}$ (right row), $t = 1.05$.

Additionally, it is possible to control the rate of change of the lift force $Flift_{wedge}$ by including differently heated layers in the energy source (compare green, purple and olive curves). It is seen that for two equally heated layers, the rate is strongly increased for smaller α_j (compare blue, orange and green curves for $Flift_{wedge}$), but by including the differently heated layers, the difference in the rate can be reduced to a greater extent, and therefore the greater the difference in α_j in the source layers (compare green, purple and olive curves).

The dynamics of the total lift forces $Flift_{total}$ for asymmetric and "reflected" asymmetric sets $\{\alpha_j\}$ are presented in Figure 14. The dynamics of $Flift_{total}$ are characterized by the same properties that were obtained for $Flift_{wedge}$: the absolute value of $Flift_{total}$ are greater for the sets $\{\alpha_j\}$ with the more heated layers, with smaller values of α_j (compare green, orange and blue curves), and it is possible to control the rate of change of $Flift_{total}$, by including the differently heated layers (compare green, purple and olive curves).

It can be emphasized that the dynamics of lift forces in Figures 13 and 14 are described by the symmetrical curves. Therefore, by replacing the asymmetric set $\{\alpha_j\}$ with its "reflected" set, it is possible to obtain the oppositely directed lift forces. Thus, lift forces can be created and controlled using a thermally stratified energy source by changing the temperature values in its layers. Notably, it is assumed that the stratified energy source is located symmetrically relative to the AD body (at zero angle of attack).

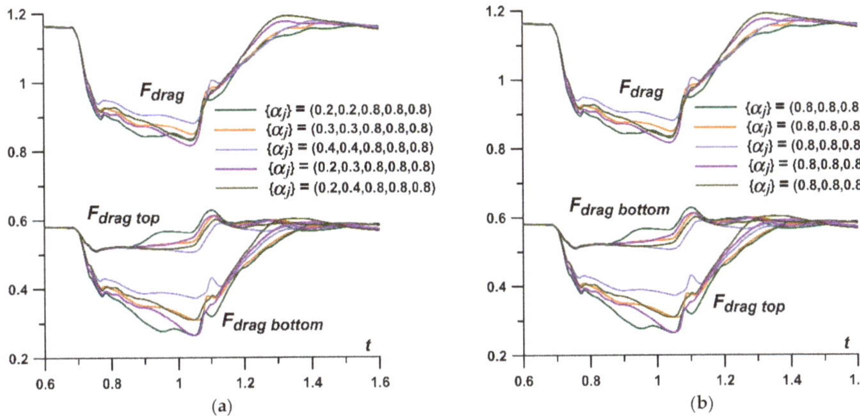

Figure 12. Dynamics of drag force for different asymmetric (**a**) and "reflected" asymmetric (**b**) sets $\{\alpha_j\}$.

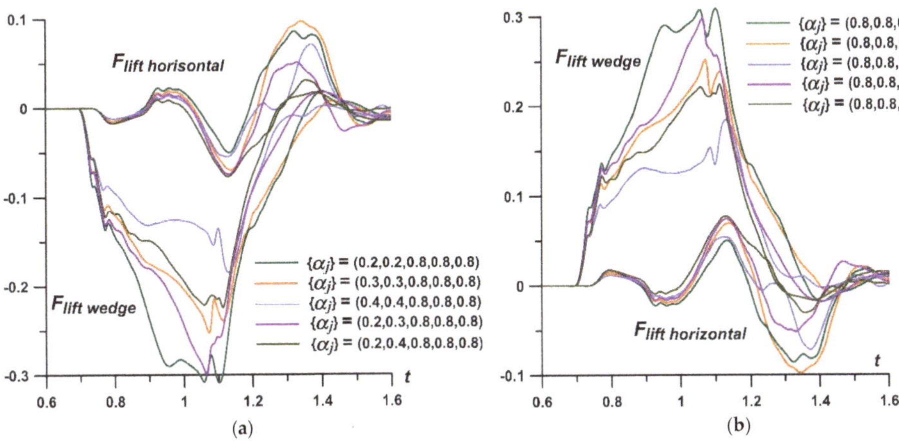

Figure 13. Dynamics of lift forces for different asymmetric (**a**) and "reflected" asymmetric (**b**) sets $\{\alpha_j\}$.

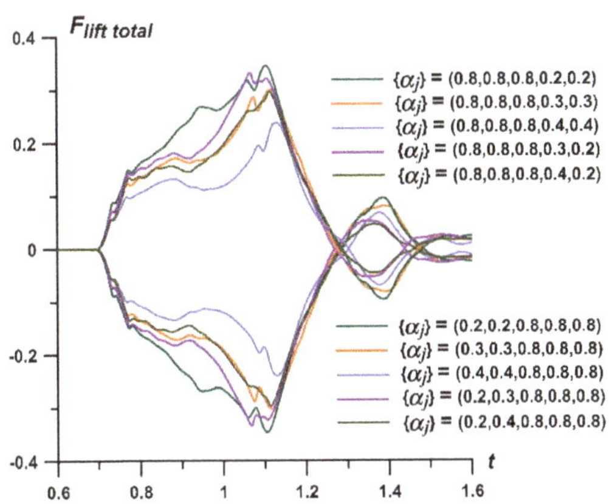

Figure 14. Dynamics of total lift forces for different asymmetric and "reflected" asymmetric sets $\{\alpha_j\}$.

5. Conclusions

The unsteady effect of a thermally stratified energy source of limited length in time on the supersonic flow past a pointed plate at M = 2 was numerically investigated. The axis of symmetry of the stratified source was supposed to coincide with the axis of symmetry of the AD body. The new results obtained are as follows:

1. Almost complete destruction of the bow shock wave in the density field was achieved due to the multiple generation of the Richtmyer-Meshkov instabilities during its interaction with a stratified energy source. The multi-vortex mechanism of the impact of a stratified energy source on the aerodynamic characteristics of an AD body was shown.
2. The principles of supersonic flow, local in time control using a stratified energy source of limited length, have been formulated, namely:

- By setting lower values of α_j (higher temperatures) in the layers and larger differences in density values, it is possible to obtain more intensive vortices accompanied the Richtmyer-Meshkov instabilities.
- By setting symmetrical sets $\{\alpha_j\}$, one can temporarily decrease the drag force of an AD body. Drag reduction is greater (and the longer action in time) the more layers are in the stratified energy source with the reduced values of α_j (or higher temperatures in the layers).
- By setting asymmetric sets $\{\alpha_j\}$, it is possible to create a temporarily lift (pitch) force (at zero angle of attack), which is greater (and the longer action in time) the more layers are in the source with the reduced values of α_j (or of higher temperatures).
- By setting "reflected" asymmetric sets $\{\alpha_j\}$, it is possible to create an oppositely directed lift (pitch) force (at zero angle of attack), which is greater (and the longer action in time) the more layers are in the source with the reduced values of α_j. Moreover, the drag forces for the "reflected" asymmetric sets $\{\alpha_j\}$ remain the same as for the asymmetric sets $\{\alpha_j\}$ used.
- It is possible to control the rate of change in the drag and lift forces. For equally heated layers, the rate is strongly increased for smaller α_j (or higher temperatures), but by including the differently heated layers (with different α_j) the rate can be reduced to a greater extent, and the greater the difference in α_j in the source layers.

Author Contributions: Conceptualization—O.A.A.; Formal analysis—O.A.A. and O.V.K.; Software—O.A.A. and O.V.K.; Visualization—O.A.A. and O.V.K.; Writing—original draft—O.A.A. All authors have read and agreed to the published version of the manuscript.

Funding: This research received no external funding.

Data Availability Statement: All obtained data underlying the conclusions made can be provided by the authors upon request.

Conflicts of Interest: The authors declare no conflict of interest.

Nomenclature

D	transverse size of the aerodynamic body
E, ε	volume kinetic energy and specific internal energy of the gas
h_j	the width of the layers in the energy source
h_x, h_y	the space steps in x- and y- directions
N	a number of layers in the energy source
M_∞	the freestream Mach number
p, ρ, T	pressure, density, and temperature of the gas
Re, Pr	the Reynolds number and the Prandtl number
T	time
$\boldsymbol{U}$	vector of the flow velocity, $\boldsymbol{U} = (u,v)$
y_0	coordinate of the body's axis of symmetry

y_{es}	coordinate of the lower boundary of the energy source
α_j	rarefaction parameter in the layers of the stratified energy source
γ	ratio of specific heats
j	parameters in the layers of the stratified energy source
n	normalizing parameters
t	parameters at the apex of the body
∞	freestream parameters

Abbreviations

AD	aerodynamic
SW	shock wave
CD	contact discontinuity

References

1. Knight, D.D. *Energy Deposition for High-Speed Flow Control*; Cambridge University Press: Cambridge, MA, USA, 2019. [CrossRef]
2. Ahmed, M.Y.M.; Qin, N. Forebody shock control devices for drag and aero-heating reduction: A comprehensive survey with a practical perspective. *Prog. Aerosp. Sci.* **2020**, *112*, 100585. [CrossRef]
3. Leonov, S.B. Review of plasma-based methods for high-speed flow control. In Proceedings of the Sixth International Conference on Fluid Mechanics, Guangzhou, China, 30 June–3 July 2011. [CrossRef]
4. Russell, A.; Zare-Behtash, H.; Kontis, K. Joule heating flow control methods for high-speed flows. *J. Electrost.* **2016**, *80*, 34–68. [CrossRef]
5. Fomin, V.M.; Tretyakov, P.K.; Taran, J.-P. Flow control using various plasma and aerodynamic approaches (short review). *Aerosp. Sci. Technol.* **2004**, *8*, 411–421. [CrossRef]
6. Georgievsky, P.Y.; Levin, V.A. Supersonic flow over bodies in the presence of external energy release. *Pis'ma v Zh. Tekh. Fiz.* **1988**, *14*, 684–687. (In Russian). Available online: http://journals.ioffe.ru/articles/viewPDF/31216 (accessed on 27 September 2022).
7. Artem'ev, V.I.; Bergel'son, V.I.; Nemchinov, I.V.; Orlova, T.I.; Smirnov, V.A.; Hazins, V.M. Changing the regime of supersonic streamlining obstacles via raising the thin channel of low density. *Izv. Akad. Nauk SSSR Meh. Židk. Gaza.* **1989**, *5*, 146–151. (In Russian)
8. Nemchinov, I.V.; Artem'ev, V.I.; Bergelson, V.I.; Khazins, V.M.; Orlova, T.I.; Rybakov, V.A. Rearrangement of the bow shock shape using a "hot spike". *Shock Waves* **1994**, *4*, 35–40. [CrossRef]
9. Kolesnichenko, Y.F.; Brovkin, V.G.; Azarova, O.A.; Grudnitsky, V.G.; Lashkov, V.A.; Mashek, I.C. Microwave energy release regimes for drag reduction in supersonic flows. AIAA 2002-0353. In Proceedings of the 40th Aerospace Sciences Meeting and Exhibit, Reno, NV, USA, 14–17 January 2002. [CrossRef]
10. Bityurin, V.; Klimov, A.; Leonov, S.; Brovkin, V.; Kolesnichenko, Y.; Popov, N.; Van Wie, D.M. Shock waves structure and velocity at propagation through non-homogeneous plasma. AIAA 2000-2571. In Proceedings of the 31st Plasmadynamics and Lasers Conference, Denver, CO, USA, 19–22 June 2000. [CrossRef]
11. Leonov, S.B.; Carter, C.D.; Hedlund, B.E.; Houpt, A.W.; Ombrello, T.; Firsov, A.A. Control of amplitude and position of reflected shock wave by stripwise plasma. AIAA 2018-0683. In Proceedings of the Aerospace Sciences Meeting, Kissimmee, FL, USA, 8–12 January 2018. [CrossRef]
12. Leonov, S.; Carter, C.; Houpt, A.; Ombrello, T. Mitigation of reflected shock wave by streamwise plasma array. In Proceedings of the 7th European Conference for Aeronautics and Space Sciences, Milan, Italy, 3–6 July 2017. [CrossRef]
13. Leonov, S.B.; Firsov, A.A.; Houpt, A.W. Suppression of reflected oblique shock wave by multi-filamentary plasma. *J. Phys. Conf. Ser.* **2018**, *1112*, 012005. [CrossRef]
14. Gan, T.; Wu, Y.; Sun, Z.; Jin, D.; Song, H.; Jia, M. Shock wave boundary layer interaction controlled by surface arc plasma actuators. *Phys. Fluids* **2018**, *30*, 055107. [CrossRef]
15. Apazidis, N.; Sembian, S.; Liverts, M. Blast wave interaction with thermal and density inhomogeneities in air. Proceeding of the 32nd International Symposium on Shock Waves, Singapore, 14–19 July 2019. [CrossRef]
16. Lapushkina, T.A.; Erofeev, A.V.; Azarova, O.A.; Kravchenko, O.V. Interaction of a plane shock wave with an area of ionization instability of discharge plasma in air. *Aerosp. Sci. Technol.* **2019**, *85*, 347–358. [CrossRef]
17. Azarova, O.A. Generation of Richtmyer-Meshkov and secondary instabilities during the interaction of an energy release with a cylinder shock layer. *Aerosp. Sci. Technol.* **2015**, *42*, 376–383. [CrossRef]
18. Azarova, O.A. Supersonic flow control using combined energy deposition. *Aerospace* **2015**, *2*, 118–134. [CrossRef]
19. Azarova, O.A.; Kravchenko, O.V.; Lapushkina, T.A.; Erofeev, A.V. Density and temperature fluctuations behind a shock wave under the influence of a stratified energy source. *Tech. Phys. Lett.* **2020**, *46*, 649–652. [CrossRef]
20. Azarova, O.A.; Krasnobaev, K.V.; Kravchenko, O.V.; Lapushkina, T.A.; Erofeev, A.V. Redistribution of energy in a viscous heat-conductive medium during the interaction of a shock wave with a temperature layered plasma region. *J. Phys. Conf. Ser.* **2020**, *1698*, 012004. [CrossRef]

21. Azarova, O.A.; Lapushkina, T.A.; Krasnobaev, K.V.; Kravchenko, O.V. Redistribution of energy during interaction of a shock wave with a temperature layered plasma region at hypersonic speeds. *Aerospace* **2021**, *8*, 326. [CrossRef]
22. Azarova, O.A.; Kravchenko, O.V. Impact of a thermally stratified energy source on the bow shock wave and aerodynamic characteristics of a body. *J. Phys. Conf. Ser.* **2021**, *1891*, 012025. [CrossRef]
23. Roache, P.J. *Computational Fluid Dynamics*; Mir: Moscow, Russia, 1980.
24. Azarova, O.A. Complex conservative difference schemes for computing supersonic flows past simple aerodynamic forms. *J. Comp. Math. Math. Phys.* **2015**, *55*, 2025–2049. [CrossRef]
25. Lapushkina, T.A.; Erofeev, A.V.; Azarova, O.A.; Kravchenko, O.V. Passage of a plane shock wave through the region of a glow gas discharge. *Tech. Phys.* **2019**, *64*, 34–41. [CrossRef]
26. Rozhdestvensky, B.L.; Yanenko, N.N. *Systems of Quasilinear Equations and Their Applications to Gas Dynamics*; Nauka: Moscow, Russia, 1978.

fluids MDPI

Article

Dynamics of Shock Structure and Frontal Drag Force in a Supersonic Flow Past a Blunt Cone under the Action of Plasma Formation †

Irina Znamenskaya [1], Vladimir Chernikov [1] and Olga Azarova [2,*]

1 Faculty of Physics, Lomonosov Moscow State University, 119234 Moscow, Russia; znamen@phys.msu.ru (I.Z.); vachernikov@rambler.ru (V.C.)
2 Federal Research Center "Computer Science and Control" of the Russian Academy of Sciences, 119333 Moscow, Russia
* Correspondence: olga_azarova@list.ru

† This paper is an extended version of our paper "Supersonic Flow past a Blunt Cone under the Action of Plasma Formation" published at the 8th European Conference for Aeronautics and Space Sciences (EUCASS), Madrid, Spain, 1–4 July 2019.

Abstract: The paper is devoted to the experimental and CFD investigation of a plasma formation impact on the supersonic flow over a body "blunt cone-cylinder". In the experiments, a series of schlieren pictures of bow shock wave–blast waves non-stationary interaction was obtained with the use of high speed shadowgraphy. The accompanying calculations are based on the system of Euler equations. The freestream Mach number is 3.1. The plasmoid is modeled by the instantaneous release of energy into a bounded volume of gas, increasing the pressure in the volume. The research of the dynamics of a shock wave structure caused by the bow shock wave and blast flow interaction has been conducted. The significant value of energy released to a supersonic flow (500J) allowed constructing a diagram of the generation and dynamics of the resulting shock waves and contact discontinuities, as well as obtaining a significant drop in the drag force and stagnation pressure (up to 80%). The dynamics of a low density and high gas temperature zone, which becomes the main factor reducing the frontal body drag force, was researched. The dynamics of the front surface drag forces have been studied for different values of the plasmoid energy as well. Qualitative agreement of the numerical flow patterns with the experiment ones has been obtained.

Keywords: supersonic flow; bow shock wave; plasmoid; blast shock wave; shock-wave structure; drag force reduction

Citation: Znamenskaya, I.; Chernikov, V.; Azarova, O. Dynamics of Shock Structure and Frontal Drag Force in a Supersonic Flow Past a Blunt Cone under the Action of Plasma Formation. *Fluids* **2021**, *6*, 399. https://doi.org/10.3390/fluids6110399

Academic Editor: Mehrdad Massoudi

Received: 22 September 2021
Accepted: 1 November 2021
Published: 4 November 2021

1. Introduction

Control of supersonic flows by means of plasma formations generated by electrical discharges, microwave energy release, and laser pulses is currently an extensive field of aerospace engineering studies (see [1] and surveys in [2–6]). A review of various applications of theoretical and experimental studies for supersonic and hypersonic flow modes in order to control drag reduction, establish the effective geometry of the vehicle, and mitigate sonic boom from it was presented in [7].

Reorganization of unsteady flow under the action of an external energy release has been researched since the second half of the last century and the beginning of this century [8–12]. In air, the effect of the external energy source produced by microwave discharge was shown to result in decreasing stagnation pressure together with the reduction in the drag force of a blunt cylinder [13]. A vortex mechanism of these phenomena was established in the calculations. Microwave energy releases, which effects the supersonic flow over a cylinder, were studied experimentally in [14,15] and numerically in [16].

Numerous experiments and simulations have been devoted to investigating laser impact on supersonic/hypersonic flow since the end of the twentieth century [17,18].

Among others there are the results of laser pulse impacts presented in [19,20] and the results of numerical simulations for laser action presented in [20–23]. The interaction of laser plasma with a bow shock wave, starting from the moment the laser spark formed until the moment when a gasdynamic perturbations developed has been considered [24]. The calculations were performed using the Navier–Stokes equations for supersonic and hypersonic air flows for a spherical blunt body and a body in the form of a double cone. The simulation results confirmed that localized energy release can be effectively used to control the bow shock configurations.

The discharge plasma effect on supersonic flow was investigated experimentally in [25] taking into account plasma, electric, and magnetic effects. In experiments, the possibility of using plasma formations (plasmoids) to change the supersonic flow near the model has been investigated in [26]. Complicated shock structures are formed in the processes of the energy sources-shock layers interaction including formation of triple-shock configurations [27,28]. The essential impact on the dynamics of the frontal drag force and the bow shock wave behaviour due to the presence of a heated area produced by energy release was obtained in all of these studies.

This paper is devoted to the experimental and numerical research of the plasmoid effect on the shock structure and frontal drag force that occurs during the supersonic streamlining a body "blunt cone-cylinder". The novelty of the work is connected with the significant value of energy released to a supersonic flow (500 J) which allowed to track the generation and dynamics of the resulting shock waves and contact discontinuities, as well as to obtain a significant drop in the drag force and stagnation pressure (up to 80%). Comparison of the experimental schlieren images and computation flow patterns is analyzed together with the frontal surface drag force dynamics and the dynamics of shock wave (SW) fronts and contact discontinuities (CD) in the developing shock-wave structures.

2. Experimental Study

2.1. Experimental Setup

A scheme of the experimental installation is shown in Figure 1a (here, the flow direction is taken from top to bottom). Setup contains the Laval nozzle (1) (which had been designed for Mach number M = 2–3.5) and a plasma generator (2). It is based on a magnetoplasma compressor of special construction. These devices are mounted inside a low-pressure chamber. High pressure at the nozzle inlet was varied from 0.2 to 0.5 from that of the magnetoplasma compressor. The pressure at the nozzle inlet was supplied by a valve via connecting the pile. Power was supplied (4) via a connector.

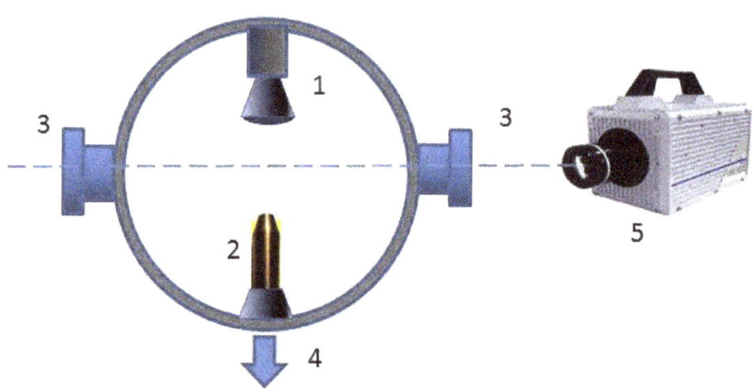

(a)

Figure 1. *Cont.*

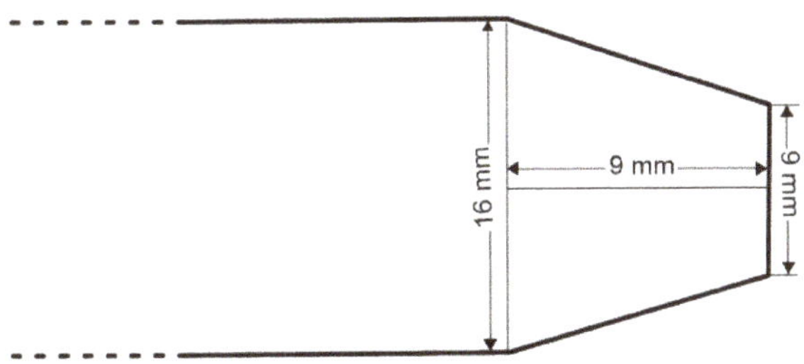

(b)

Figure 1. Experimental installation (schematic) (**a**): 1—Laval nozzle; 2—electrical discharger; 3—windows for visualization; 4—to power supply; 5—high speed camera; (**b**) model dimensions.

When a high-voltage pulse was fed to the spark gap, the supply voltage was fed to the magnetoplalma compressor, and it then was discharged with the generation of a plasma jet. Electric current time duration was ~100 μs, maximal current was ~12 kA, and voltage drop across the discharge was 700V; the average electron density is 10^{15}–10^{16} cm^{-3}. The freestream Mach number tested was 3.1. The diameter of the cylinder part of the body is $D = 1.6{\cdot}10^{-2}$ m; the diameter of a frontal surface of the body is $D_f = 9{\cdot}10^{-3}$ m (Figure 1b).

The classical Tepler shadow scheme was used for flow visualization, including a parallel light beam passing through the windows (3). A digital recording system with the high temporal and spatial resolution was employed. The high-speed digital camera (5) has the exposure time of a frame about 1 μs. Recording regimes with 150,000 frames/s and 325,000 frames/s were used; the interval between the frames was about 7 and 4 μs, respectively. Films with a duration up to 3 s including all the stages of the plasma initiation, relaxation, and gasdynamic processes up to stationary streamlining have been recorded.

A scheme of the supersonic flow past the model is presented in Figure 2.

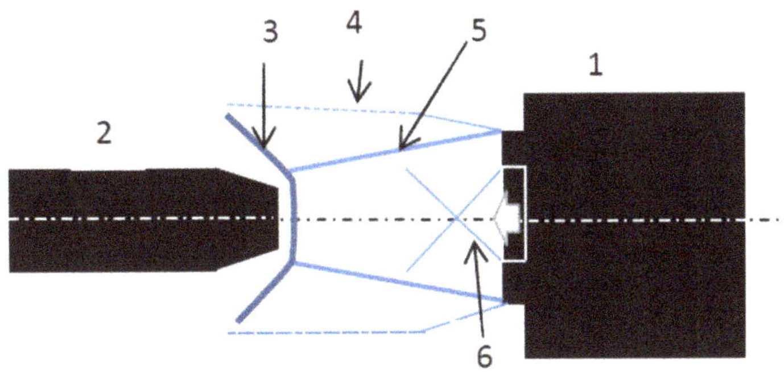

Figure 2. Scheme of the supersonic flow field: 1—supersonic nozzle; 2—plasma formation generator in the flow; 3—bow shock wave; 4—outer boundary of the mixing layer; 5—inner boundary of the mixing layer; 6—suspended shocks.

2.2. Experimental Results

High-speed shadow imaging showed that due to the short time of plasma energy release, a blast wave arises from the discharge area. Its dynamics controls the non-steady stage of the bow shock wave structure evolution. In front of the streamlined body, the shock layer was shown to be reconstructed, with the value of the bow shock wave standoff on the axis of symmetry increasing significantly: approximately as long as the value of the plasmoid diameter (4 cm). Schlieren images of the plasmoid impact on the supersonic flow are presented in Figure 3. Steady streamlining is established at 150–200 μs after nozzle launching; some fractures of the bow shock wave are a result of the method of supersonic flow organization by means of the nozzle.

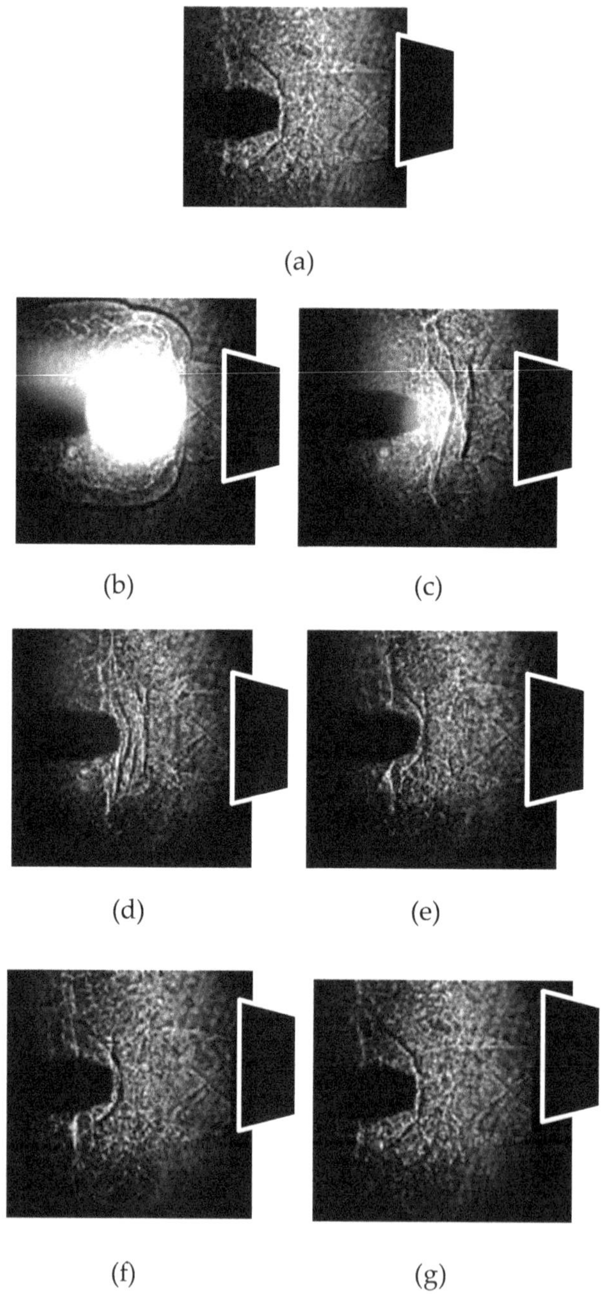

Figure 3. Schlieren images: (**a**) t = 0; (**b**) t = 66 μs; (**c**) t = 107 μs; (**d**) t = 140 μs; (**e**) t = 166 μs; (**f**) t = 173 μs; (**g**) t = 193 μs.

After the establishment of steady flow mode, the plasma energy release begins (Figure 3b), thus causing a strong reconstruction of the entire flow. In the experiment, the moment of time of the discharge inclusion is accepted as the initial time. Several shock waves are visualized (Figures 2d and 3c), which interact with each other (Figure 3e,f), and finally they form a new bow shock when the flow becomes steady again sometime after the end of the impact of the energy deposition area (Figure 3g).

3. Numerical Simulations

3.1. Methodology, Statement of the Problem, and Grid Convergence

Supersonic flow over a body "blunt cone-cylinder" under the impact of an energy release was studied at M = 3.1. The simulation is based on the Euler system of equations for perfect inviscid gas in curvilinear orthogonal coordinates with the ratio of specific heats $\gamma = 1.4$.

$$(\mathbf{U}r)_t + (\mathbf{F}r)_x + (\mathbf{G}r)_r = \mathbf{H}, \quad (1)$$

$$\begin{aligned} \mathbf{U} &= (\rho, \rho u, \rho v, E)^T, \mathbf{F} = \left(\rho u, p + \rho u^2, \rho u v, u(E+p)\right)^T, \\ \mathbf{G} &= \left(\rho v, \rho u v, p + \rho v^2, v(E+p)\right)^T, \mathbf{H} = (0, 0, p, 0)^T, \end{aligned} \quad (2)$$

$$E = \rho(\varepsilon + 0.5(u^2 + v^2)). \quad (3)$$

Here, the r-coordinate is directed on the radius of a body. The state equation for a perfect gas is used:

$$\varepsilon = p/(\rho(\gamma - 1)),$$

where ρ, p, u, and v are the gas density, pressure, and velocity of the x-components and y-components, and ε is the specific internal energy.

The problem is solved in dimensionless variables. Dimensionless quantities for time, spatial variables, components of sound velocity and velocity, gas density, pressure, and temperature are expressed with the dimensional ones (marked with the index "dim") as follows.

$$\begin{gathered} t = \frac{t_{dim}}{t_n}, x = \frac{x_{dim}}{l_n}, r = \frac{r_{dim}}{l_n}, u = \frac{u_{dim}}{u_n}, \\ v = \frac{v_{dim}}{u_n}, c = \frac{c_{dim}}{u_n}, \rho = \frac{\rho_{dim}}{\rho_n}, p = \frac{p_{dim}}{p_n}, \\ T = \frac{T_{dim}}{T_n}. \end{gathered} \quad (4)$$

Here, the following scales for the parameters are accepted:

$$\rho_n = \rho_\infty, p_n = p_\infty, l_n = k_l^{-1} D, T_n = T_\infty, \quad (5)$$

$$u_n = (p_\infty / \rho_\infty)^{0.5}, t_n = \frac{l_n}{u_n}.$$

where k_l is the dimensionless value of D.

A domestic code based on the complex conservative difference schemes of the second approximation order in space and in time is used in the simulations [29]. The body's boundaries are introduced into the calculation area without breaking the space-time conservation properties in it. For this purpose, the boundaries of the body are approximated by stepped lines, and discrete conservation laws are written for each resulting configuration. This allows calculations to be carried out conservatively in the entire computational domain, including the boundaries of the body [29]. The position of the angular part of the body on the grid in an enlarged form is shown in Figure 4. In the calculations, the staggered numerical grids are used with the distance between the nodes at each time level equal to $2h_x$ and $2h_y$ (h_x and h_y are the space steps in x-directions and y-directions).

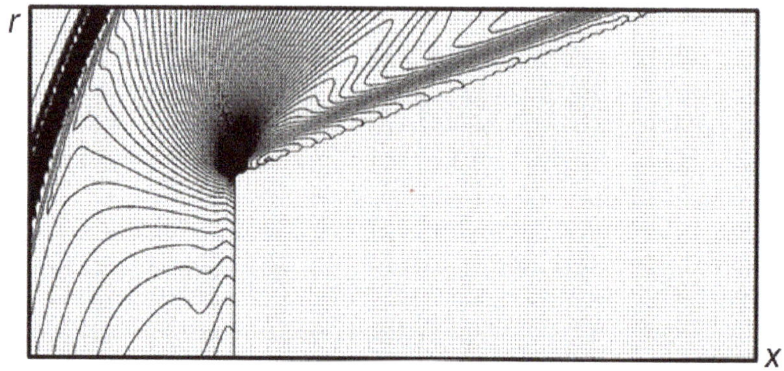

Figure 4. The position of the angular part of the body on a computational grid (enlarged).

Grid convergence analysis was conducted for three difference grids (Table 1, $t = 0.6$). Here, the characteristics of the selected grids are presented and are chosen by taking into account the flow symmetry. The analysis of the stagnation parameters obtained using these grids is presented as well as the relative errors. Figure 5 demonstrates the flow fields in isochores (Figure 5a) and the dynamics of the parameters at the stagnation point (Figure 5b) obtained by using these three difference grids. It can be observed that despite the fact that the grids differ significantly (the numbers of nodes of Grid1 and Grid3 differ by 16 times; Grid1 and Grid2 differ by four times), the values at the stagnation point differ from their theoretical values from 0.5% for the stagnation pressure (Grid1) to 5.5% for the stagnation density (Grid3); the relative errors are smaller for finer grids. In addition, the positions of the bow shock wave almost coincide (see Figure 5a).

Table 1. Characteristics of grids and analysis of grid convergence.

Grid	Steps $h_x = h_y$	Sizes	Relative Error, p_t	Relative Error, ρ_t
Grid1	0.0005	2000 × 1000	0.469%	1.873%
Grid2	0.001	1000 × 500	1.336%	4.477%
Grid3	0.002	500 × 250	2.182%	5.536%

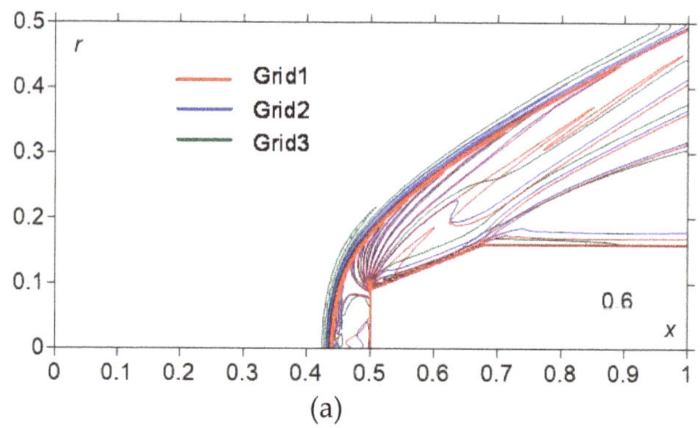

Figure 5. *Cont.*

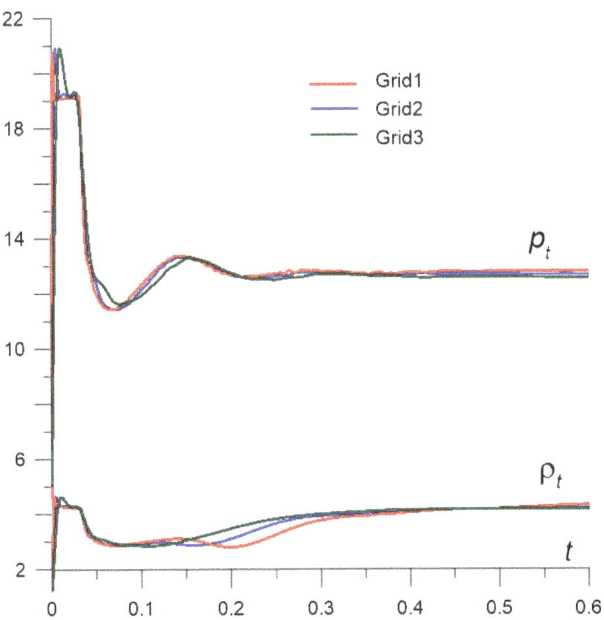

(b)

Figure 5. Analysis of the grid convergence on three different grids: (**a**)—density fields (superposed); (**b**)—dynamics of the stagnation pressure p_t and density ρ_t.

Thus, all of these factors indicate that the grid convergence takes place. For the simulation, we used Grid1, which contains 10^6 working nodes, and $\approx 10^3$ nodes are located on the diameter of a cylinder part of the body (2*R*). In the simulation, the symmetrical flow picture is considered by regarding the experimental one, which is connected with the possibilities of the using software.

Initial conditions for the problem are the fields of gas parameters in a converged supersonic steady flow past the body (Figure 6), $t = 0.6$. At this time moment, the stagnation pressure and density differ from their theoretical values by 0.469% and 1.873%, accordingly. The boundary conditions have a sense of the absence of normal flows for the corresponding parameters on the body surfaces and the absence of the reflection in the normal directions at the exit flow boundaries.

The energy source is supposed to have a spherical shape. It is assumed to arise instantly in the steady flow at the time moment t_i; the coordinate of its center x_0 was chosen from the experiment. The radius of the energy source is chosen so that the volume to be located in front of the bow shock wave. The pressure in the energy source p_i is supposed to be larger than in the surrounding flow while density and velocity remain the same (so the temperature in the energy source is increased in comparison with the surrounding flow). Thus, the model of the instant explosion of a bounded gas volume is used for energy deposition. The pressure value p_i in the energy source is defined from the following relation.

$$\eta E_0 = 4/3\pi r_i^3 (p_i - p_\infty)/(\gamma - 1). \quad (6)$$

Here, η is the part of the discharge energy spent to the expansion of a gas, $E_0 = 500$ J (from the experiment). The value of η was estimated from the results of numerical modeling from the conditions of qualitative proximity of the processes occurring relative to the experiment. In the simulations, η was set to 0.07, i.e., it was assumed that 7% of the energy was consumed in gas expansion. It should be noted that in [22], an estimate of 0.1 was obtained for the value of η. The defining flow parameters and the normalizing coefficients used in the simulations are presented in Table 2.

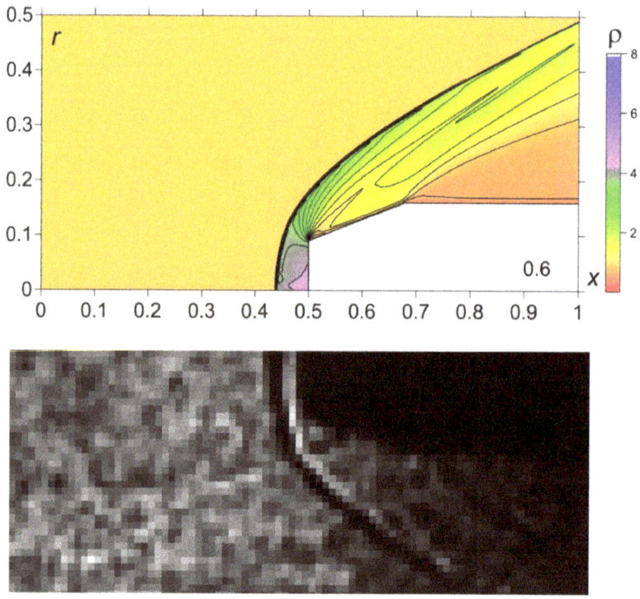

Figure 6. Steady flow: comparison with the experiment; **upper**—calculations; **bottom**—experiment (enlarged and rotated).

Table 2. Defining flow parameters and normalizing coefficients accepted in the simulations.

Parameter	Dimensional Value	Dimensionless Value	Normalizing Coefficient
Mach number of the incoming flow M_∞		3.1	
Ratio of specific heats γ		1.4	
Initial gas pressure p_∞	2 atm	1.0	p_n = 2 atm = 2 × 1.01325 × 10^5 Pa
Initial gas density ρ_∞	4.71 kg/m^3	1.0	ρ_n = 4.71 kg/m^3
Initial gas temperature T_∞	150 K	1.0	T_n =150 K
Pressure in the energy supply zone p_i	66.42 atm	33.2124	p_n = 2 atm = 2 × 1.01325 × 10^5 Pa
Radius of the energy supply zone r_i	8 × 10^{-3} m	0.16	l_n = 5 × 10^{-2} m
Energy spent on the expansion of the gas	35 J	1.3817	$E_n = l_n{}^3 p_n$
Length		1	l_n = 5 × 10^{-2} m
Velocity		1	$u_n = (p_n/\rho_n)^{0.5}$ = 207.4258 m/s
Time		1	$t_n = l_n/u_n$ = 2.4105 × 10^{-4} s = 241 µs
The time of switching on the energy source	144.8 µs	0.601	t_n = 241 µs
Interaction start time	145.9 µs	0.6053	t_n = 241 µs

3.2. Results of the Simulations

The interaction of the energy source with the shock layer was shown to cause the change of the entire flow. The gas in the energy source moves from its center to the periphery and moves towards the bow shock as well. The shape of the energy source becomes asymmetrical. The process of the instant explosion is accompanied by the break of a shock on the boundary of the compressed gas, which can be described by the solution of the Riemann problem for the decay of an arbitrary discontinuity. As a result of gas expansion, the shock wave and contact discontinuity moving from the center are originated together with a rarefaction wave moving to the volume center. For a particular set of the parameters, a weak shock wave also can be generated from a boundary of the rarefaction wave. In this manner, an area of heated gas is formed in the internal region of expanding gas. The impact of this area was shown to be a reason for the front drag force reduction under the action of energy release [21,22].

In Figure 7, the initial stage of the dynamics of energy source-shock layer interaction is presented (dimensionless time instants are indicated in the lower right corner). This stage is associated with the creation of a hot area due to energy release and its impact on the body. At the beginning of the interaction (t = 0.6053), a strong source shock wave and contact discontinuity are generated (t = 0.606). An area of compressed gas with increased pressure is formed together with a strong shock structure that is caused by the source's shock wave and contact discontinuity interacting with the bow shock (t = 0.62). Then, the pressure in the area of the compressed gas decreases and a heated gas region is formed at the central part of the expanding energy source (t = 0.64). The right boundary of this heated gas area gives rise to the weak shock wave from which a modified new bow shock will be formed (t = 0.66). Later, it comes to the less hot left area between the contact discontinuity and the source shock wave (t = 0.68) and is strengthened there (t = 0.7). This modified bow shock can be interpreted as a bow shock formed by the streamlining of an energy source. It moves to the blast shock wave (t = 0.7, 0.7174). A boundary of the heated area is clearly observed (for t = 0.62–0.80 in Figures 7 and 8), which is a contact discontinuity with different values of the flow density. This discontinuity (see image for t = 0.7174) can be the nearest shock to the body in the experimental image in Figure 3c. Multiple bow shock diffractions are observed during this stage of the interaction (t = 0.62–0.68), with the formation of a triple configuration in the upper part of the flow (t = 0.66–0.7) [27,28].

In Figure 8 the middle stage of the interaction is presented. This stage is associated with the movement of the hot area to the surface of the body and passing it behind the body. The modified bow shock moves to the source blast wave, becomes weaker, and later merges with it (t = 0.74, 0.76). At the same time, a new shock wave (SW1) is formed at the heated area boundary (t = 0.76). This shock wave is moving to the left towards the blast wave (t = 0.78). Another shock wave is initiated as well (SW2) (t = 0.80). After the hot area passing behind the body, a less heated gas (located between the contact discontinuity and the blast wave) is affecting the body's surface. Thus, during the movement of the perturbation area to the body, a series of the shock waves was generated (up to three or even four ones) near the front surface of the body (t = 0.76–0.8165). They merge together, originating a strong shock wave that interacts with the left fragment of the source wave forming a new bow shock (t = 0.84–0.88). The area of compressed gas with high pressure near the body front surface at this time was observed.

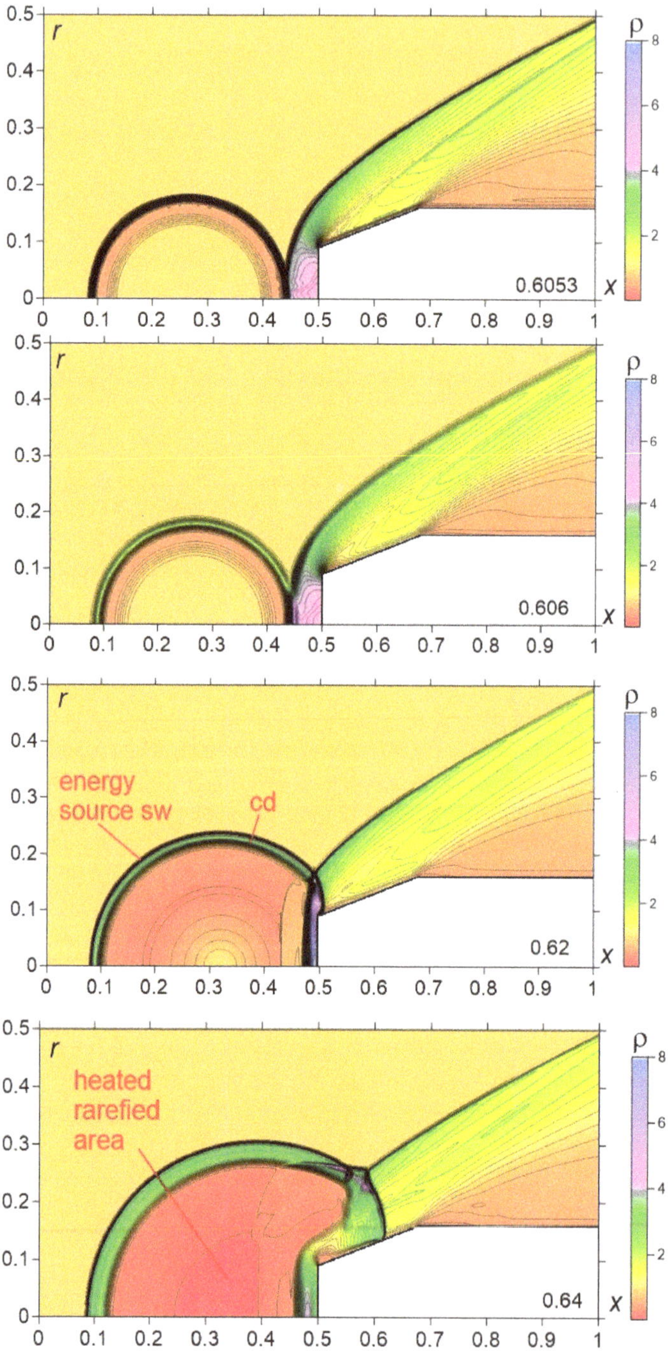

Figure 7. *Cont.*

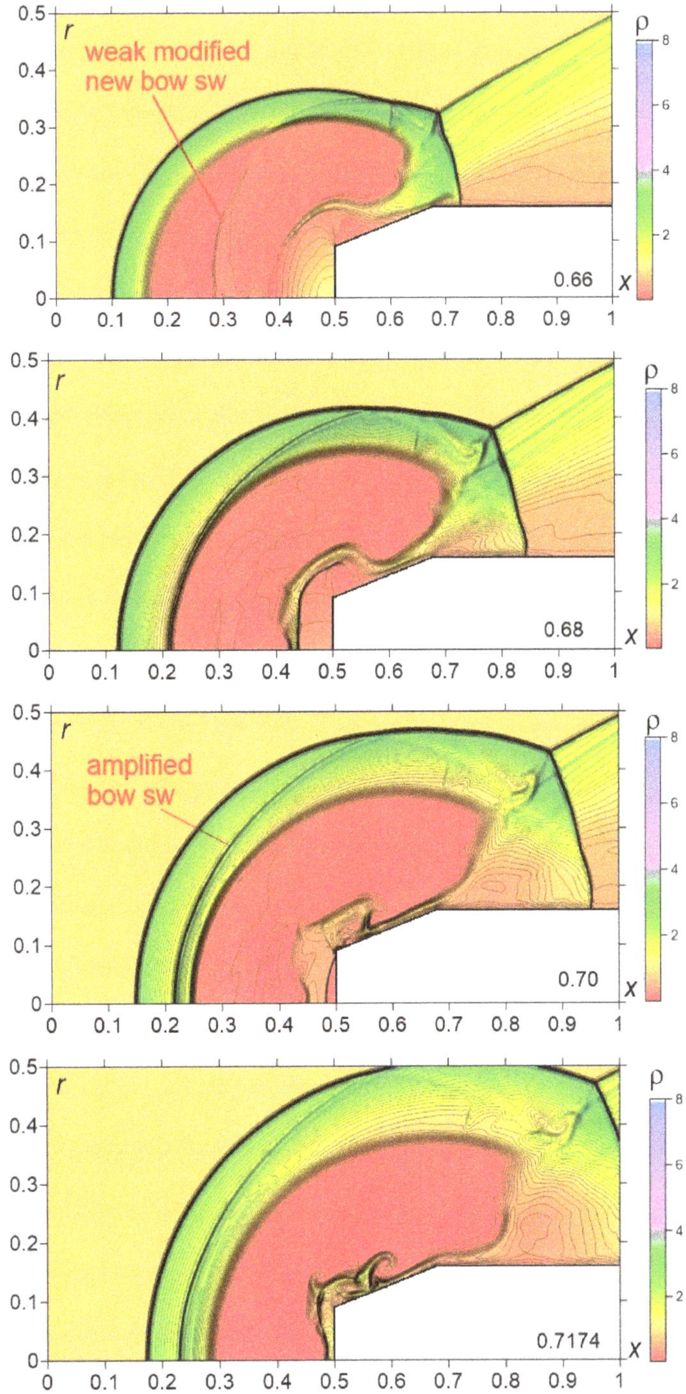

Figure 7. Initial stage of energy source-shock layer interaction dynamics.

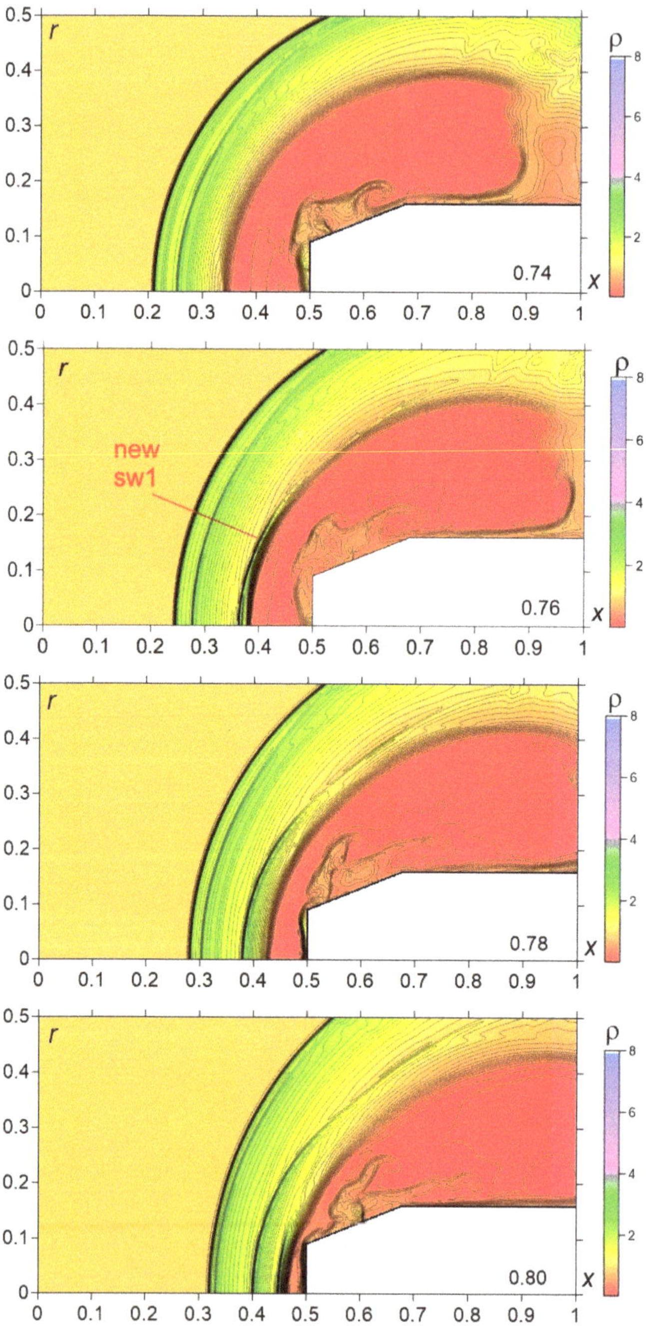

Figure 8. *Cont.*

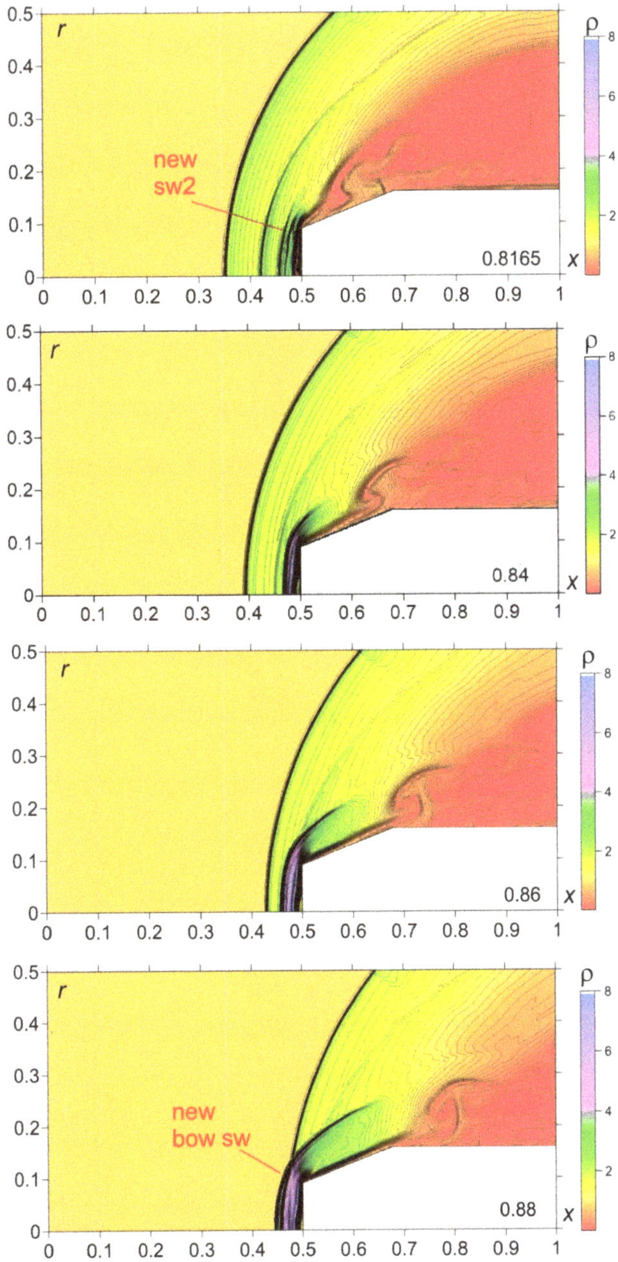

Figure 8. Middle stage of energy source-shock layer interaction dynamics.

Figure 9 demonstrates the final stage of the interaction. This stage is associated with the dynamics of a new bow shock wave and setting a stationary flow mode. The new bow shock is moving to the left accompanied by another shock, which is the remainder part of the source shock wave (t = 0.9–0.94). Then, it stops and begins to move towards the body

(t = 0.94–1.0). The area of compressed gas near the front surface gradually decreases and disappears. Finally, flow returns to the initial steady state (t = 1.2–1.6).

It should be noted that the separation of the process at stages is conditional and reflects qualitatively ongoing processes and new details forming during these stages.

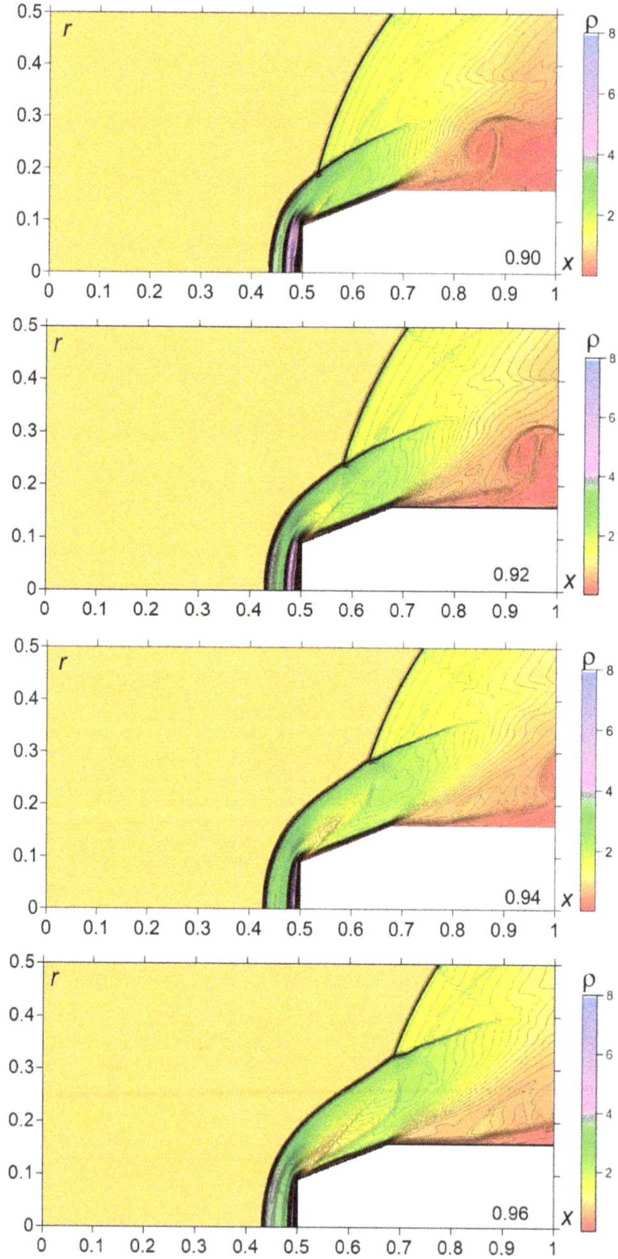

Figure 9. *Cont.*

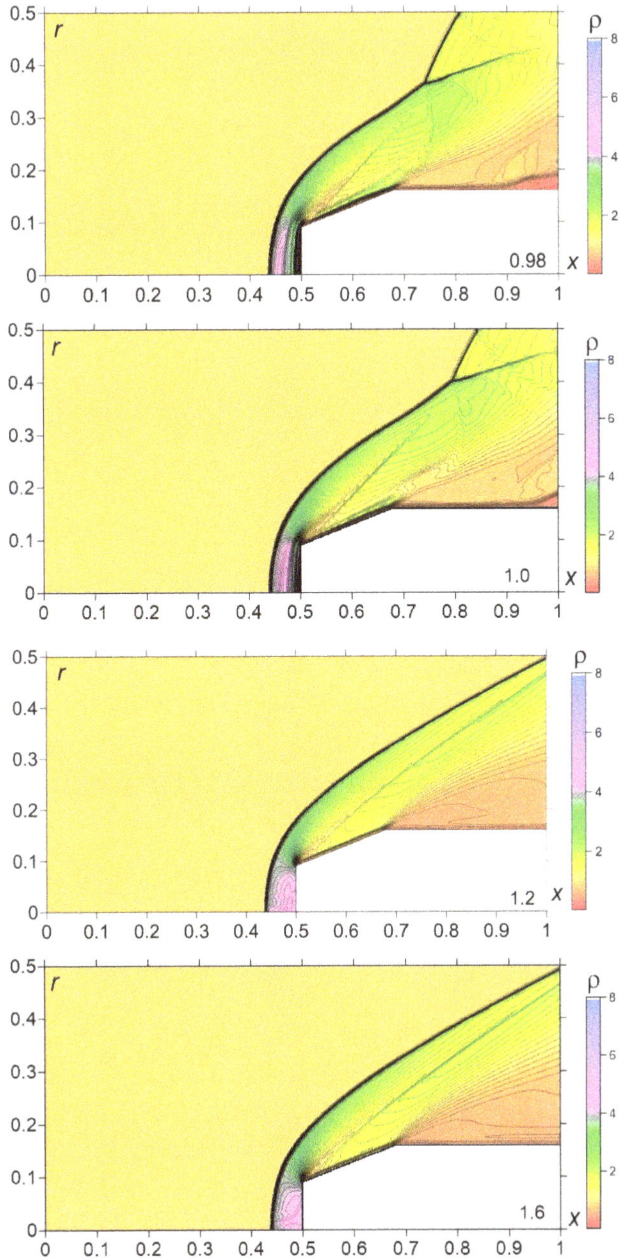

Figure 9. Final stage of energy source-shock layer interaction dynamics.

The trajectories of the main resultant shocks are presented in Figure 10. It should be noted that the considered interactions are complicated and characterized by multiple generations of additional shock waves and discontinuities. In this situation, it is not entirely clear which shock wave has to be considered as the bow shock wave (since the leftmost wave is the blast wave). We consider a bow shock wave to be a shock wave that occurs as a result of forming or a fusion with the main initial bow wave (and has the form of a bow shock wave).

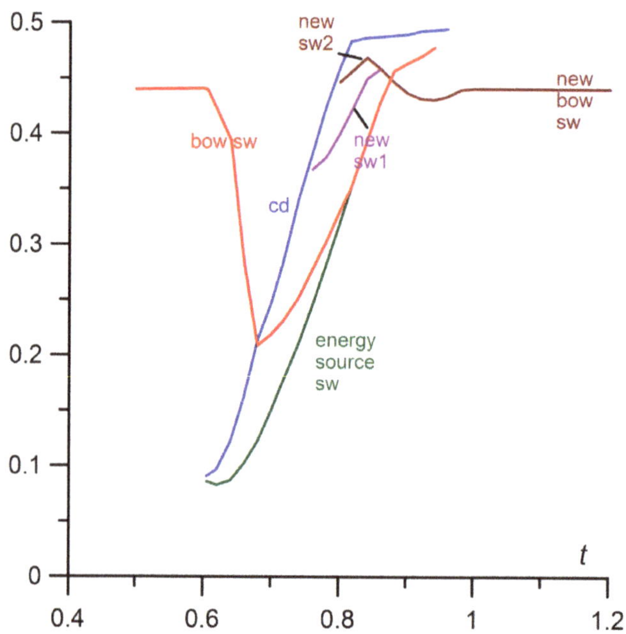

Figure 10. Dynamics of shock waves during energy source-shock layer interaction, $p_i = 33.212$.

The impact of the energy deposition is a reason for the essential front drag force reduction that occurs simultaneously with the movement of the bow shock away from the body. Figure 11 shows the dynamics of the relative stagnation pressure (Figure 11a) and the relative frontal drag force F/F_0 (Figure 11b) for two values of pressure in the energy source. Here, we have the following:

$$F = \int_0^R p_G r_G dr_G, \tag{7}$$

where p_G and r_G are the pressure value at the point with r_G—coordinate at the frontal and conical parts of the body's boundary, and R is the radius of the cylinder part of the body. F_0 in Figure 11b is the value of F without energy deposition.

The first pick of the pressure (and drag force) reflects the impact of the source blast shock wave on the bow shock; the following decreasing front drag force is caused by the action of the heated gas area [22]. It can be observed that due to the action of the heated gas area, a significant local frontal drag force reduction (up to 80%) occurs (here $\min(F/F_0) = 0.189385$, $t = 0.67$). This is caused by the action of the central heated zone's impact upon the body front surface (see Figure 7). At this time, the local stagnation pressure decrease according to the Euler approach used in the simulations is more than 90% (Figure 11a).

In Figure 11a,b the relative stagnation pressure and frontal drag force are also presented for initial pressures in the energy source $p_i = 19.407$ ($\eta = 0.04$) when only 4% of the discharge energy has been spent on gas expansion. Nevertheless, it can be observed that the stagnation pressure drop and drag force reductions are significant as well.

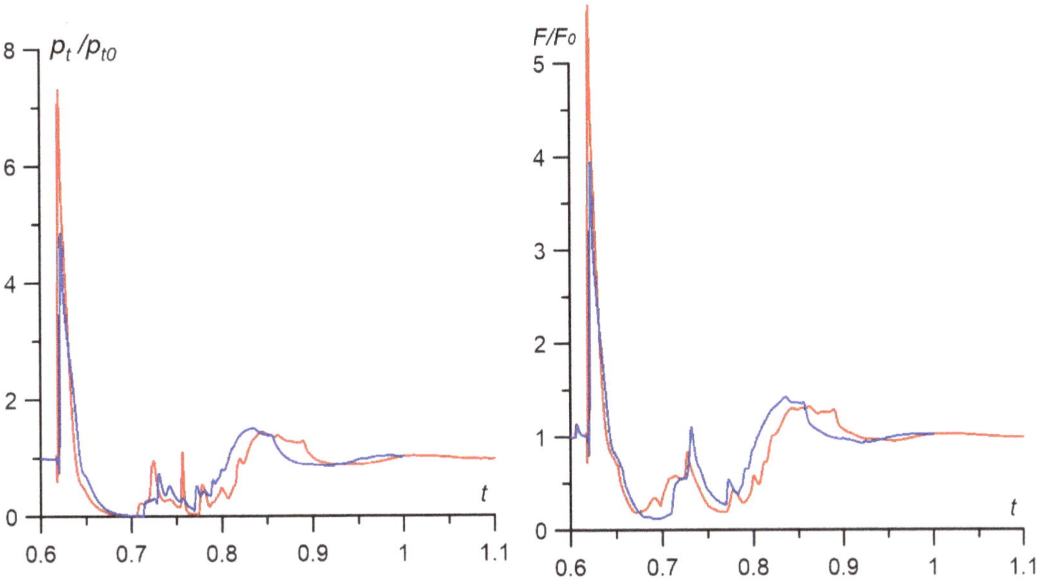

(a) (b)

Figure 11. Dynamics of relative stagnation pressure (**a**) and relative front surface drag force (**b**) during energy source-shock layer interaction: *red curve*—p_i = 33.212 (η = 0.07), *blue curve*—p_i = 19.407 (η = 0.04).

4. Discussion

It should be underlined that the simulations based upon the Euler system of equations with the use of the instant explosion as the model of an energy deposition provide only a qualitative understanding of the considered phenomena. Additionally, in the experiment, the heated area is of a pulsing nature in time, which can be connected with complicated plasma processes needed to be described by using a non-equilibrium approach. Moreover, the time for the formation of the plasma area is not registered in the experiment.

Nevertheless, the qualitative flow features obtained in the conducted simulations are in agreement with the experimental results:

1. Steady flow with the close values of numerical and experimental standoff of the bow shock wave and the close numerical and experimental shapes of the bow shock waves were obtained (Figure 6).
2. The generation of three shocks (two shock waves and a contact discontinuity—a boundary of the heated area) in the region between the left part of the source shock wave and the body at the initial and middle stages of the interaction has been obtained numerically and recorded at the schlieren pictures (Figure 3c vs. flow image in Figure 7 for t = 0.7174) (Figure 12a).
3. The generation of a series of shock waves (up to three) in the vicinity of the body at the middle stage of the interaction was obtained numerically and recorded experimentally (Figure 3d vs. flow image in Figure 8 for t = 0.8165) (Figure 12b).
4. The formation of a new bow shock from this shock wave which is accompanied by the pulsation of this new bow shock (during the steady flow establishing at the final stage of the interaction). This result can be observed in the experimental flow images (Figure 3e–g) and in the numerical flow patterns in Figures 8 and 9 for t = 0.86–1.2 (Figure 12c, t = 0.88).

It should be noted that in this paper we chose such a degree of depth of analysis of the generation and dynamics of the discontinuities (shock waves and contact discontinuities), which allowed us to draw a diagram of discontinuities (Figure 10). This construction essentially was the purpose of this work. Straightly speaking, the generation of discontinuities is associated with the solution in the local domain of the corresponding (two-dimensional) Riemann problems of the decay of an arbitrary discontinuity. Drag change reflects the dynamics of the generation and the dynamics of the discontinuities. Therefore, the considerations stated above also apply to the dynamics of the drag force.

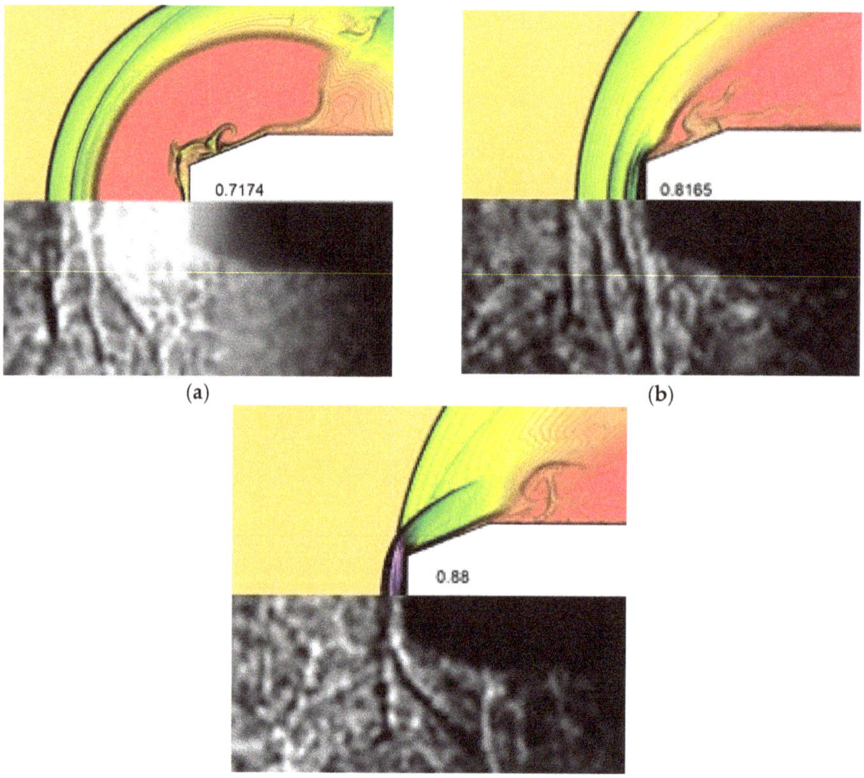

(a) (b) (c)

Figure 12. Qualitative agreement of numerical and experimental flow patterns. (**a**) $t = 0.7174$; (**b**) $t = 0.8165$; (**c**) $t = 0.88$.

5. Conclusions

High speed flow schlieren images have been obtained experimentally for the process of the impact of the plasma area (plasmoid) on the supersonic layer past a body "blunt cone-cylinder" at Mach number 3.1. The images showed that the bow shock wave standoff on the axis of symmetry increases significantly upwards, as long as the diameter of the plasma formation. The dynamics of a complicated shock wave structure, which could include up to three additional shock waves being generated, resulting from the impact of the plasma zone have been visualized. This flow structure includes a series of new shock waves and new bow shock formation after the plasmoid action.

Numerical simulations on the base of the Euler system of equations have been conducted. The model of an energy release as an instant explosion in a gas was used. The simulations provided the qualitative understanding of the considered phenomena and showed sufficient agreement between the numerical flow patterns and the experimental shadow images. In the simulations, a steady flow with the shape of the bow shock wave and the value of its standoff close to the experimental ones were obtained. The generation of up to three shock waves and a contact discontinuity in the region between the left fragment of the blast shock wave and the body at the initial and middle stages of the interaction was predicted numerically; it is in agreement with the experimental images. The calculated numerical series for the other shock waves were also experimentally visualized. The formation of a new bow shock, accompanied by its pulsation, can be observed both in numerical flow patterns and in the experimental flow images. Additionally, it was observed that, due to the action of the heated gas area, a significant local stagnation pressure decrease and local frontal drag force reduction (up to 80%) occurred. In the future, the consideration of chemical reactions relative to the model of plasma formation is planned to be included.

Author Contributions: Conceptualization—I.Z.; formal analysis—I.Z., V.C. and O.A.; software—O.A.; visualization—I.Z., V.C. and O.A.; writing—original draft—I.Z. and O.A. All authors have read and agreed to the published version of the manuscript.

Funding: This research received no external funding.

Acknowledgments: The study was partly carried out within the framework of the Development Program of the Interdisciplinary Scientific and Educational School of Moscow State University "Photonic and Quantum Technologies: Digital Medicine".

Conflicts of Interest: The authors declare no conflict of interest.

Nomenclature

M	freestream Mach number
γ	ratio of specific heats
p_∞, ρ_∞, u_∞, v_∞	freestream pressure, density and velocity components
R	diameter of a cylinder part of a body
t_i	time moment of an energy source arising
p_i	pressure in an energy source
r_i	radius of an energy source
x_0	distance between the center of an energy source and a frontal surface of a body
η	part of the discharge energy spent to the expansion of a gas
p_G, r_G	pressure and r-coordinate at the body's boundary
SW	shock wave
CD	contact discontinuity

References

1. Knight, D.D. *Energy Deposition for High-Speed Flow Control*; Cambridge University Press: Cambridge, UK, 2019; 450p. [CrossRef]
2. Knight, D. Survey of aerodynamic drag reduction at high speed by energy deposition. *J. Propuls. Power.* **2008**, *24*, 1153–1167. [CrossRef]
3. Russell, A.; Zare-Behtash, H.; Kontis, K. Joule heating flow control methods for high-speed flows. *J. Electrost.* **2016**, *80*, 34–68. [CrossRef]
4. Leonov, S.B. Review of plasma-based methods for high-speed flow control. In Proceedings of the Sixth International Conference on Fluid Mechanics AIP Conference Proceedings, Guangzhou, China, 30 June–3 July 2011; Volume 1376, pp. 498–502. [CrossRef]
5. Starikovskiy, A.Y.; Aleksandrov, N.L. Gasdynamic flow control by ultrafast local heating in a strongly nonequilibrium pulsed plasma. *Plasma Phys. Rep.* **2021**, *47*, 148–209. [CrossRef]
6. Ahmed, M.Y.; Qin, N. Forebody shock control devices for drag and aero-heating reduction: A comprehensive survey with a practical perspective. *Prog. Aerosp. Sci.* **2020**, *112*, 100585. [CrossRef]
7. Shneider, M.N.; Macheret, S.O.; Zaidi, S.H.; Girgis, I.G.; Miles, R.B. Virtual Shapes in Supersonic Flow Control with Energy Addition. *J. Propuls. Power* **2008**, *24*, 900–915. [CrossRef]
8. Georgievsky, P.Y.; Levin, V.A. Supersonic flow over bodies in the presence of external energy input. *Pis'ma Zhurnal Tekh. Fiziki.* **1988**, *14*, 684–687. (In Russian)
9. Artem'ev, V.I.; Bergel'son, V.I.; Nemchinov, I.V.; Orlova, T.I.; Smirnov, V.A.; Hazins, V.M. Changing the regime of supersonic streamlining obstacles via raising the thin channel of low density. *Izv. Akad. Nauk SSSR Meh. Židk. Gaza.* **1989**, *5*, 146–151. (In Russian)
10. Nemchinov, I.V.; Artem'ev, V.I.; Bergel'son, V.I.; Hazins, V.M.; Orlova, T.I.; Rybakov, V.A. Rearrangement of the bow shock shape using a "hot spike". *Shock. Waves* **1994**, *4*, 35–40. [CrossRef]

11. Riggins, D.; Nelson, H.; Johnson, E. Blunt-body wave drag reduction using focused energy deposition. *AIAA J.* **1999**, *37*, 460–467. [CrossRef]
12. Miles, R.B. Flow Control by Energy Addition into High-Speed Air. In Proceedings of the 38th AIAA Aerospace Sciences Meeting and Exhibit, AIAA, Denver, CO, USA, 19 June–22 June 2000; pp. 1–14, Paper AIAA-2000-2324.
13. Kolesnichenko, Y.F.; Brovkin, V.G.; Azarova, O.A.; Grudnitsky, V.G.; Lashkov, V.A.; Mashek, I.C. Microwave energy release regimes for drag reduction in supersonic flows. In Proceedings of the 40th AIAA Aerospace Sciences Meeting & Exhibit, Reno, NV, USA, 14 January–17 January 2002; pp. 1–12, Paper AIAA-2002-0353.
14. Knight, D.D.; Kolesnichenko, Y.F.; Brovkin, V.G.; Khmara, D.; Lashkov, A.V.; Mashek, I.C. Interaction of microwave-generated plasma with a hemisphere cylinder at Mach 2.1. *AIAA J.* **2009**, *47*, 2996–3010. [CrossRef]
15. Knight, D.D.; Kolesnichenko, Y.F.; Brovkin, V.G.; Khmara, D.; Lashkov, A.V.; Mashek, I.C. Interaction of microwave-generated plasma with hemisphere-cone-cylinder. In Proceedings of the 48th AIAA Aerospace Sciences Meeting Including the New Horizons Forum and Aerospace Exposition, Orlando, FL, USA, 4 January 2010–7 January 2010; pp. 1–16, Paper AIAA-2010-1005.
16. Azarova, O.A.; Knight, D.D. Numerical prediction of dynamics of microwave filament interaction with supersonic combined cylinder bodies. In Proceedings of the 53rd AIAA Aerospace Sciences Meeting, Kissimmee, FL, USA, 5–9 January 2015; pp. 1–12, Paper AIAA-2015-0581.
17. Myrabo, L.N.; Raizer, Y.P. Laser-induced air spike for advanced transatmospheric vehicles. In Proceedings of the 25th Plasmadynamics and Lasers Conference, Colorado Springs, CO, USA, 20 June–23 June 1994. Paper AIAA-1994-2451.
18. Tretyakov, P.K.; Fomin, V.M.; Yakovlev, V.I. New principles of control of aerophysical processes—research development. In Proceedings of the International Conference on the Methods Aerophysical Research, Novosibirsk, Russia, 29 June–3 July 1996; The Institute of Theoretical and Applied Mechanics: Novosibirsk, Russia, 1996; pp. 210–220.
19. Adelgren, R.G.; Yan, H.; Elliott, G.S.; Knight, D.D.; Beutner, T.J.; Zheltovodov, A.A. Control of Edney IV interaction by pulsed laser energy deposition. *AIAA J.* **2005**, *43*, 256–269. [CrossRef]
20. Zheltovodov, A.A.; Pimonov, E.A.; Knight, D.D. Energy deposition influence on supersonic flow over axisymmetric bodies. In Proceedings of the 45th AIAA Aerospace Sciences Meeting and Exhibit, Reno, NV, USA, 8 January–11 January 2007; pp. 1–31, Paper AIAA-2007-1230.
21. Mortazavi, M.; Knight, D.D.; Azarova, O.A.; Shi, J.; Yan, H. Numerical simulation of energy deposition in a supersonic flow past a hemisphere. In Proceedings of the 52nd Aerospace Sciences Meeting, National Harbor, MD, USA, 13–17 January 2014; pp. 1–18, Paper AIAA-2014-0944.
22. Azarova, O.A.; Knight, D.D. Interaction of microwave and laser discharge resulting "heat spots" with supersonic combined cylinder bodies. *Aerosp. Sci. Technol.* **2017**, *43*, 343–349. [CrossRef]
23. Kim, S.; Lee, H.J. Influence of laser energy deposition conditions on the drag of a sphere in supersonic flow. *Energies* **2019**, *16*, 3914. [CrossRef]
24. Alberti, A.; Munafo, A.; Pantano, C.; Panesi, M. Supersonic and hypersonic non-equilibrium flow control using laser energy deposition. In Proceedings of the AIAA Aviation 2019 Forum, Dallas, TX, USA, 17–21 June 2019; pp. 1–27, Paper AIAA 2019-2867.
25. Lapushkina, T.A.; Erofeev, A.V. Supersonic flow control via plasma, electric and magnetic impacts. *Aerosp. Sci. Technol.* **2017**, *69*, 313–320. [CrossRef]
26. Znamenskaya, I.A.; Naumov, D.S.; Sysoev, N.N.; Chernikov, V.A. Analysis of dynamic processes occurring during generation of plasmoid formations in a supersonic flow. *Tech. Phys.* **2019**, *64*, 802–806. [CrossRef]
27. Azarova, O.A.; Gvozdeva, L.G. Control of triple-shock configurations in high speed flows over a cylindrically blunted plate in gases at different Mach numbers. *Proc. Inst. Mech. Eng. Part G J. Aerosp. Eng.* **2018**. [CrossRef]
28. Azarova, O.A.; Gvozdeva, L.G. Control of triple-shock configurations and vortex structures forming in high speed flows of gaseous media past AD body under the action of external energy sources. *Aerospace* **2017**, *4*, 9. [CrossRef]
29. Azarova, O.A. Complex conservative difference schemes for computing supersonic flows past simple aerodynamic forms. *Comput. Math. Math. Phys.* **2015**, *55*, 2025–2049. [CrossRef]

fluids

MDPI

Article

Flow Characterization at Heated Air Supersonic Facility SBR-50

Philip Andrews [1,*], Philip Lax [1], Skye Elliott [1], Alexander Firsov [2] and Sergey Leonov [1,*]

[1] Department of Aerospace and Mechanical Engineering, Institute for Flow Physics and Control, University of Notre Dame, South Bend, IN 46637, USA; plax@nd.edu (P.L.); selliot4@nd.edu (S.E.)
[2] Joint Institute for High Temperature RAS, 125412 Moscow, Russia; alexander.a.firsov@gmail.com
* Correspondence: pandrew2@nd.edu (P.A.); sleonov@nd.edu (S.L.)

Abstract: The supersonic wind tunnel facility SBR-50 at the University of Notre Dame was built in 2015 for experimental research related to shock wave (SW) interactions with obstacles and boundary layers (BL) as well as supersonic combustion and a plasma-based flow control. Currently, the facility provides the following range of flow parameters with a test section area at the nozzle exit of 76.2 × 76.2 mm: Mach number $M = 2$ and 4, total pressure $p_0 = 1$–4 bar, stagnation temperature $T_0 = 300$–775 K, and typical duration of the steady-state flow $t = 0.5$–2 s. One distinct feature of the facility is the Ohmic gas heater installed in a long plenum section. Objective of this study is to characterize flow in the SBR-50 facility, specifically the dynamics of the gas temperature. Two measuring methods were applied for collection of a detailed dataset: thermocouple measurements and schlieren-based thermal mark (laser spark) velocimetry. The experimental data are compared with 3D Navier–Stokes modelling of the gas parameters over the entire flowpath. Particularly, this study proves that the original facility schematics (the concept of a virtual piston in the plenum) allow for a longer operation with a constant stagnation temperature compared to a constant plenum volume with adiabatic cooling of the stored gas.

Keywords: blowdown wind tunnel; Ohmic heating; supersonic thermocouple probe; laser spark velocimetry (LSV); RANS simulation

Citation: Andrews, P.; Lax, P.; Elliot, S.; Firsov, A.; Leonov, S. Flow Characterization at Heated Air Supersonic Facility SBR-50. *Fluids* **2022**, *7*, 168. https://doi.org/10.3390/fluids7050168

Academic Editor: Olga A. Azarova

Received: 19 April 2022
Accepted: 8 May 2022
Published: 11 May 2022

1. Introduction

High-speed wind tunnels (WT) are typically designed for ground testing where the major requirements are to match the flow Mach number M, the flow Reynolds number Re, and the pressure P to flight conditions. If supersonic Mach number similarity is achieved by application of an appropriate nozzle, Reynolds number congruence requires proper selection of test model size, gas pressure, and to a lesser degree temperature. Continuously operating WTs with high velocities (supersonic and hypersonic) require powerful equipment for gas heating which makes them impractical for cost-sensitive university-based testing. Among short-duration, cost-effective, high-speed test facilities, the most commonly used configurations include Ludwieg tubes and blowdown tunnels. Both of them have a well-known benefits and drawbacks and here readers are referred to classic works [1–3] and more recent manuscripts [4,5], each consisting of a more than comprehensive list of available publications.

For hypersonic engine testing, the requirements are different and include predefined flow velocity, pressure, and temperature similarities [6–8]. The facility operation time should be about $t > 0.1$ s due to a relatively long time of chemical reactions coupled to the flow structure. One more important limitation is the oxygen concentration in the working gas and a diminishing of chemical pollutants, which makes some air heating techniques problematic, such as a pre-combustion or arc heating. The application of "clean" heaters, such as Ohmic heaters or heat exchangers, is far more preferable [9]. These additional limitations make specialized facilities development and implementation more complex, especially in educational laboratories [10–14]. The University of Notre Dame Supersonic

Test Rig SBR-50 blowdown facility was designed in 2014 and was in operation starting from 2015 for experimental studies of active flow control techniques, scramjet/dual mode flameholding patterns, and development of active flameholding control systems.

A general photograph of the SBR-50 is shown in Figure 1. It consists of a high-pressure tank, plenum/air heater, nozzle, test section, diffuser, and low-pressure/vacuum tank. The test section can operate as a supersonic combustor, with the fuel injectors and electrical discharge generator flush-mounted on a plane wall or in cavity flameholder geometries [15,16]. The SBR-50 is also used for an active flow control research [17,18].

The SBR-50 facility features an Ohmic air heater installed in the plenum section. The air (or other gas if needed) is heated at stagnation pressure p_0 before the run for several minutes up to a maximum of $T_0 = 775$ K. At the beginning of the run, when the gate valve opens, gas rarefaction leads to temperature reduction in the plenum and test Section [19]. To overcome this, additional valves open on the back side of the plenum, compressing the air in the plenum to maintain a constant pressure and ensure that test section temperature remains constant. The boundary between cold and hot air moves along the plenum section during the run working, similar to a virtual piston. The plenum section is designed long enough to provide a run time up to 1 s at flow Mach number $M = 2$. To some extent, such a configuration could be treated as a combination of a blowdown scheme with a Ludwieg tube. For the SBR-50 facility, it is not known a priori how the plenum cold-hot air mixing at the virtual piston boundary affects the flow parameters' stability.

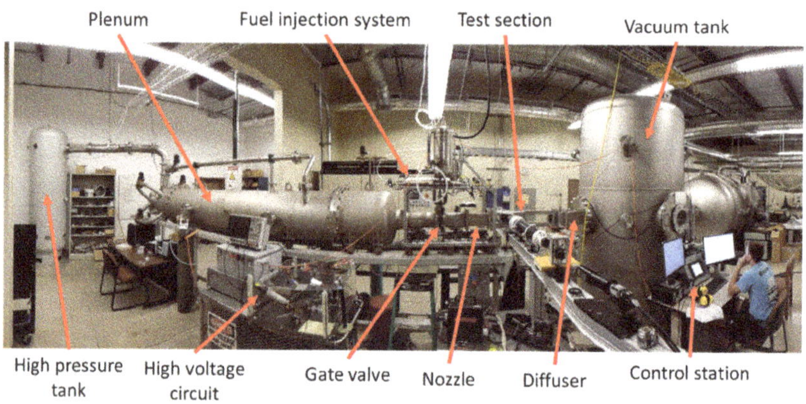

Figure 1. Overall view of the SBR-50 facility with major components labeled.

A variety of methods are employed for the characterization of stagnation temperature in supersonic flows. The most simple direct method is a stagnation temperature probe consisting of a thermocouple near the end of a metal tube with side vents to increase the recovery factor. However, the use of this type of probe is challenging in short duration supersonic flows because the response time is long, typically on the order of second(s), and the recovery factor must be found by empirical calibration. A stagnation point heat flux probe [20] has a much faster response time, down to the order of microseconds, but computing stagnation temperature from the heat flux of a semi-infinite body requires detailed calculations. Additionally, the method requires that the stagnation temperature be significantly higher than the initial temperature of the probe in order for appreciable heat flux to occur.

Stagnation temperature may also be indirectly obtained by measuring freestream static temperature or velocity along with Mach number, and then computing stagnation temperature using isentropic relations. Coherent anti-Stokes Raman scattering (CARS) is a commonly used method [21–24] for measuring both rotational and vibrational temperature in a wide variety of flows. However, the method is difficult to implement in most cases and requires specialized (and expensive) laser equipment. A multitude of techniques exist for measuring gas velocity, including laser Doppler velocimetry (LDV) [25], particle image velocimetry (PIV) [26], schlieren image velocimetry (SIV) [27], femtosecond laser electronic excitation tagging (FLEET) [22,23,28] and krypton tagging velocimetry (KTV) [29]. However, PIV techniques in supersonic flows suffer from issues with seeding and particle slip. The laser-based tagging techniques posses insufficient accuracy and require specialized equipment and proper optical access. Since conventional schlieren methods are spanwise integrated, cross-correlation based SIV techniques are complicated by the need to distinguish the freestream velocity U_∞ from the convective velocity U_c in the side wall boundary layers, where $U_c/U_\infty < 1$. An alternative and readily applied velocimetry method for freestream flow is laser spark velocimetry (LSV) [30,31], in which a laser-induced plasma (laser spark) is generated in the flow and convects downstream. The plasma luminescence may be tracked directly [30], or the hot gas kernel created by the plasma may be tracked using schlieren visualization [31].

In the excitation of the laser spark, the high electric field at the focal point of the laser ionizes the neutral gas, which then absorbs a fraction of the laser energy, creating a hot plasma that expands outward with a strong concomitant shock wave [32,33]. Due to the elongated shape of the laser beam waist, the hot plasma and the resulting SW is at first elongated rather than spherical. The dynamics of the shock wave and hot gas kernel can be approximated using the Sedov-Taylor [34,35] self-similarity solution for a strong explosion. The Sedov-Taylor solution assumes a strong shock and so is only valid while the Mach number of the shock wave is $M > 2$ [32]. In the weak shock limit an extended blast wave solution is required, such as those of Refs. [36,37], while the blast wave continues expanding into the gas medium as a progressively weakening shock wave, the hot gas kernel reaches a final radius when the hot gas pressure equals that of the surrounding gas. As the pressure behind the shock wave decreases, the pressure gradient of the hot kernel is inverted and the kernel collapses to a degree, with the hot gas on the laser axis continuing to move towards the laser source and the surrounding hot gas forming a turbulent vortex ring around the jet [38].

The objective of this work is perform SBR-50 flow characterization at two Mach numbers, $M = 2$ and $M = 4$, with variable stagnation pressures and temperatures. Figure 2 provides a basic summary of the facility. The flow temperature is measured by two different methods: direct measurements by a thermocouple and indirect measurements through LSV. A stagnation temperature probe is used to obtain qualitative stagnation temperature and its dynamics data. Complementary, an LSV method with schlieren tracking is used to obtain quantitative data on freestream velocity. Mach number is independently measured using a standard Pitot rake and the Rayleigh Pitot tube equation. The stagnation temperature is then computed using isentropic relations.

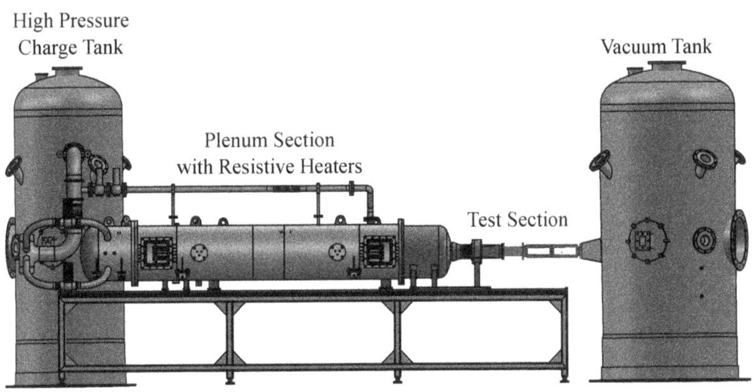

Figure 2. Overview of the SBR-50 facility with major components labeled.

2. SBR-50 Description

The SBR-50 facility consists of a 1.9 m^3 high pressure charge tank, a 0.94 m^3 plenum, and a 5.6 m^3 vacuum tank. The plenum is double walled, with direct connection between the outer plenum section and the charge tank for filling the plenum and a spiral path connecting the outer and inner plenum sections which opens at the rear of the plenum. The plenum is wheeled and supported by two tracks on a steel frame so that the entire plenum section can by moved along its axis when not secured to the test section. A fast gate valve separates the downstream plenum head from a transition region and the nozzle section, which consists of two interchangeable 2D planar nozzle halves. To minimize large scale rotation when adding air from the charge tank to the plenum, the charge tank is connected to the plenum in four branches, each offset 90 degrees from the next with a 90 degree inlet to the plenum. Within the plenum are three sets of hexacomb flow straighteners to reduce vorticity and two sets of Ohmic heater banks with 12 heater elements each, total electrical power 67 kW.

Four valves, one on each connecting branch, control the time sequence when gas from the charge tank is flown into the plenum. Over the text, the wording is used when "back valves on" means that the back valves between the charge tank and the plenum section open at 0.1 s after the opening of the main valve, and remain open during the entire operation. The phrase "back valves off" means that back valves was not opening and no additional cold gas is supplied. In the "back valves on" operation, higher pressure gas from the charge tank pressurizes the hot gas in the plenum in a virtual piston configuration aiming for better stabilization of the pressure and temperature in the test section over the course of each run. This virtual piston concept is an alternative approach to a mechanical piston in the hope that adiabatic cooling due to volumetric expansion is reduced. In other words with a lesser similarity, it could be compared to a contact boundary in a Ludwieg tube configuration with a much larger gas volume involved.

2.1. *Test Section*

The nozzle is followed immediately by the test section, which has a cross section of 76.2 $\times$ 76.2 mm at the nozzle exit and a 1 degree half angle expansion on the top and bottom walls to account for boundary layer growth. Four 5 $\times$ 12 inch quartz side windows provide optical access to the test section. Two top and two bottom stainless steel wall inserts have 16 static pressure ports each. The inserts are removable to allow for the insertion of a variety of specialized test section articles. These removable inserts as well as a full schematic of the test section is provided in Figure 3. The test section is connected to the vacuum tank through a diffuser, which has a 4 1/2 inch flanged fused silica window directly opposite the test section for optical access along the test section centerline. The vacuum tank is connected to a vacuum pump. Two 2 m long 95 mm square aluminum rails are joined and orientated orthogonal to the test section for the mounting of cameras and other diagnostic instruments.

2.2. *Schlieren Visualization*

Density gradients are visualized using a conventional refractor-based schlieren system. A high-power white LED (Luminus Devices CFT-90-WCS-X11-VB600) is powered at 40 A by a pulsed diode driver (PicoLAS LDP-V 240-100 V3.3) with external Peltier cooler (TE Technology CP-065 and TC-24-10). The broadband white light is collected and focused by an aspherical condenser lens (Thorlabs ACL50832U-A) and achromatic doublet (Edmund Optics 49-289-INK) and passed through an iris diaphragm. Two refractor lenses (Celestron Omni XLT 120) collimate and refocus the light. A vertical knife edge is place at the second focal point to visualize density gradients. The image is recorded by a high speed camera (Phantom v1611) with relay lens (Nikon 200mm f/4 AI). The high speed camera and diode driver are controlled and synchronized by a pulse generator (Berkeley Nucleonics 577). The LED optical pulse width is 100 ns with a repetition rate of 200 kHz. The exposure time of the high speed camera was setup to the minimum value of 300 ns: the LED optical pulse was triggered to be within this window.

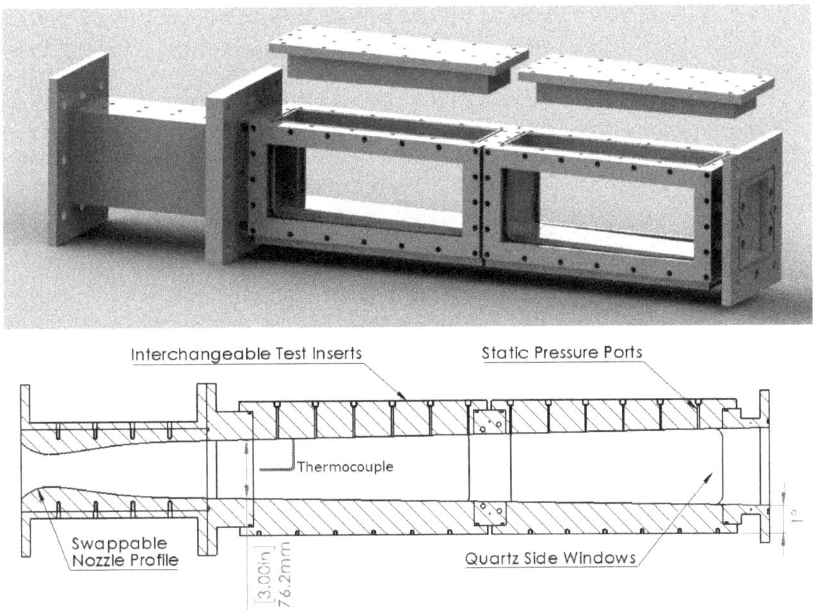

Figure 3. Detailed view of SBR-50 test section: the thermocouple and Pitot tube are posed at same location.

3. Thermocouple Measurements

In an effort to characterize the temperature conditions inside the wind tunnel test section over the course of a run, a thermocouple probe was installed mid-stream. Temperature and pressure data was collected simultaneously over a variety of conditions in order to characterize the flow parameters and compare differing cases. For thermocouple measurements, the matrix of test conditions involved 300 K, 500 K, and 700 K nominal temperatures, Mach 2 and 4, higher and lower pressures $P_0 = 1.3$ and 3.2 bar, and charge tank back valves on and off. Table 1 displays exact information regarding pressure conditions used.

Table 1. Summary of different Mach number and pressure test conditions.

Mach Number	Plenum Pressure (Bar)	Charge Tank Pressure (Bar)
2	1.6	2.6
2	3.2	4.5
4	1.6	2.6
4	3.2	4.5

The difficult task of measuring flow temperature using a thermocouple involves recovering the fluid temperature from a measurement of the thermocouple junction's temperature. When trying to measure flow temperature, the total error can be divided into velocity error, conduction error, and radiation error. Velocity error refers to the fact that the probe cannot recover all of the kinetic energy of the flowing gas as thermal energy. The ratio of kinetic energy recovered as thermal energy can be expressed as the recovery factor $\alpha = \frac{T_j - T_s}{T_0 - T_s}$ where T_j is the indicated temperature at the junction, T_s is the free stream static temperature, and T_0 is the actual stagnation temperature. Choosing the entrance to vent area ratio of the probe determines the velocity of gas over the probe tip, and can greatly impact the recovery factor and major source of error. For example, decreasing the velocity may reduce velocity error, but it also reduces convective heat transfer coefficients and thus leads to a larger conduction error. With regard to radiation errors, the thermocouple receives radiation from both the outer probe shield as well as some fraction from the test section walls as governed by the Stefan Boltzmann law $P_{rad} = \epsilon\sigma A(T_{env}^4 - T_{probe}^4)$. In this setting radiation error is negligible compared to other errors, but by setting the probe tip further inside the shielding, the area of the colder tunnel walls that the tip sees is small compared to area of the hotter walls of the probe shielding. All these considerations were taken into account when designing the thermocouple probe as well as balancing robustness and practicality [39].

At the expected temperature ranges inside the test section a type K thermocouple provides a very close to linear response. The size of the thermocouple was selected to balance the response time with robustness. In this case, a bead-welded thermocouple with a bead diameter of 1.2 mm was selected to provide the fastest response times possible without having to worry about the probe breaking under harsh flow conditions. The probe was mounted inside aluminum tubing. The inside of this tubing was coated with insulating paint near the probe tip to prevent shorting out the probe. For the details on the design and mounting of this probe see Figure 4. The probe tip was set back inside the tubing by about 4mm and a small notch was cut in one side of the tubing behind the probe tip to act as a vent and increase the recovery factor.

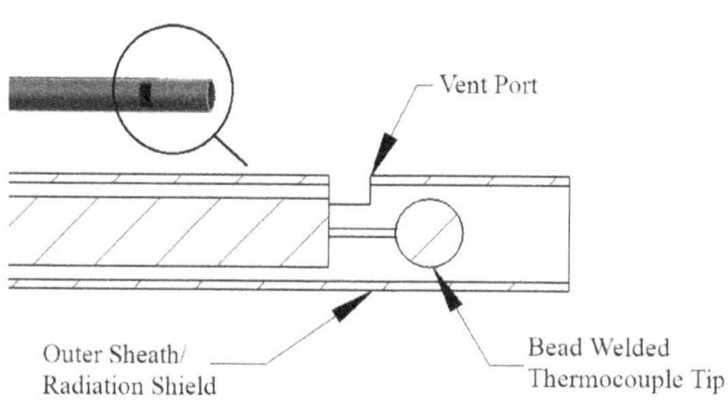

Figure 4. Schematic of thermocouple probe design.

Voltage data was collected with a Teledyne LeCroy HDO6034A-MS HDO6000A High Definition Oscilloscope and the known room temperature was used as the reference temperature. Since the thermocouple voltage amplitude is only a few millivolts, the voltage data is naturally quite noisy and was processed using a series of filters to mitigate this and generate a smooth calculated temperature time series. First a 60 Hz notch filter was applied with the goal of removing noise generated by surrounding electronics. Then, a third order digital Butterworth lowpass filter with a cutoff frequency of 0.0003 half-cycles/sample or with our 0.5 MHz sampling rate, 75 Hz. Last a rolling window average was applied over each 0.1 s. The voltage data was converted to temperature using a Type K thermocouple voltage response reference table and assuming a linear response in the operating region according to $T(V) = RV + T_r$ where R is the response coefficient and T_r is the reference temp which was room temperature [40]. Note that none of computed temperature data presented in this section is strictly quantitative without proper calibration in well-certified flow conditions, but results provide a qualitative match and are generally representative of the actual temperature dynamics. Processed data is presented in Figures 5–7.

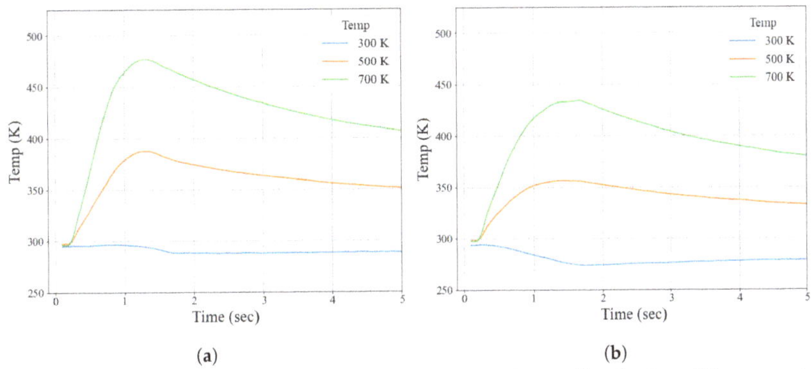

Figure 5. Thermocouple temperature measurements taken at $P_0 = 1.6$ bar for three different nominal plenum temperatures at Mach 2: (**a**) measured temperature with valves on; (**b**) measured temperature with valves off.

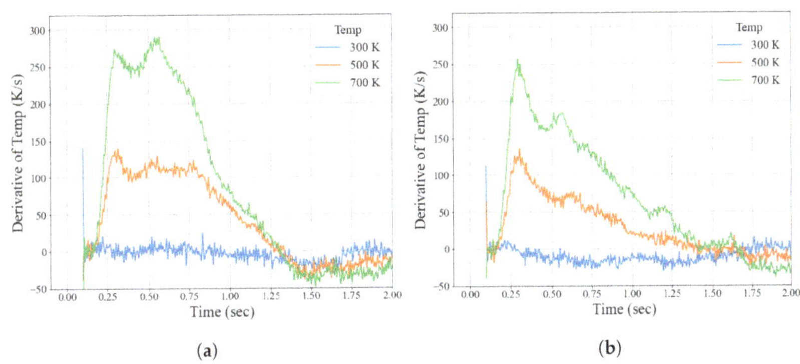

Figure 6. Comparing measured temperature derivative for $P_0 = 1.6$ bar and $M = 2$: (**a**) valves on; (**b**) valves off.

Comparing runs with the back valves on versus back valves off demonstrates that, in general, the facility is operating as designed. Examination of Figures 5–8 demonstrates that keeping the valves closed introduces cooling due to the pressure drop whereas using the valves allows the plenum to push air out at a consistent temperature. According to Figure 7, temperatures are higher at higher pressure runs, however, this discrepancy is largely due to the increase in thermal recovery factor of the probe at higher pressure due to increased convective heat transfer. The pressure jump at t = 1.7 s corresponds to the gate valve closure and supersonic-to-subsonic flow transition. Examination of Figure 8 reveals the cooling and overall lower flow temperatures that occur when the back valves are off, while data is only presented for Mach 2, additional tests indicated that turning the back valves off introduced a more severe cooling effect at Mach 2 compared to Mach 4 which is because the mass flowrate is significantly lower at $M = 4$.

For simplicity, the output of thermocouple was assumed to be dictated by thermal transfer from the gas to the probe as subject to Newton's law of heating/cooling, $\dot{T} = r(T_{env} - T(t))$ where r is the coefficient of heat transfer. Solving this initial value problem for a step change in temperature and accounting for recover factor α yields the measured temperature model

$$\frac{T(t)}{\alpha} = T_{env} + (T(0) - T_{env})e^{-rt}$$

Without an empirical calibration under specific flow conditions, the recovery factor of this probe is unknown and thus is assumed to be unity during an initial data processing. This assumption in the model leads to measured temperatures that are lower than real values but still provide useful qualitative results. The temperature data were fit with this model using a thermocouple time constant of 0.7 s to examine over what time period this exponential fit matches the data. This gives information on how long the tunnel can hold a constant stagnation temperature as well as a rough extrapolation what this temperature is. Figure 9 shows that flow temperature is roughly constant for about 0.7 s. Figure 9 also shows that with the valves on the recorded temperatures are higher and a stable temperature is maintained in the test section for slightly longer. Predicted flow temperatures from extrapolation are always lower than the actual tunnel stagnation temperature. In part this is due to the imperfect recovery factor of the probe, but also these predictions agree with a lower than nominal value of T_0 calculated by laser spark experiments presented in the following section.

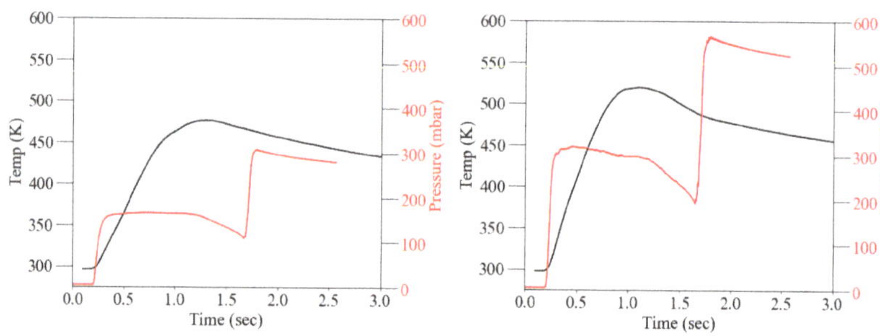

(a) (b)

Figure 7. Comparing temperature at Mach 2 operation for (**a**) Lower pressure condition, 1.6/2.6 bar for Plenum/Charge Tank (**b**) Higher pressure condition, 3.2/4 bar for Plenum/Charge Tank.

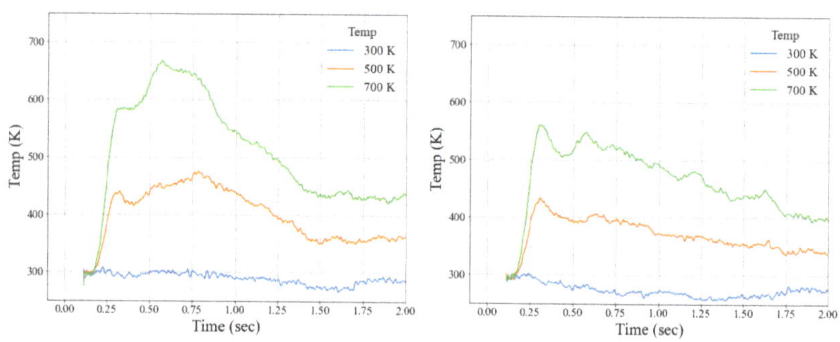

(a) (b)

Figure 8. Calculated total temperature using $T_{env} = T(t) + \frac{\dot{T}}{r}$ at $P_0 = 1.6$ bar for (**a**) Valves on (**b**) Valves off.

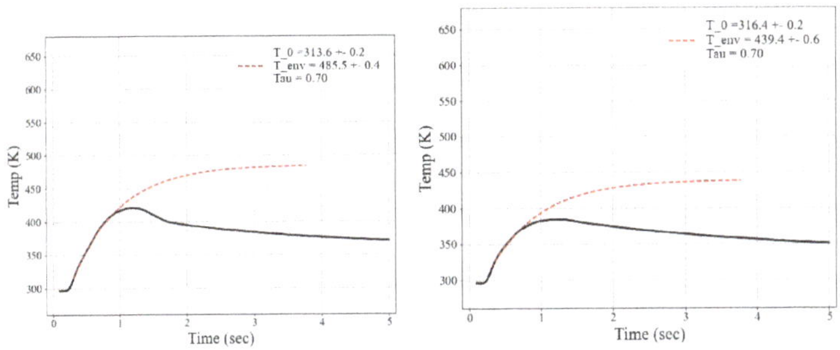

(**a**) (**b**)

Figure 9. Extrapolated total temperature at 500 K plenum setting and $P_0 = 1.6$ bar for (**a**) Valves on (**b**) Valves off.

4. Laser Spark Dynamics

For Mach 2 flow, the fundamental wavelength of a ns-pulsed 100 Hz Nd:YAG laser (Solar Laser Systems LQ 629-100) is frequency doubled to 532 nm, expanded from a beam diameter of 4.6 mm to 12.3 mm, and focused into the test section using an $f = 75$ mm fused silica lens. The 532 nm pulse energy is 70 mJ/pulse as measured by a thermopile power meter (Ophir 50A-PF-DIF-18). The post-laser spark hot gas kernel is visualized using the high-speed schlieren system discussed above with 200 kHz framerate, 100 ns optical pulse width, and a vertical knife edge. For increased energy deposition in the lower density Mach 4 flow, the fundamental wavelength of 1064 nm (170 mJ/pulse) is used and is focused into the test section with an $f = 50$ mm fused silica lens. Representative image sequences for schlieren visualization of the hot gas kernel in Mach 2 and Mach 4 flow are presented in Figure 10, where $t = 0$ and $x = 0$ correspond to the time and x-location, respectively, of the initial breakdown. Figure 10a,b each contain 17 consecutive schlieren images of a hot gas kernel as it convects downstream in Mach 2 and Mach 4 flow, respectively, with flow from left to right.

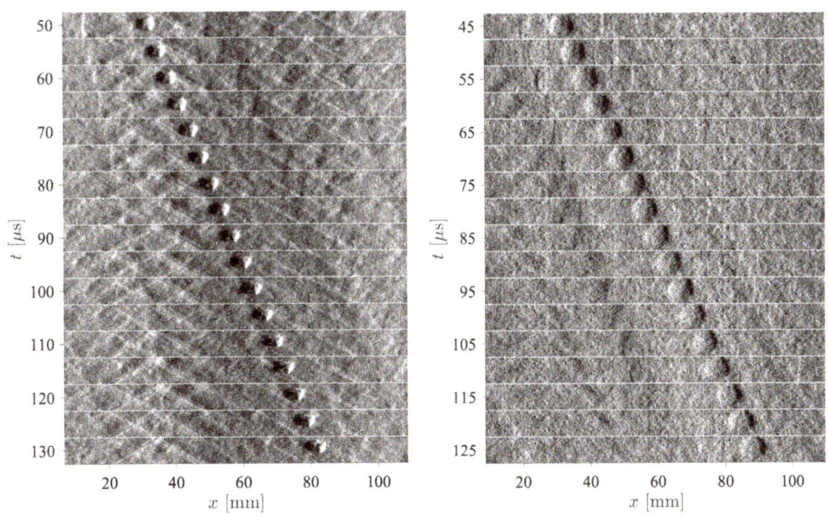

(**a**) (**b**)

Figure 10. Representative compiled schlieren images of the hot gas kernel. Flow is left to right. (**a**) Mach 2 flow with T_0 = 500 K and P_0 = 2.6 bar. (**b**) Mach 4 flow with T_0 = 350 K and P_0 = 4 bar.

The hot gas kernel is tracked as it convects downstream using a cross correlation based algorithm. Each individual schlieren image (the sub-images in Figure 10) is summed along vertical pixels, and the resulting 1D signal is cross correlated with a function consisting of a single sawtooth wave. The cross correlation peak is determined with sub-pixel interpolation using a second order polynomial. This process is repeated for every schlieren image containing the hot gas kernel, and an x–t plot is constructed using the fixed time between images of 5 µs. The velocity of the hot gas kernel is then simply the slope of the x–t plot. Representative results are shown in Figure 11, where Figure 11a is the x–t plot for the convecting hot gas kernel shown in Figures 10a and 11b–d are the cross correlation results for the 90 µs sub-image in Figure 10a. Since the R-squared value for the x–t plot is typically $R^2 = 0.99999 \approx 1$, the error in measured convective speed is small and can be neglected.

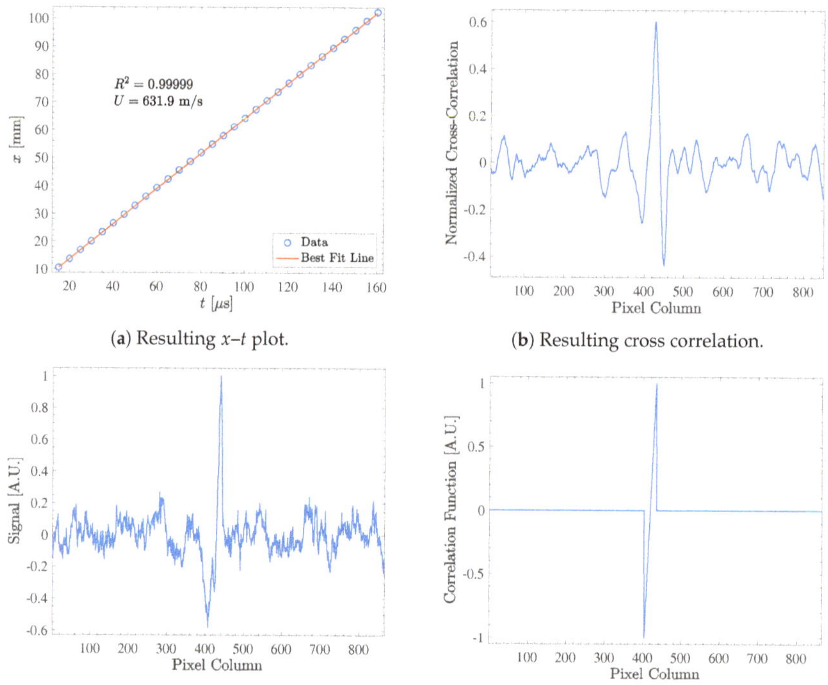

(**a**) Resulting x–t plot. (**b**) Resulting cross correlation.

(**c**) Summed schlieren data. (**d**) Correlation function.

Figure 11. Representative cross correlation results for the hot gas kernel in Figure 10a.

Mach number is calculated from static pressure and Pitot tube pressure collected at 800 Hz by a multi-channel pressure scanner (Scanivalve MPS4264) using the Rayleigh Pitot tube equation. The results are presented in Figure 12. Stagnation temperature is computed using isentropic relations, and is presented in Figure 13a,b for Mach 2 flow and in Figure 13c for Mach 4 flow, where $t = 0$ corresponds to the time of tunnel start. For greater accuracy at high temperatures, the results at nominal stagnation temperatures of 600 K and 700 K in Figure 13a,b are five run averages, with the standard deviation computed using the data of all five runs. Due to the low gas densities at Mach 4, the laser spark is formed only sporadically at high stagnation temperatures when the gas density is lowest. Therefore, fewer data points are included in the high temperature Mach 4 results, and the nominal stagnation temperature is limited to $T_0 \leq 500$ K in Figure 13c.

In Figure 13, it is observed that stagnation temperature is relatively constant throughout the steady-state runtime of the facility, which is about 400–900 ms for Mach 2 flow and

400–1200 ms for Mach 4 flow. However, there are fluctuations in stagnation temperature that increase as the nominal stagnation temperature increases, which are thought to be due to incomplete mixing of gas within the plenum. Additionally, the average calculated stagnation temperature shown in Figure 14a deviates from the nominal value as the nominal stagnation temperature is increased. It is hypothesized that this reduction in the calculated stagnation temperature is caused by the mixing of a relatively colder gas into the hot gas supplied by the plenum. This colder gas may originate from either a zone of high wall heat transfer or an imperfect seal between the plenum core and the helical cooling channel surrounding the plenum. The magnitude of the difference between the nominal and measured stagnation temperature is

$$\Delta T_0 = a(T_0 - T_\infty), \tag{1}$$

where T_∞ is the ambient temperature and $a = 0.20$ for Mach 2 flow and $a = 0.33$ for Mach 4 flow.

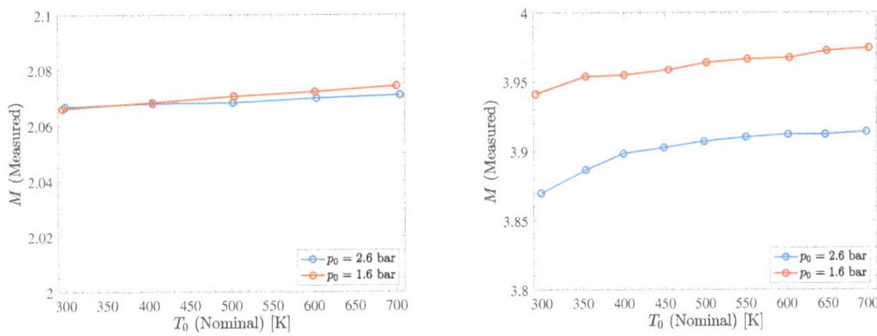

(a) Mach 2 nominal flow.

(b) Mach 4 nominal flow.

Figure 12. Average steady-state Mach number.

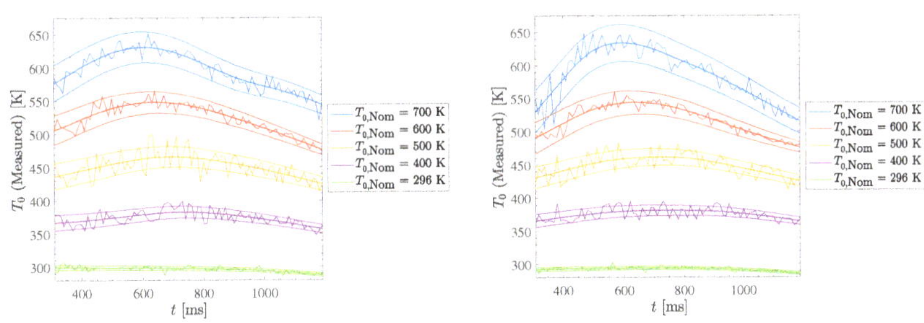

(a) $M = 2$, P_0 = 2.6 bar.

(b) $M = 2$, P_0 = 1.6 bar.

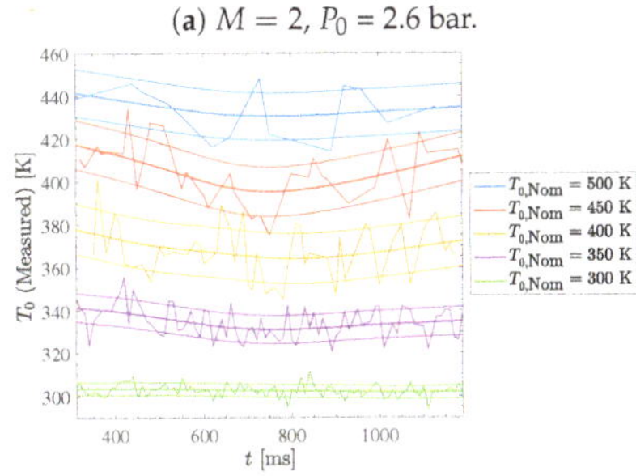

(c) $M = 4$, P_0 = 4 bar.

Figure 13. Results for T_0. Curved line is a cubic smoothing spline, and error bars represent one standard deviation.

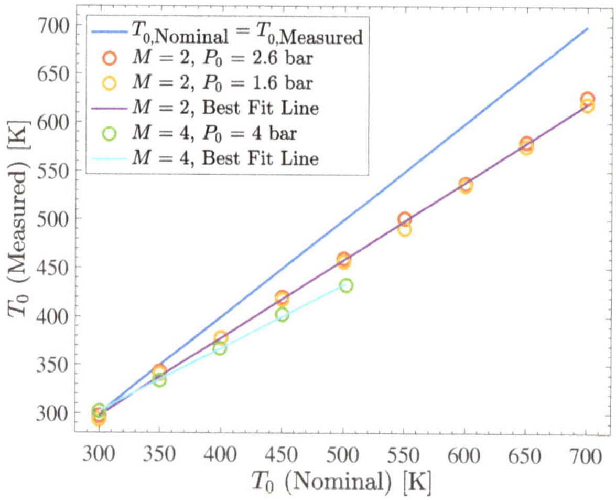

Figure 14. Average steady-state results for T_0 for Mach 2 and Mach 4 flow.

5. Numerical Navier–Stokes Simulation

Numerical simulation of SBR-50 operation in different modes was performed using FlowVision 3.12.04 CFD software with the purpose of understanding the process of cold-hot air mixing at the virtual piston boundary and how the plenum configuration affects the stability of flow parameters. The simulation was based on the solution of the three-dimensional unsteady Reynolds averaged Navier–Stokes equations accompanied by the $k - \varepsilon$ turbulence model. The geometry of the simulation test section corresponds to the experimental one and the calculation domain includes the either the full setup including high-pressure charge tank, plenum, honeycombs and nozzle as in Figure 15 or the symmetric half part of plenum, honeycombs and nozzle for more detailed temperature dynamics analysis as in Figure 16. The full geometry was used to analyze the influence of slight deviations from symmetry, the real solenoid valve opening, and for determining the total pressure drop between charge tank and plenum. The second geometry was used for the simulation of SBR-50 operation with the virtual piston. In the half geometry, the symmetry condition was set to the vertical symmetry plane. No-slip conditions, adiabatic conditions, and wall functions were used on all other walls. The outlet boundary of nozzle was set to a free supersonic flow exit. The operation of the gate valve and solenoid valves were simulated using the "moving bodies" function of FlowVision, and in this approach the mesh around moving bodies is rebuilt at each time step. In this simulation, the gate valve opens with constant velocity from 0 to 0.2 ms, and solenoid valves open with constant velocity from 0.05 to 0.08 ms. Additionally, the temperature dynamics at a point directly at the center of the nozzle exit are compared with measured thermocouple data recorded from a point slightly downstream of the nozzle as shown in Figure 17. In the used 3D URANS numerical method in conjunction with the k-ε-model the grid independence was previously tested for similar tasks. The grid near the walls was based on the y^+ appropriate for k-ε-model ($30 < y^+ < 150$). The grid in the flow volume allows for all large-scale features of the flow to be resolved. The software routine used allows the grid to be changed at any calculation step using adaptation (one cell splits to 8 at one level of adaptation and the number of levels is limited only by available RAM). The short period, when cold air is injected from charge tank, and the time period, when jet formation is possible inside the plenum, were simulated at different levels of adaptation to ensure the independence of large-scale features in the flow from the grid size or adaptation level.

Simulation results indicate that the boundary between hot and cool air inside the plenum experiences significant distortion as is seen in Figure 16. These distortions could lead to unpredictable fluctuations in the stagnation temperature during a tunnel run, but expecially near the end of the run. When the back valves are open, air from the charge tank at room temperature pushes heated air out of the plenum but tends to form cold air jets near the centerline of the plenum as shown in Figure 15. The presence of these cold air jets inside the plenum due to the introduction of cold higher pressure air from the charge tank back valves is a possible explanation of observed temperature fluctuation in stagnation temperature computed from laser spark measurements. As it is seen from numerical simulation, the delay between the first portion of cold air in the nozzle and main onset of cold flow could be about 0.25 s. One reason for the cold jet formation in the plenum in front of the main hot-cold boundary is the high speed of injected air coming from the charge tank. It is expected that further optimization of the cold air supply system could prevent such cold air jets and provide an increase in the duration of stable flow parameters from 0.45 s to up to 0.7 s at discussed operational parameters.

Figure 17 compares the simulation result with the data acquired by thermocouple and by the laser spark velocimetry. In both simulation and experiment, runs with the back valves closed lead to immediate expansion cooling after the gate valve opens, and runs with the back valves on lead to slight heating from initial over-compression. The data for the valves on operation mode prove the concept of the virtual piston. Additionally, on one side, data provides reasonable validation for these simulations, as key flow behaviors are matched by experimental results. On another side, an obvious discrepancy has to be discussed. A reasonable explanation is that the temperature data were extracted from indirect measurement datasets.The thermocouple data is computed with an assumed recovery factor of 0.9 in accordance with available literature suggesting that total temperature probes of this design in similar conditions demonstrated recovery factors ranging from 0.89 to 1.02 [41]. The temperature values were recalculated by a differentiation procedure. Taking into account that the thermocouple time constant is close to the run duration, some error could be assumed. For the laser spark dataset, the gas temperature is recalculated from the direct gas velocity measurements. Because T is proportional to v^2, a small fluctuations in v lead to clearly visible fluctuations in T. In simulations the gas temperature in the plenum distributes itself uniformly, while in the real facility, the near-gate valve portion of the plenum is not heated. This leads to significant difference in simulated and measured temperature values at t = 0.3–0.4 s.

A deviation in the geometry of internal elements in the plenum, such as an installation of a blocking disk near the plenum axes, affects the hot-cold gas mixing and an axial cold jet appearance in a significant degree. This opens a window for a further optimization of the flow parameters, including the duration of a steady state of the flow. In general, a manipulation of the timing of the back valves and the charge tank pressure allows for the ability to generate a flow field with time-variable predefined parameters, introducing an additional flexibility to the facility operation.

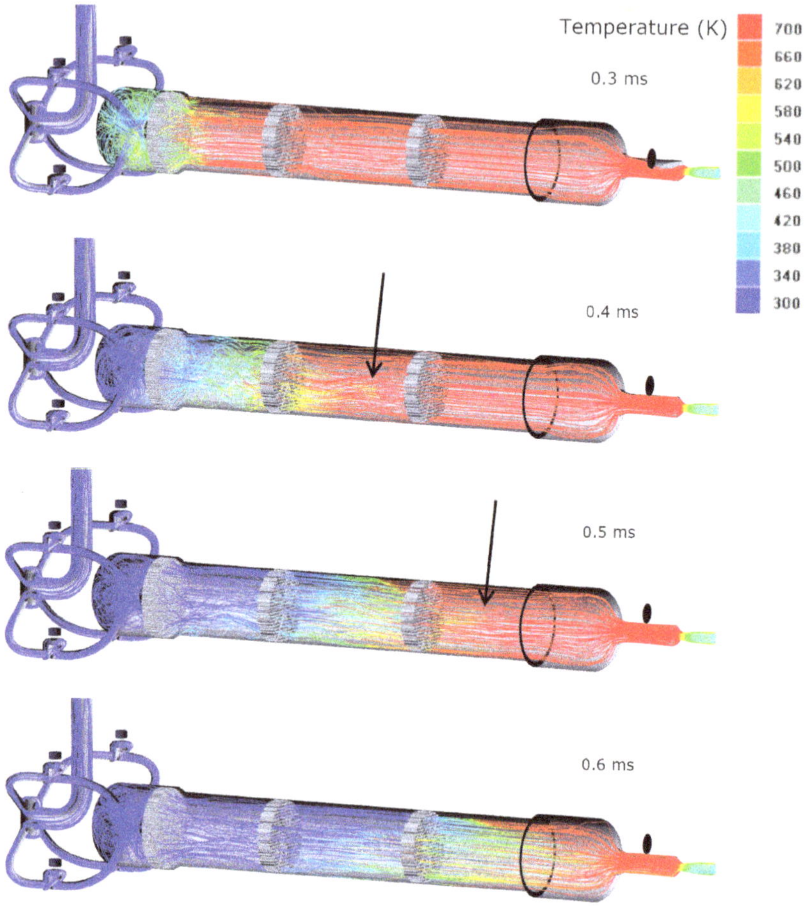

Figure 15. Full domain numerical NS simulation of time evolution of hot-cool air boundary inside plenum section.

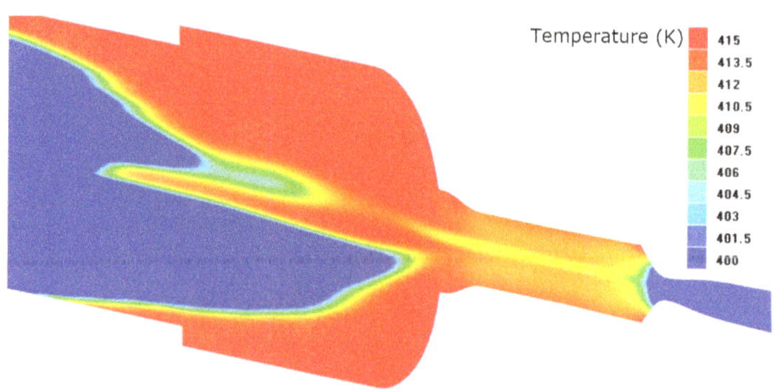

Figure 16. Example of hot-cold boundary distortion inside plenum during operation with back valves on.

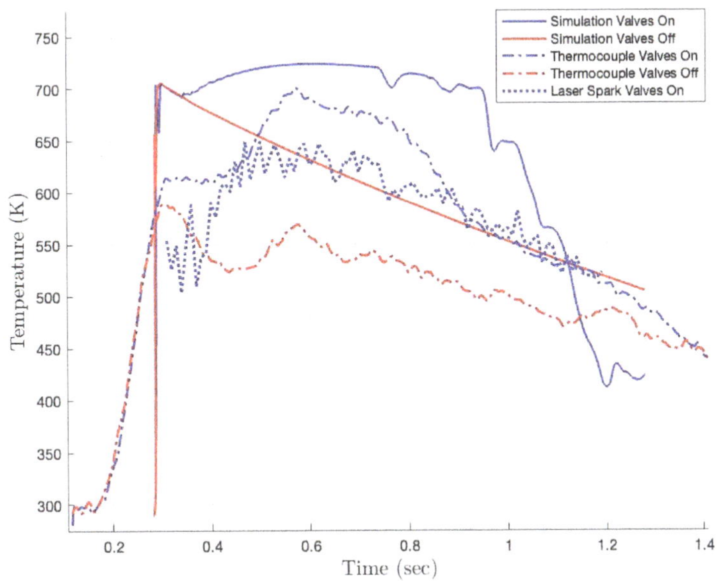

Figure 17. Simulated temperature dynamics for half geometry domain at $P_0 = 1.6$ bar and $T_0 = 700$ K compared to laser spark data and thermocouple probe measured temperature with assumed recovery factor of r = 0.9.

6. Conclusions

The supersonic facility SBR-50 at the University of Notre Dame is used for research efforts studying supersonic combustion and plasma-based flow control. The facility provides Mach number M = 2 and 4 flow with the total pressure $P_0 = 1$–4 bar, stagnation temperature $T_0 = 300$–775 K and typical duration of the steady-state flow $t = 0.5$–2 s. For the temperature control, an Ohmic gas heater is installed in a long plenum section. This manuscript describe some results of the flow characterization, specifically the dynamics of the gas temperature. Two measuring methods were applied for collection of a detailed dataset: thermocouple measurements, and schlieren-based thermal mark (laser spark) velocimetry.

The general conclusion resulting from these measurements is that the facility original schematics (virtual piston in the plenum concept) allows for a longer operation with a relatively constant stagnation temperature compared to a constant plenum volume with adiabatic cooling of the stored gas. At the same time, results demonstrate a significant level of temperature perturbation, which needs additional clarification. Another effect observed and under further analysis is a notably lower stagnation temperature measured at M = 2 and M = 4 than the one measured in the plenum. This effect is attributed to an uneven gas temperature distribution over the plenum section.

The numerical simulation indicates that the gas temperature in the test section could potentially equal the plenum gas temperature for as long as t = 0.45 s. Additionally, it shows that an optimization of the plenum geometry and the gas premixing in the plenum could resolve some of the issues with the gas temperature dynamics. These results also demonstrate that the virtual piston concept implemented by injecting air through the back valves stabilizes pressure and temperature values (variations less than 5%) in the test section. It was noted that further optimization of the cold air supply system could extend the stable window of operation from 0.45 s to up to 0.7 s under explored conditions if the cold jet formed by the charge tank air supply is suppressed.

Author Contributions: P.A. contributed to the test arrangement, data collection, data analysis, and the manuscript composition. P.L. contributed to the laser spark measurements and data analysis. S.E. contributed to the facility adjustment and the data acquisition. A.F. performed the numerical simulations. S.L. contributed to the conceptualization, project oversight, and the data analysis. All authors have read and agreed to the published version of the manuscript.

Funding: This work was supported by the University of Notre Dame, AME Department and Institute of Flow Physics and Control.

Institutional Review Board Statement: Not applicable.

Informed Consent Statement: Not applicable.

Data Availability Statement: Data is available upon request.

Conflicts of Interest: The authors declare no conflict of interest.

References

1. Lukasiewicz, J. *Experimental Methods of Hypersonics*, 3rd ed.; M. Dekker: New York, NY, USA, 1973.
2. Pope, A.; Goin, K.L. *High-Speed Wind Tunnel Testing*; John Wiley and Sons: Hoboken, NJ, USA, 1965.
3. Hucho, W.H.; Hannemann, K.; Schramm, J.M.; Williamson, C.H.K. Aerodynamics. In *Springer Handbook of Experimental Fluid Mechanics*; Tropea, C., Yarin, A.L., Foss, J.F., Eds.; Springer: Berlin/Heidelberg, Germany, 2007; Chapter 16, pp. 1043–1146. [CrossRef]
4. Gu, S.; Olivier, H. Capabilities and limitations of existing hypersonic facilities. *Prog. Aerosp. Sci.* **2020**, *113*, 100607. [CrossRef]
5. Marren, D.; Lu, F. *Advanced Hypersonic Test Facilities*; AIAA: Reston, VA, USA, 2002. [CrossRef]
6. Gildfind, D.E.; Morgan, R.G.; Jacobs, P.A.; McGilvray, M. Production of High–Mach–Number Scramjet Flow Conditions in an Expansion Tube. *AIAA J.* **2014**, *52*, 162–177. [CrossRef]
7. Donohue, J.M. Dual-Mode Scramjet Flameholding Operability Measurements. *J. Propuls. Power* **2014**, *30*, 592–603. [CrossRef]
8. Marren, D.; Lafferty, J. The AEDC Hypervelocity Wind Tunnel 9. In *Advanced Hypersonic Test Facilities*; Marren, D., Lu, F., Eds.; AIAA: Reston, VA, USA, 2012; Chapter 17, pp. 467–478. [CrossRef]
9. Mitani, T.; Hiraiwa, T.; Sato, S.; Tomioka, S.; Kanda, T.; Tani, K. Comparison of Scramjet Engine Performance in Mach 6 Vitiated and Storage-Heated Air. *J. Propuls. Power* **1997**, *13*, 635–642. [CrossRef]
10. Fotia, M.L.; Driscoll, J.F. Isolator-Combustor Interactions in a Direct-Connect Ramjet-Scramjet Experiment. *J. Propuls. Power* **2012**, *28*, 83–95. [CrossRef]
11. Karimi, M.; Ochs, B.; Liu, Z.; Ranjan, D.; Sun, W. Measurement of methane autoignition delays in carbon dioxide and argon diluents at high pressure conditions. *Combust. Flame* **2019**, *204*, 304–319. [CrossRef]
12. Baccarella, D.; Liu, Q.; Lee, T.; Hammack, S.D.; Do, H. The Supersonic Combustion Facility ACT-2. In Proceedings of the 55th AIAA Aerospace Sciences Meeting, Kissimmee, FL, USA, 8–12 January 2018 . [CrossRef]
13. Sánchez-González, R.; Srinivasan, R.; Hofferth, J.; Kim, D.Y.; Tindall, A.J.; Bowersox, R.D.W.; North, S.W. Repetitively Pulsed Hypersonic Flow Apparatus for Diagnostic Development. *AIAA J.* **2012**, *50*, 691–697. [CrossRef]
14. Segal, C. A Unique, Mach 6 Enthalpy, Non-Vitiated Facility for Hypersonic Aerodynamics Research. In Proceedings of the 17th AIAA International Space Planes and Hypersonic Systems and Technologies Conference, Washington, DC, USA, 13–17 June 2016; Paper No. 2011-2278. [CrossRef]
15. Leonov, S.B. Electrically Driven Supersonic Combustion. *Energies* **2018**, *11*, 1733. [CrossRef]
16. Leonov, S.B.; Elliott, S.; Carter, C.; Houpt, A.; Lax, P.; Ombrello, T. Modes of plasma-stabilized combustion in cavity-based M = 2 configuration. *Exp. Therm. Fluid Sci.* **2021**, *124*, 110355. [CrossRef]
17. Watanabe, Y.; Elliott, S.; Firsov, A.; Houpt, A.; Leonov, S. Rapid control of force/momentum on a model ramp by quasi-DC plasma. *J. Phys. D Appl. Phys.* **2019**. *52*, 444003. [CrossRef]
18. Elliott, S.; Hasegawa, M.; Sakaue, H.; Leonov, S. Shock-Dominated Flow Control by Plasma Array: Pressure Analysis Including Pressure-Sensitive Paint Visualization. *Exp. Therm. Fluid Sci.* **2021**, *131*, 110522. doi: [CrossRef]
19. Vuković, D.S.; Damljanović, D.B. Thermal effects influencing measurements in a supersonic blowdown wind tunnel. *Therm. Sci.* **2016**, *20*, 2101–2112. [CrossRef]
20. Grossir, G.; Paris, S.; Rambaud, P.; Hove, B.V. Design of Static Pressure Probes for Improved Free-Stream Characterization in Hypersonic Wind-Tunnels. In Proceedings of the 52nd Aerospace Sciences Meeting, National Harbor, MD, USA, 13–17 January 2014; Paper No. 2014-1410. [CrossRef]
21. Montello, A.; Nishihara, M.; Rich, J.W.; Adamovich, I.V.; Lempert, W.R. Picosecond CARS measurements of nitrogen rotational/translational and vibrational temperature in a nonequilibrium Mach 5 flow. *Exp. Fluids* **2012**, *54*, 1422. [CrossRef]
22. Dogariu, A.; Dogariu, L.E.; Smith, M.S.; McManamen, B.; Lafferty, J.F.; Miles, R.B. Velocity and Temperature Measurements in Mach 18 Nitrogen Flow at Tunnel 9. In Proceedings of the AIAA Scitech 2021 Forum, Nashville, TN, USA, 11–15 & 19–21 January 2021; Paper No. 2021-0020. [CrossRef]

23. Dogariu, L.E.; Dogariu, A.; Miles, R.B.; Smith, M.S.; Marineau, E.C. Non-Intrusive Hypersonic Freestream and Turbulent Boundary-Layer Velocity Measurements in AEDC Tunnel 9 Using FLEET. In Proceedings of the 2018 AIAA Aerospace Sciences Meeting, Kissimmee, FL, USA, 8–12 January 2018; Paper No. 2018-1769. [CrossRef]
24. Cutler, A.; Magnotti, G.; Cantu, L.; Gallo, E.; Danehy, P.; Rockwell, R.; Goyne, C.; McDaniel, J. Dual-Pump CARS Measurements in the University of Virginia's Dual-Mode Scramjet: Configuration "A". In Proceedings of the 50th AIAA Aerospace Sciences Meeting including the New Horizons Forum and Aerospace Exposition, Nashville, TN, USA, 9–12 January 2012; Paper No. 2012-114. [CrossRef]
25. Ecker, T.; Brooks, D.R.; Lowe, K.T.; Ng, W.F. Development and application of a point Doppler velocimeter featuring two-beam multiplexing for time-resolved measurements of high-speed flow. *Exp. Fluids* **2014**, *55*, 1819. [CrossRef]
26. Beresh, S.; Kearney, S.; Wagner, J.; Guildenbecher, D.; Henfling, J.; Spillers, R.; Pruett, B.; Jiang, N.; Slipchenko, M.; Mance, J.; et al. Pulse-burst PIV in a high-speed wind tunnel. *Meas. Sci. Technol.* **2015**. *26*, 095305. [CrossRef]
27. Biswas, S., Schlieren Image Velocimetry (SIV). In *Physics of Turbulent Jet Ignition: Mechanisms and Dynamics of Ultra-lean Combustion;* Springer International Publishing: Cham, Switzerland, 2018; pp. 35–64. [CrossRef]
28. Gopal, V.; Palmquist, D.; Maddalena, L.; Dogariu, L.; Dogariu, A. FLEET velocimetry measurements in the ONR-UTA arc-jet wind tunnel. *Exp. Fluids* **2021**, *62*, 212. [CrossRef]
29. Zahradka, D.; Parziale, N.J.; Smith, M.S.; Marineau, E.C. Krypton tagging velocimetry in a turbulent Mach 2.7 boundary layer. *Exp. Fluids* **2016**, *57*, 62. [CrossRef]
30. Rennie, M.R.; Nguyen, M.; Gordeyev, S.; Jumper, E.; Cain, A.B.; Hayden, T., Windspeed and Flow Angle Measurement by Tracking of a Laser-Induced Breakdown Spark. Proceedings of the In 33rd AIAA Applied Aerodynamics Conference, Dallas, TX, USA, 22–26 June 2015; [CrossRef]
31. Lester, L.; Gragston, M. 10 kHz laser-induced schliere anemometry for velocity, Mach number, and static temperature measurements in supersonic flows. *Appl. Opt.* **2021**, *60*, 8644–8650. [CrossRef]
32. Campanella, B.; Legnaioli, S.; Pagnotta, S.; Poggialini, F.; Palleschi, V. Shock Waves in Laser-Induced Plasmas. *Atoms* **2019**, *7*, 57. [CrossRef]
33. Harilal, S.S.; Brumfield, B.E.; Phillips, M.C. Lifecycle of laser-produced air sparks. *Phys. Plasmas* **2015**, *22*, 063301. [CrossRef]
34. Sedov, L.I. Propagation of strong blast waves. *Prikl. Mat. Mekhanika* **1946**, *10*, 241–250. (In Russian)
35. Taylor, G.I. The formation of a blast wave by a very intense explosion I. Theoretical discussion. *Proc. R. Soc. Lond. Ser. Math. Phys. Sci.* **1950**, *201*, 159–174. [CrossRef]
36. Raga, A.C.; Cantó, J.; Rodríguez, L.F.; Velázquez, P.F. An analytic model for the strong-/weak-shock transition in a spherical blast wave. *Mon. Not. R. Astron. Soc.* **2012**, *424*, 2522–2527. [CrossRef]
37. Kandula, M.; Freeman, R. On the interaction and coalescence of spherical blast waves. *Shock Waves* **2008**, *18*, 21–33. [CrossRef]
38. Dors, I.G.; Parigger, C.G. Computational fluid-dynamic model of laser-induced breakdown in air. *Appl. Opt.* **2003**, *42*, 5978–5985. [CrossRef]
39. Bontrager, P.J. Development of Thermocouple-Type Total Temperature Probes in the Hypersonic Flow Regime. 1969. Available online: https://apps.dtic.mil/sti/pdfs/AD0681489.pdf (accessed on 10 March 2022).
40. Omega. *Revised Thermocouple Reference Tables, Type K Reference Tables*; N.I.S.T. Monograph 175 Revised to ITS-90; 1993. Available online: https://cn.omega.com/temperature/Z/pdf/z204-206.pdf (accessed on 10 November 2022).
41. Albertson, C.W.; Bauserman, W.A. Total Temperature Probes for High-Temperature Hypersonic Boundary-Layer Measurements. NASA, 1993. Available online: https://ntrs.nasa.gov/api/citations/19930012627/downloads/19930012627.pdf (accessed on 10 March 2022).

fluids

MDPI

Article

Thrust Vectoring of a Fixed Axisymmetric Supersonic Nozzle Using the Shock-Vector Control Method

Emanuele Resta, Roberto Marsilio and Michele Ferlauto *

Department of Mechanical and Aerospace Engineering, Politecnico di Torino, Corso Duca degli Abruzzi 24, 10129 Torino, Italy; emanuele.resta@polito.it (E.R.); roberto.marsilio@polito.it (R.M.)
* Correspondence: michele.ferlauto@polito.it

Abstract: The application of the Shock Vector Control (SVC) approach to an axysimmetric supersonic nozzle is studied numerically. SVC is a Fluidic Thrust Vectoring (FTV) strategy that is applied to fixed nozzles in order to realize jet-vectoring effects normally obtained by deflecting movable nozzles. In the SVC method, a secondary air flow injection close to the nozzle exit generates an asymmetry in the wall pressure distribution and side-loads on the nozzle, which are also lateral components of the thrust vector. SVC forcing of the axisymmetric nozzle generates fully three-dimensional flows with very complex structures that interact with the external flow. In the present work, the experimental data on a nozzle designed and tested for a supersonic cruise aircraft are used for validating the numerical tool at different flight Mach numbers and nozzle pressure ratios. Then, an optimal position for the slot is sought and the fully 3D flow at flight Mach number $M_\infty = 0.9$ is investigated numerically for different values of the SVC forcing.

Keywords: thrust vectoring; shock-vector control method; active flow control; supersonic nozzle

Citation: Resta, E.; Marsilio, R.; Ferlauto, M. Thrust Vectoring of a Fixed Axisymmetric Supersonic Nozzle Using the Shock-Vector Control Method. *Fluids* **2021**, *6*, 441. https://doi.org/10.3390/fluids6120441

Academic Editor: Olga A. Azarova

Received: 8 November 2021
Accepted: 2 December 2021
Published: 7 December 2021

1. Introduction

Thrust Vectoring (TV) consists of the modulation of the thrust vector in a variable direction other than the axial direction, thus introducing an additional control variable in the equation of motion of the aircraft. In so doing, otherwise inoperable flight regimes, such as maneuvers at low airspeeds and very high angles of attack, or even stalled conditions, can be handled safely [1–3]. Moreover, designers can explore new supersonic/hypersonic aircraft configurations with lower sonic-boom signatures [4], and with Short Take-Off and Landing (STOL) capabilities and augmented maneuverability [1]. Nowadays, thrust vectoring control is a required feature for any advanced tactical fighter [3,5]. Practical application of thrust vectoring is obtained by turning the nozzle mechanically to point in different directions. The same effect can be obtained without actuated mechanical hardware, by forcing and manipulating the flow inside a nozzle of fixed geometry. This second approach, namely the Fluidic Thrust Vectoring (FTV), uses a secondary air bleed in order to actively manipulate and control the primary air-stream of the nozzle. The injected fluid creates variable "artificial" nozzle boundaries and the perturbations generated by the forcing of the secondary flow make the nozzle wall pressure distribution asymmetric. The resulting effect is a side-force on the nozzle that can be seen as the lateral component of the thrust vector. Experimental testing and numerical simulations of simple two-dimensional nozzle configurations have shown significant thrust deflections and a dynamic response faster than that obtained by mechanical systems [6–8]. Moreover, with respect to mechanical thrust vectoring, the FTV approach does not increase the aircraft weight significantly.

Following the renewed interest in space exploration and the trans-atmospheric supersonic flight, FTV concepts are being intensively investigated [9–13]. The key point for the fluidic approach is the identification of manipulation techniques that gradually modulate the nozzle wall pressure symmetry-breaking effect within an acceptable range of deterioration of the nozzle performances [14]. Several control strategies have been investigated in the literature including techniques such as Shock Vector Control (SVC), Counter-Flows (CF), Throat Shifting (TS) and supersonic Dual-Throat Nozzle (DTN) [14]. The effectiveness of the mentioned FTV techniques has been investigated numerically and tested experimentally by several research groups [4,7,15–23]. The integration of FTV models in engine and aircraft dynamic models is an on-going research project [24–26]. Most current experimental and numerical investigations on FTV rely on fundamental aspects and are based on two-dimensional nozzle models [14]. Two-dimensional flows are easier to visualize experimentally and less expensive from a computational point of view, while retaining most of the essential features of the phenomenon under investigation. More commonly, however, aircraft and rocket nozzles are axisymmetric and this fact has an important consequence: after breaking the flow symmetry, for example, by local secondary air-injection, the nozzle flow becomes fully three-dimensional. Therefore, the applications of FTV to real nozzle geometries must deal with the analysis and/or simulation of fully three-dimensional flows.

In the present work, the FTV performances of an axisymmetric nozzle under shock vector control are investigated numerically. In order to support our analysis with experimental data, the reference nozzle geometry is deduced from the experimental work of Carlson and Lee at Nasa LaRC [27]. Five different nozzle geometries were tested in that reference. Each nozzle geometry represented a different flight condition and power setting of a variable-geometry axisymmetric nozzle designed for a variable-cycle engine of a supersonic aircraft. The nozzle was installed in a nacelle with a forebody and was tested at external flow Mach numbers ranging from 0.6 to 1.3 [27]. The nozzle configuration 2 at Mach flight number $M_\infty = 0.6$, $M_\infty = 0.9$ and $M_\infty = 1.2$ was selected for the present analysis. The outline of the papers is as follows: the nozzle setup and working conditions are described first; then the mathematical model and numerical approaches are illustrated. After that, the numerical investigations follow with increasing complexity. The numerical tools are validated with respect to experimental data [27] in the axisymmetric case for different values of the Nozzle Pressure Ratio NPR and different Flight Mach numbers. As a second step, the sensitivity of the flow to the slot position and secondary mass flow is investigated, retaining the axisymmetric assumption. The SVC approach is then finally applied to the fully three-dimensional case and the nozzle FTV effectiveness and performances are investigated numerically for different values of the secondary mass flow.

2. Nozzle Setup and Geometry

The nozzle geometry and working conditions simulated in the present numerical study are derived from the experimental testing carried out at Nasa LaRC by Carlson and Lee [27]. In that work, five models of an axisymmetric, convergent-divergent nozzle with circular-arc throat contour and conical divergent sections were investigated experimentally. The nozzle geometries refer to different configurations of the variable-geometry axisymmetric nozzle designed for a supersonic cruise aircraft equipped with a variable-cycle engine. The five configurations represent the nozzle setup for different flight conditions and power settings ranging from a subsonic cruise, dry power configuration, characterized by a low expansion ratio and a high nozzle boattail angle, to supersonic acceleration, with maximum afterburning, having a high expansion ratio and a low nozzle boattail angle. The experimental investigations were conducted at nozzle pressure ratios (NPR) from jet-off to about NPR $= 10$ and for freestream Mach numbers ranging from 0.60 to 1.30. A detailed description of the nozzle geometries and testing conditions is given in Ref. [27]. Numerical simulations of the flow are also available for comparison [28,29]. The nozzles were attached to an axisymmetric nacelle with a forebody. The general arrangement of the nacelle model and support system is shown in Figure 1. Nozzle Configuration 2 (Conf-2) of Ref. [27] has been selected for the numerical tests reported here. The configuration corresponds to the supersonic cruise setting in dry condition, that is, without afterburning. The adaptation pressure ratio of the nozzle in Conf-2 is NPR$^* = 21.23$. Nozzle geometry and relevant design parameters are presented in Figure 2.

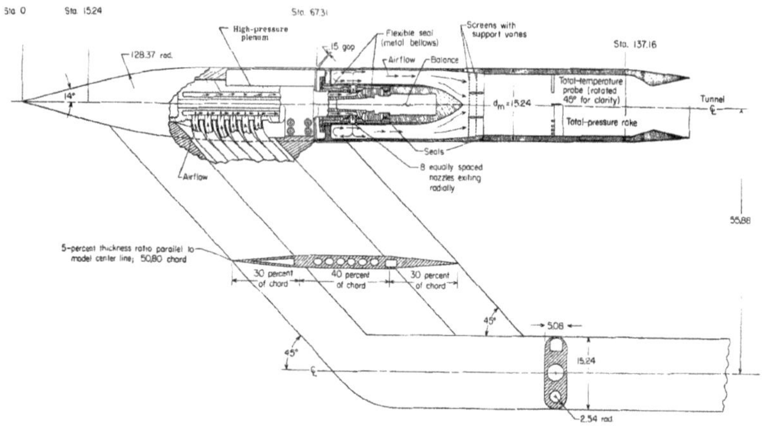

Figure 1. General arrangement of the nacelle model and support system (adapted from Ref. [27]).

Parameter	Value
A_e/A_t	3.0
A_t/A_m	0.25
d_t/d_m	0.5
d_e/d_m	0.866
l_c/d_m	0.286
l/d_m	0.979
L_{div}	11.87 cm
β	3.82^o
δ	13.18^o
ϑ	42.35^o
NPR*	21.23

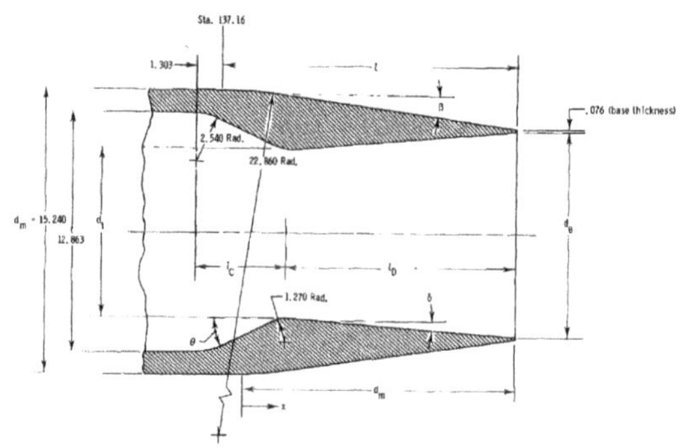

Figure 2. Nozzle Configuration-2 (adapted from Ref. [27]). Geometry and e list of relevant design parameters.

3. Mathematical and Numerical Modelling

The problem is numerically investigated solving the compressible Unsteady Reynolds-Averaged Navier–Stokes (URANS) equations by using the commercial CFD solver STAR-CCM+ [30]. The one-equation model of Spalart–Allmaras (S-A) [31,32] is used for the turbulence modelling, as it has shown a good agreement with the experimental data for the case of unsteady flow in nozzles [7,33]. The numerical approach adopted is almost equivalent to that used in the in-house 2D numerical framework developed for the unsteady simulation of the vectored nozzle in open and closed-loop conditions, which was tested and validated for the FTV with both continuous or pulsating blowing [6,34].

Briefly, the flow governing equations are represented by the compressible Unsteady Reynolds Averaged Navier–Stokes equations (URANS), written in the compact integral form:

$$\frac{\partial}{\partial t}\int_{\mathcal{V}} \vec{W} d\mathcal{V} + \int_{\mathcal{S}} \vec{F}_I \cdot \hat{n} d\mathcal{S} + \int_{\mathcal{S}} \vec{F}_V \cdot \hat{n} d\mathcal{S} = \int_{\mathcal{V}} \vec{H} d\mathcal{V}, \quad (1)$$

for an arbitrary volume $\mathcal{V}$ enclosed in a surface $\mathcal{S}$. With the usual conventions, $\vec{W} = \{\rho, \rho\vec{q}, E, \tilde{\nu}_t\}^T$ is the hyper-vector of conservative variables, $\vec{F}_I$ and $\vec{F}_V$ are tensors containing the inviscid and the viscous fluxes, respectively.

$$\vec{F}_I = \left\{\rho\vec{q}, p\bar{\bar{I}} + \rho\vec{q} \otimes \vec{q}, (E+p)\vec{q}, \tilde{\nu}_t\vec{q}\right\}^T, \tag{2}$$

$$\vec{F}_V = \left\{0, -\bar{\bar{\tau}}, -\kappa\nabla T - \bar{\bar{\tau}} \cdot \vec{q}, -\frac{\nu + \tilde{\nu}_t}{\sigma}\nabla\tilde{\nu}_t\right\}^T. \tag{3}$$

$\vec{q} = \{u, v, w\}^T$ is the velocity vector, E the total energy per unit volume, γ is the ratio of the specific heats and $\bar{\bar{I}}$ is the unit matrix. The term $\vec{H}$

$$\vec{H} = \left\{0, 0, 0, c_{b1}\tilde{S}\tilde{\nu}_t + \frac{c_{b2}}{\sigma}(\nabla\tilde{\nu}_t)^2 - c_{w1}f_w\left(\frac{\tilde{\nu}_t}{d}\right)^2\right\}^T \tag{4}$$

contains turbulence model source terms. The viscous stresses terms can be written as:

$$\tau_{ij} = (\mu + \mu_t)\left[\frac{\partial q_j}{\partial x_i} + \frac{\partial q_i}{\partial x_j} - \frac{2}{3}(\nabla \cdot \vec{q})\delta_{ij}\right]. \tag{5}$$

The laminar viscosity μ is computed via the Sutherland's law, whereas the turbulent viscosity $\mu_t = \rho\nu_t$ is defined according to the Spalart–Allmaras (S-A) model [31,32]. Inviscid fluxes are evaluated through the AUSM+ flux–vector splitting scheme [35], based on the upwind concept, which applies to both the convective and pressure parts of the inviscid flux. The computation takes into account local flow characteristics for a correct propagation of physical information inside the domain. A second order discretization in space and time has been chosen for the solution of the set of equations, by means of an implicit, dual time-stepping solver. The implicit solver features a preconditioned pseudo-time derivative term within the equation system. This additional term vanishes as convergence is reached in the inner loop, and the solution at the next physical time level is computed. A local pseudo time-step is used in steady state simulations and for inner loop iterations of transient simulations.

Computational Domain and Boundary Conditions

The domain geometry is chosen in such a way to capture the relevant interactions between the free flowing air and the forebody, as well as interactions between the external flow and the nozzle jet. The computational domain consists of a mesh of about 170 k quadrangular elements for the 2D axisymmetric analysis, whereas for 3D calculations, a grid of about 3.5 M cells has been generated by revolution around the symmetry axis. The boundary conditions for the problem were chosen according to the nature of the flow and the available experimental information. For the external flow inlet, a free-stream boundary condition has been used to impose Mach number, pressure and temperature. A pressure outlet boundary is utilized to impose an outflow condition with specified static ambient pressure, while values for velocity or temperature are extrapolated from the interior of the domain. A total inlet condition is adopted for the nozzle flow, in order to prescribe values of total pressure and total temperature for the jet, whereas for the injection of the secondary flow, a mass flow rate boundary condition has been used. For solid surfaces a no-slip wall condition is utilized. For the 2D domain a symmetry axis has been prescribed in order to perform the axisymmetric simulations, whereas for the 3D simulation a symmetry plane condition is utilized to account for the other half of the physical domain. A representation of the computational grid and a visualization of the boundaries can be found in Figure 3.

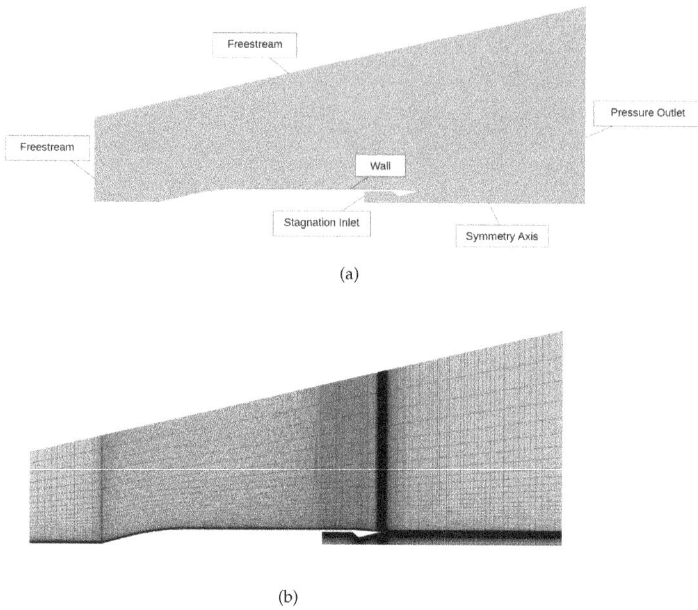

Figure 3. Computational domain with boundary conditions (**a**) and grid (**b**).

4. Numerical Results

4.1. Grid Independence Analysis

The number of nodes for the axisymmetric mesh has been defined after a grid dependence study on three different levels of refinement. A coarse grid of 73 k cells, a medium grid of 170 k cells, and a fine grid of 325 k cells have been selected for this analysis. The results, in terms of temperature, mach number and pressure profiles obtained along the x-axis, that is, along the nozzle axis of symmetry on the different grids, are presented in Figure 4. A small mismatch can be appreciated between the temperature profile obtained on the coarse mesh with respect to that obtained on the medium and fine meshes (Figure 4a). The latter two temperature profiles match perfectly instead. The coarse grid also tends to slightly under-predict the Mach number values, as visible in Figure 4b. Nevertheless, the pressure distributions along the x-axis evaluated on the three meshes are almost indistinguishable (Figure 4c). The x-axis is also the region of inner nozzle computational domain with the lower resolution. A refinement of the mesh has been performed to increase the resolution of the results near walls and other regions of interest as, for instance, at the exit section of the nozzle and in the vicinity of the secondary flow injection opening. The grid has also been stretched near the walls in order to obtain $y^+ = 1$, and in such a manner that no wall functions were needed to correctly model turbulence at the boundary layer level. Although the coarsest mesh provided more than adequate results for the wall pressure distributions, the medium grid was chosen for all subsequent analyses for its greater resolution inside the nozzle duct and the increased shock capturing accuracy.

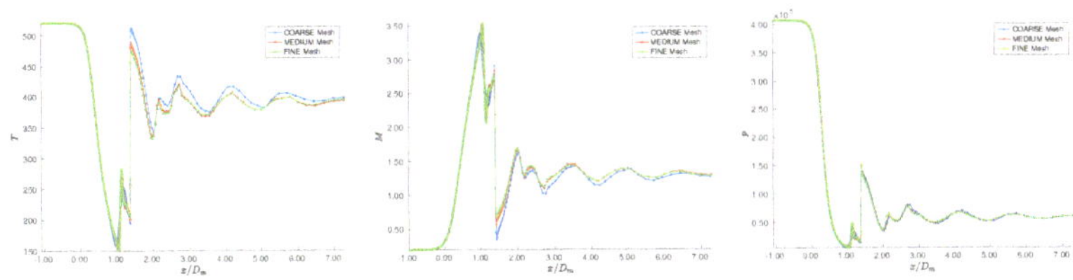

(a) (b) (c)

Figure 4. Temperature (**a**), Mach number (**b**) and pressure (**c**) distributions along the nozzle axis of symmetry at different grid levels (coarse mesh = 73 k cells; medium mesh = 170 k cells; fine mesh = 325 k cells).

4.2. Simulation of the Axisymmetric Nozzle Flow

The flowfield in an axisymmetric propulsive nozzle is very complex and is characterized by a highly nonlinear response to incoming perturbations. An extensive validation of the computational framework used must be carried out before going toward the simulations of fluidic forcing. As mentioned in previous sections, the wide database of experiments reported in Carlson and Lee [27] offers very useful test-cases for this kind of validation, including the effects of external flow interference at different nozzle pressure ratios. In this section, the validation carried out for nozzle configuration 2 (Conf-2) of Ref. [27], without SVC forcing, is described with details. Three different flight Mach number conditions have been selected for the validation, specifically $M_\infty = 0.6, 0.9$, and 1.2, in order to obtain a fairly complete set of working conditions for the nozzle, ranging from relatively low flight Mach numbers to supersonic cruise conditions, and across a variety of nozzle pressure ratios. For validation purposes, the selected sets of parameters (M_∞, NPR) match the working conditions tested in the NASA LaRC experiments [27].

A full view of the simulated flowfield inside and around the nozzle system is presented in Figure 5. This flowfield represents a typical nozzle off-design condition of interest. The nozzle is over-expanded (the pressure ratio is $\text{NPR} = 11$) at high transonic flight ($M_\infty = 0.9$). As is visible, an adequate resolution of the nozzle plume is obtained and no significant flow perturbations arise from or reach the boundaries, other than the jet flow. Moreover, the interference generated by the forebody does not propagate up to the nozzle region. Therefore, the jet flow issuing from the nozzle is influenced only by the external pressure and by the flow conditions in the aft-body region in general.

The results of the complete sets of simulations at M_∞= 0.6, M_∞= 0.9 and M_∞= 1.2 are presented in Figures 6–8, respectively. For each pressure ratio, the flowfield is represented in terms of both Mach number and density contour maps. The NPRs values are not the same for each flight mach number M_∞ in order to adhere to that reported in the experiments [27]. In all these figures, the different levels of nozzle over-expansion, its effect of the shock induced separation and on the mach disk position is clearly visible. Focusing on Figure 6, once the nozzle flow becomes supersonic, shock structures begin to form, as is already visible for $\text{NPR} = 2.53$, where oblique shock waves and the Mach disk are first generated. As the NPR is increased, the region of separated flow shrinks in size, the Mach disk is shifted forward, and wave reflections become more visible. For even higher NPR, supersonic shocks and reflections become increasingly distinguishable in the jet plume too. The density field follows the same considerations, but its values mirror those of the Mach field, decreasing during expansions where the Mach number increases and vice versa through shocks. Very similar results are obtained for external Mach numbers $M_\infty = 0.9$ and $M_\infty = 1.2$, as illustrated in Figures 7 and 8.

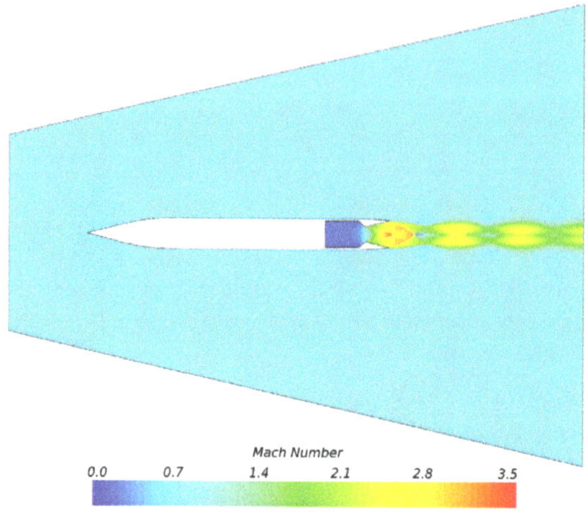

Figure 5. Overview of the flowfield inside and around the nozzle Configuration 2 at $M_\infty = 0.9$ and NPR = 11.

The plots of numerical results against available experimental data are given in Figure 9, for both the pressure coefficient on the boat-tail c_{p_β}, and pressure distribution on the internal wall, normalized with the jet total pressure p_{tj}. A good agreement between numerical and experimental data is achieved with the chosen model, as the point of separation was accurately predicted in all cases. The best results are obtained at higher nozzle pressure ratios, in particular for NPR > 4, where the separated regions are at first limited and then eliminated. The external pressure coefficient plots also show good agreement with experimental values, following the curve compression and expansion features for all three flight Mach numbers. The higher mismatch between numerical and experimental data is obtained at $M_\infty = 1.2$ and NPR $= 2.02$ in Figure 9c. At that conditions, the numerical prevision shows a choked flow inside the nozzle, whereas the experimental data do not collapse in a single curve after chocking as expected and the flow seems to exhibit a weak separation instead of an abrupt shock induced separation. After that, the location of the flow separation and the exit conditions are matched correctly.

4.3. Application of Secondary Injection to the Axisymmetric Nozzle Flow

In this section, a study of the flow forcing effectiveness is carried out for identifying an adequate location of the secondary massflow injection point required for the SVC based thrust vectoring. Obviously, under the axisymmetric flow assumption, we do not expect any jet flow deflection. It was assumed, anyway, that the flow forcing may produce local effects that are similar in the axisymmetric and fully three-dimensional case. It is well-known that the SVC technique depends on a flow structure generated by the interaction between the secondary blowing, normal to the wall, and the main flow. In the supersonic region upwind the flow injection a fluid ramp is formed and an oblique shock is generated [20,36]. We are interested in the numerical estimation of parameters such as the shock distance from the injection slot and the shock inclination, in order to rapidly assess the effects on the main flow and to identify a suitable position of the injection slot for the more costly 3D simulations. Naturally, multiple parameters other than the position of the slot alone can influence the effect of the injection on the main flow and its structures; for instance, the secondary mass flow ratio and the opening area have been shown to have a direct influence on the deflection of the flow [37]. In this preliminary analysis, the secondary mass flow-rate has been kept constant and equal to 3% of the main mass flow-rate, while the injection slot area in the axisymmetric case is naturally the lateral surface of a truncated cone. The resulting mass flux per unit area is thus considerably smaller than that in the case of a secondary injection of the same mass flow-rate through a reduced area. The opening has been sequentially positioned at 70%, 80% and 90% of the length of the diverging part of the nozzle until the position of the fluidic ramp that would be generated was considered satisfactory for the subsequent fully three-dimensional analysis. The modification of the grid close to the injection slot and the required grid refinement are shown in Figure 10.

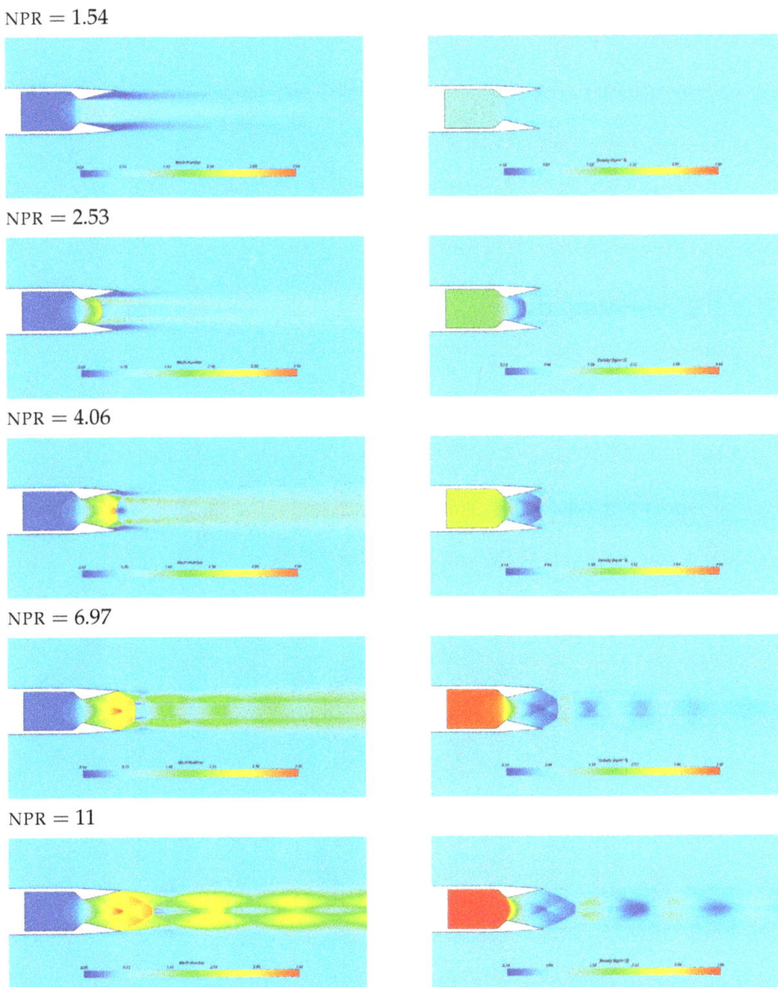

Figure 6. Mach number (**left**) and density (**right**) fields for different NPR's for $M_\infty = 0.6$.

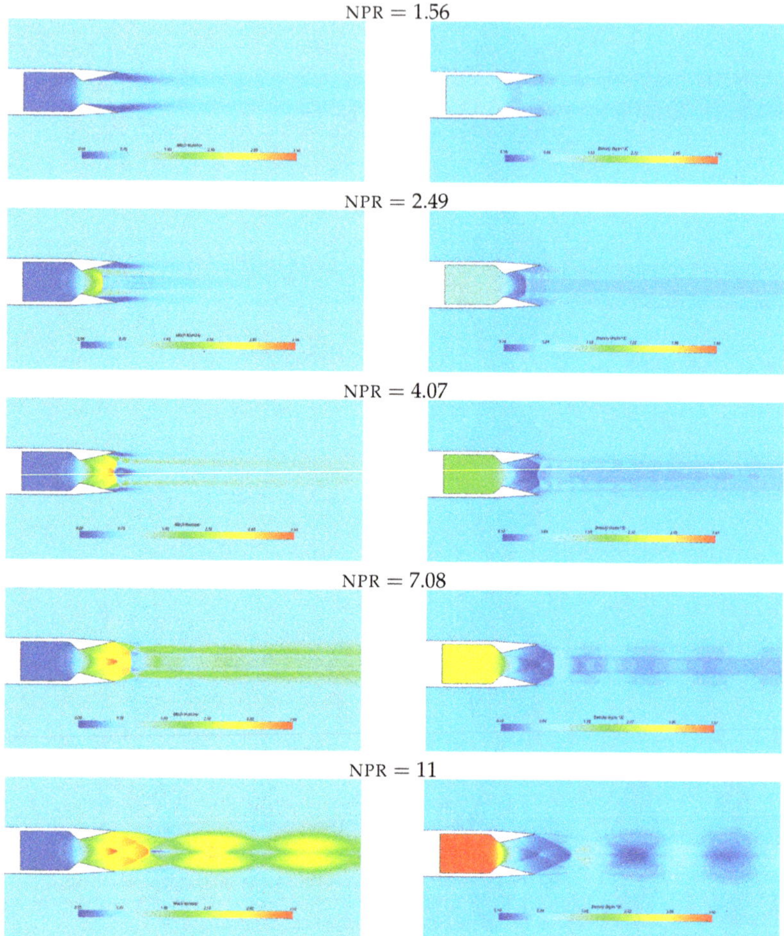

Figure 7. Mach number (**left**) and density (**right**) fields for different NPR's and $M_\infty = 0.9$.

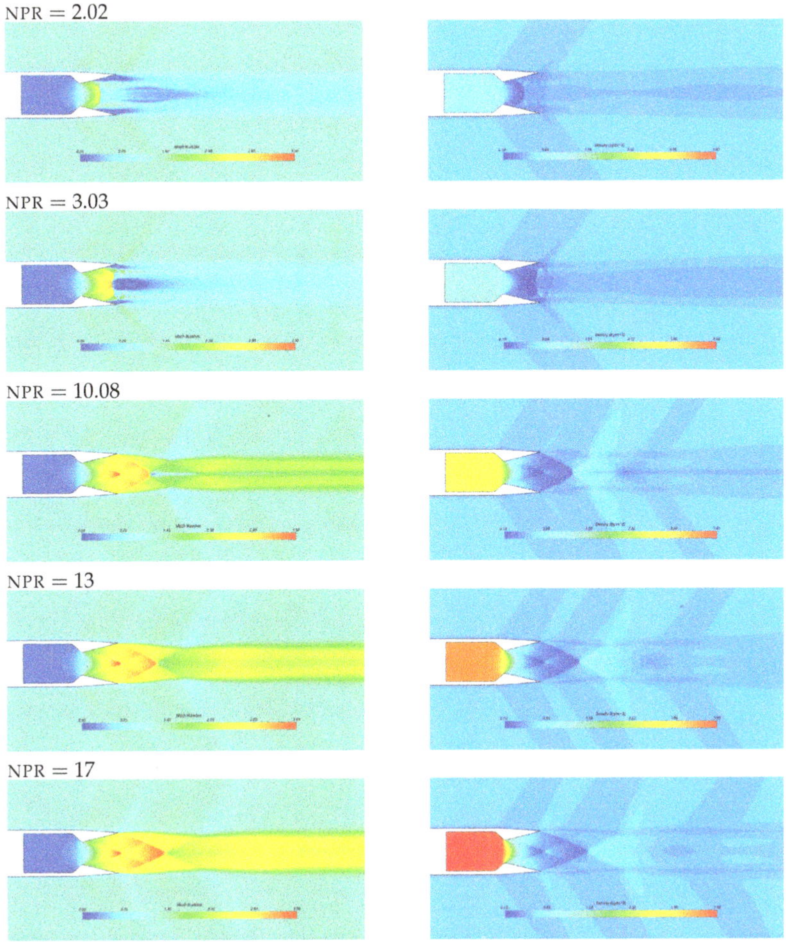

Figure 8. Mach number (**left**) and density (**right**) fields for different NPR's and $M_\infty = 1.2$.

As can be seen in Figure 11, the recession of the injection slot position moves the first oblique shock backwards, and expands the separated flow region near the slot. When the opening position is too far upstream a reattachment of the flow can be seen for this value of the mass flux per unit area, however this problem is eliminated when either the injection area is restricted or the mass flux per unit area is increased, or both. A comparison between the perturbation of the main flow in axisymmetric conditions and in three dimensions, for the same position of the slot and the same value of mass flux is shown in Figure 12. The chosen mass flux for the comparison is that corresponding to a secondary injection of 6% of total mass flow rate in the three dimensional case. The increase in value of the mass flux eliminates the aforementioned problem of reattachment of the flow for the axisymmetric case and, as can be seen in the figure, the effect of the injection in these conditions is more pronounced than that in the three dimensional case. With these considerations in mind, the slot position at 70% of the nozzle length has been considered the most adequate for the purposes of the subsequent three dimensional analysis.

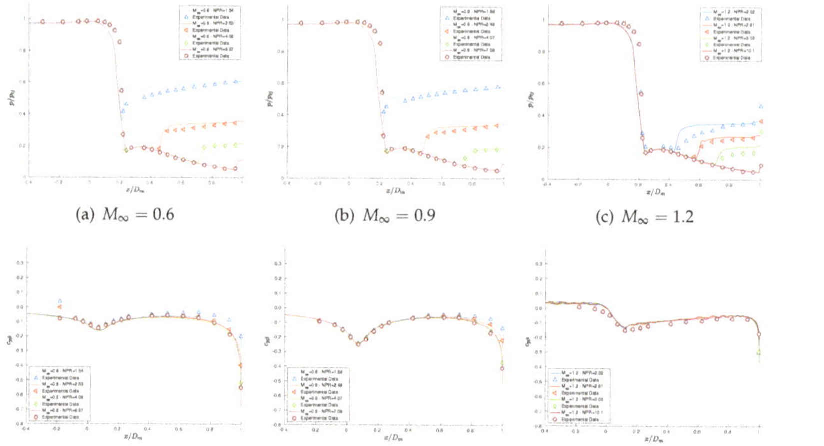

(a) $M_\infty = 0.6$ (b) $M_\infty = 0.9$ (c) $M_\infty = 1.2$

(d) $M_\infty = 0.6$ (e) $M_\infty = 0.9$ (f) $M_\infty = 1.2$

Figure 9. Internal nozzle walls static pressure distribution (**a**–**c**), and boat-tail pressure coefficient distribution (**d**–**f**). Comparison with experimental data.

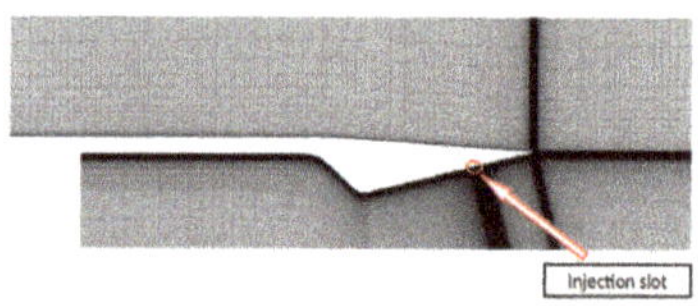

Figure 10. Mesh detail close to the injection zone.

4.4. *SVC Thrust Vectoring of the Axisymmetric Nozzle in 3D*

In this section, the analysis based on the fully three-dimensional numerical simulations of the nozzle under SVC forcing is discussed. In previous SVC vectored nozzle studies [19,20,36,38] performances are analyzed and characterized by varying the NPR and the secondary mass flow ratio SMF for a jet-flow efflux in calm air. External flow interactions have been accounted for mainly by CFD approaches in two dimensions [39]. In Section 4.2, the NASA test-case on boat-tail nozzle flow has been used for validating the numerical tools and a good agreement for the simulations of the interaction between the nozzle plume and the external flow has been obtained for the case of co-flowing streams at different asymptotic Mach numbers M_∞. Now, we follow the same path and numerically analyze the SVC effects of secondary mass flow injection on the nozzle side force, also in the presence of an external flow. In particular, at this stage, we focused on the study of the effectiveness of the FTV for different levels of forcing, based on different SMF values, at the nozzle pressure ratio NPR $= 7.08$ and with an external flow Mach number $M_\infty = 0.9$.

The 3D grid has been generated by rotational extruding of the axisymmetric mesh by an angle of 180° around the symmetry axis while retaining symmetry on the meridian plane. Only the half nozzle has therefore been simulated because of symmetry. A 3D grid of about 3.5 M cells is thus obtained. A sketch of the nozzle configuration is shown in Figure 13. Based on the analysis described in the previous section, the injection slot, with a 2 mm width in the axial direction, has been positioned at 70% of the axial extension of the diverging part of the nozzle. The slot has been extended by a $\vartheta = 45$ degree angle along the circumferential coordinate. Due to the symmetry assumption, the actual circumferential extension ϑ_R of the slot is twice, that is $\vartheta_R = 90°$. Figure 13 also shows the injection slot and its location on the nozzle walls.

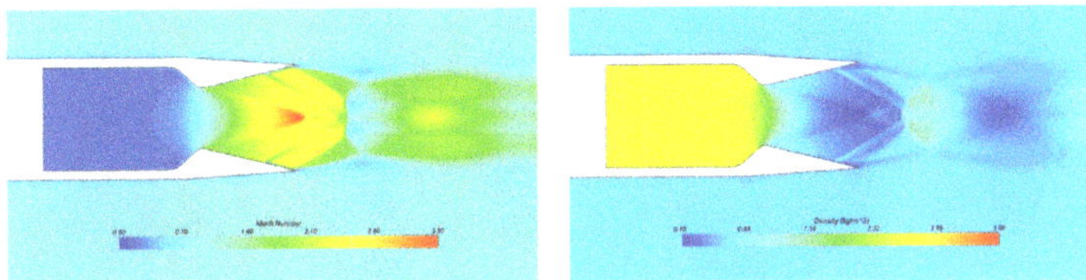

$x_s = 70\%\ L_{div}$

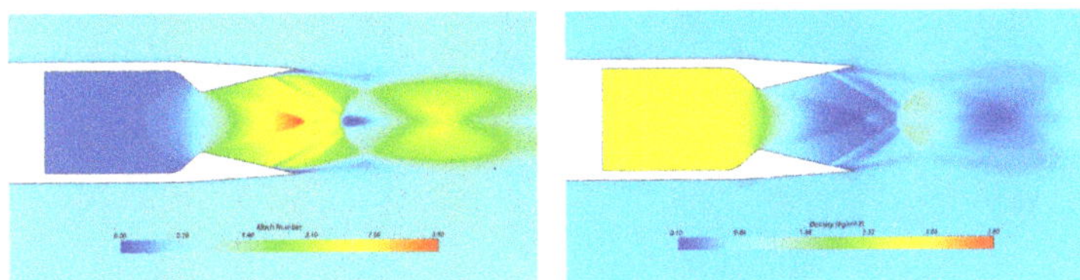

$x_s = 80\%\ L_{div}$

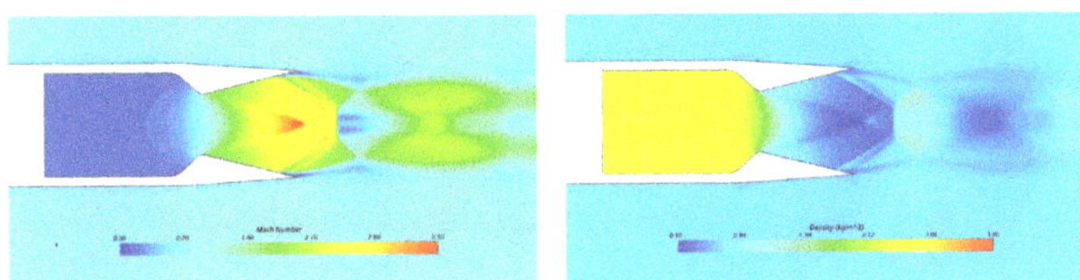

$x_s = 90\%\ L_{div}$

Figure 11. Mach number (**left**) and density (**right**) fields for axisymmetric injection at different positions in the nozzle divergent.

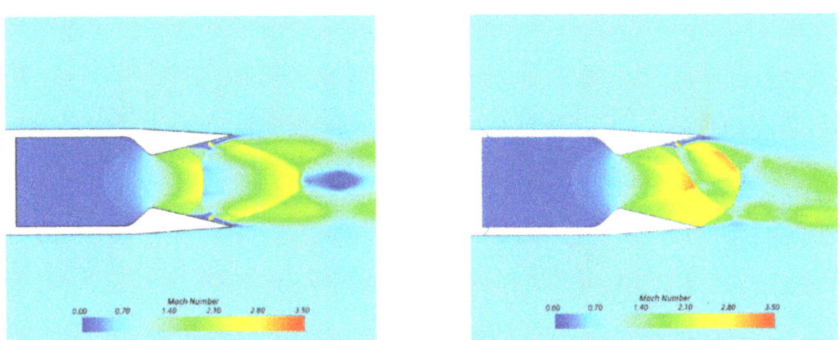

(a) (b)

Figure 12. Comparison between the axisymmetric (**a**) and three dimensional (**b**) solution of the simulation with secondary injection.

Various secondary mass flow rates SMF have been considered, ranging from 3% to 10% of the main flow rate computed through the nozzle throat. The FTV performances have been evaluated as is done in [11]. In particular, the thrust components along the axes have been computed and the ratio of lateral force component Fy to the axial component F_x gives a measure of the vectoring effectiveness of the secondary flow injection. Results obtained by the numerical simulations are reported in Figure 14 where the values of the pitch thrust-vector angle $\delta = \tan^{-1}(F_y/F_x)$ at different SMF are shown.

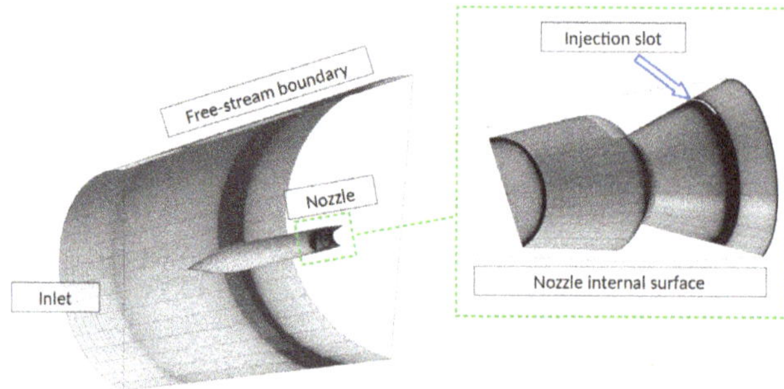

Figure 13. Three-dimensional (3D) mesh configuration: global view and nozzle detail.

The plot of Figure 14 shows that the increase of secondary mass flow rate induces a nonlinear increase in the value of δ, up to about 4.5° for SMF $= 10\%$, following an almost parabolic curve [11]. As a consequence, the lateral force F_y rises as well, increasing from 3.5% to about 7.5% of the axial force component, F_x. At the same time, the axial force component is reduced with respect to the symmetric condition of just about 3%, passing from about 1280 N for SMF $= 0\%$ to 1245 N for SMF $= 10\%$. The increase in lateral force is due to the separation induced by the injection of the secondary flow, which in turn generates a separation zone on the nozzle walls, thus changing the pressure distribution as is well highlighted in Figure 15 for the 6% and 10 % of secondary mass flow cases. Such separation is shown in Figure 16, where it is very clear that a fluidic ramp is formed ahead of the injection point, generating an oblique shock that induces the flow to separate, changing the pressure distribution and generating a force imbalance, ($F_y > 0$). Naturally, the separation zone increases as the SMF rises; however, this phenomenon does not seem to overly affect the axial force ($\Delta F_x \leq 3\%$). This last consideration gives us expectation that the method can be applied efficiently to vector the thrust of propulsion systems.

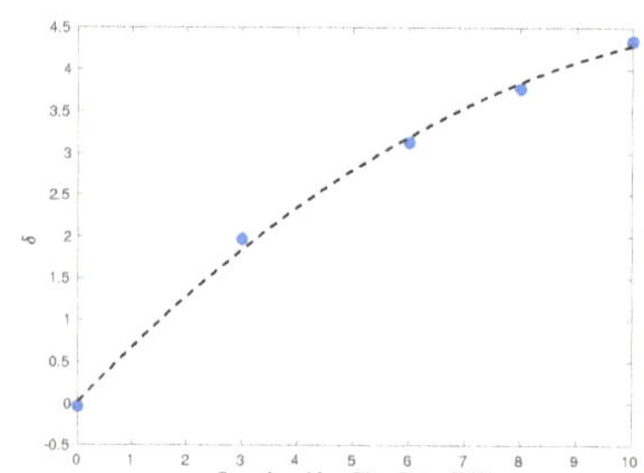

Figure 14. Diagram of pitch thrust-vector angle δ against the secondary mass-flow rate, SMF.

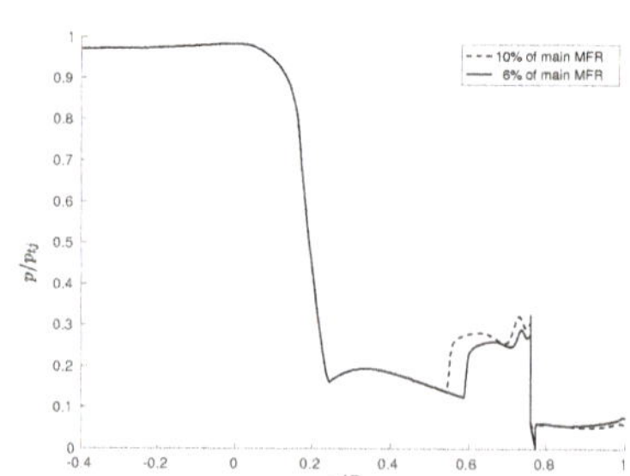

Figure 15. Wall pressure distribution at different secondary mass-flow injections rates (SMF).

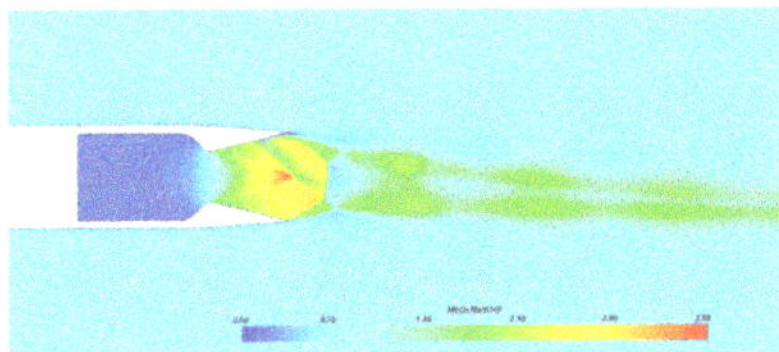

(a) SMF = 3%

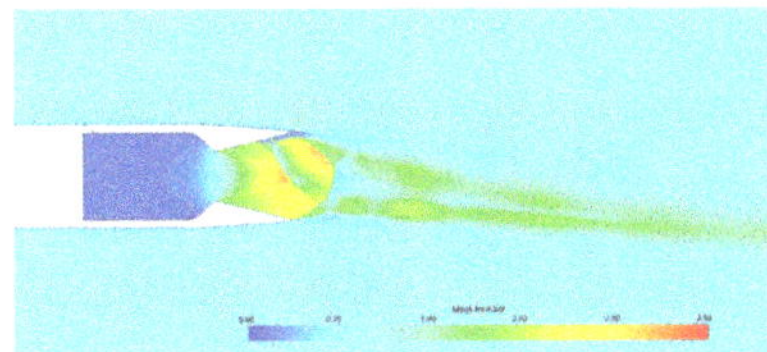

(b) SMF = 6%

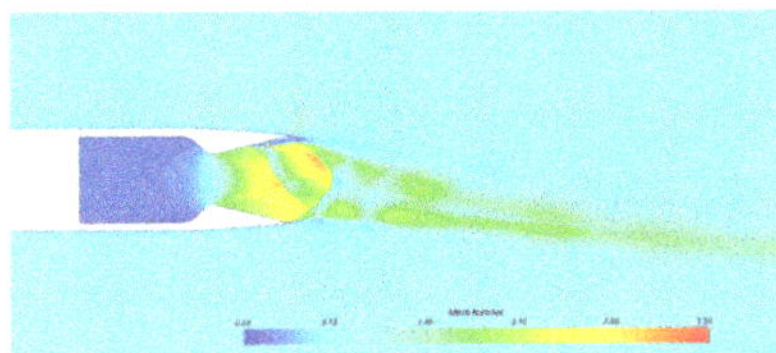

(c) SMF = 8%

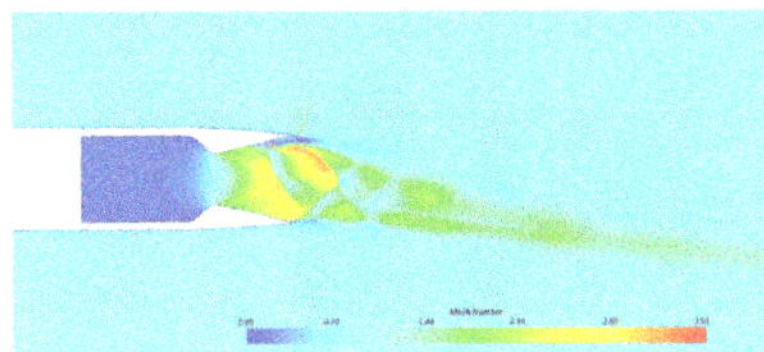

(d) SMF = 10%

Figure 16. Flowfield in the nozzle symmetry-plane. Mach contour at different secondary mass-flow injection rates (SMF).

5. Conclusions

Numerical investigations of the active flow control of an axisymmetric nozzle by using SVC at flight mach number $M_\infty = 0.9$ have been carried out. With respect to the large part of the works available in the open literature, the present study involves an axisymmetric nozzle geometry, which makes the jet-flow fully three-dimensional, and the presence of a non-negligible effect of the external flow. While fluidic thrust vectoring experiments and simulations of axisymmetric nozzles discharging in calm air are readily available in the literature (even from the present research group [12]); to the authors' knowledge there are no similar experiments including external flow interactions.

The present study has therefore been conceived as a design of experiment by applying the SVC thrust vectoring approach to an actual nozzle geometry, tested experimentally at NASA LaRC for non-vectored performances at several flight Mach numbers [27]. The analysis of the 3D flow was the last part of the following three-step process: (i) validation of the numerical framework against the NASA LaRC experimental data set [27]; (ii) numerical investigation of the nozzle sensitivity to SVC flow forcing; (iii) fully 3D verification of the SVC vectoring approach applied to the aforementioned nozzle.

The availability of experimental data on the external–internal flow interaction allowed for the validation of the numerical code against more realistic nozzle flow patterns, as opposed to just the efflux in calm air. The computations of the non-vectored performances in the axisymmetric case have been validated for several nozzle pressure ratios NPR and at flight Mach numbers $M_\infty = 0.6,\ 0.9$ and 1.2. The numerical results and experimental data were in good agreement. The computed flowfields illustrated with detail the flow topology resulting from the interactions between the internal and external flows, and the differences between subsonic and supersonic external air-streams. Before going towards three-dimensional simulations, the sensitivity of the nozzle system to SVC forcing for different locations of the injection point has been investigated in the axisymmetric case. In fact, we assumed that local effects of fluid forcing are similar in the axisymmetric and fully three-dimensional case, and that this might provide useful insight for a suitable placement of the injection slot. Finally, the effectiveness of the SVC forcing of the nozzle in three-dimensions has been investigated numerically for different secondary mass flows at the intermediate flight Mach number $M_\infty = 0.9$ and at the NPR $= 7.08$. The results show a significant turning of the thrust-vector caused by the force imbalance in the y-direction, with very limited losses of axial force F_x.

Author Contributions: Conceptualization, E.R., R.M. and M.F.; methodology, E.R., R.M. and M.F.; software, E.R., R.M. and M.F.; validation, E.R., R.M. and M.F.; formal analysis, E.R., R.M. and M.F.; investigation, E.R., R.M. and M.F.; resources, E.R., R.M. and M.F.; data curation, E.R., R.M. and M.F.; writing—original draft preparation, E.R., R.M. and M.F.; writing—review and editing, E.R., R.M. and M.F.; visualization, E.R., R.M. and M.F. All authors have read and agreed to the published version of the manuscript.

Funding: This research received no external funding.

Institutional Review Board Statement: Not applicable.

Informed Consent Statement: Not applicable.

Data Availability Statement: Not applicable.

Acknowledgments: Computational resources were provided by hpc@polito.it, a project of Academic Computing within the Department of Control and Computer Engineering at Politecnico di Torino (http://www.hpc.polito.it).

Conflicts of Interest: The authors declare no conflict of interest.

Nomenclature

C_{p_β}	pressure coefficient on the boat-tail, $2(p - p_\infty)/(\rho_\infty U_\infty^2)$
F_x	nozzle axial force
F_y	nozzle normal force
L_{div}	length of nozzle divergent part, $x_{TE} - x_t$
M_∞	external flow Mach number, flight Mach number
p_∞	external flow static pressure
p_{tj}	primary flow total pressure
U_∞	external flow velocity
w_p	primary mass flow rate
w_s	secondary mass flow rate
x_s	position of the injection slot
x_t	position of nozzle throat
x_{TE}	position of nozzle trailing edge
δ	pitch thrust-vector angle, $tan^{-1}(F_y/F_x)$
ρ_∞	external flow density
NPR	nozzle pressure ratio, p_{tj}/p_∞
SMF	secondary mass flow rate, $w_s/w_p \cdot 100$

References

1. Asbury, S.; Capone, F. High-Alpha Vectoring Characteristics of the F-18/HARV. *J. Propuls. Power* **1994**, *10*, 116–121. [CrossRef]
2. Wilde, P.; Crowther, W.; Buonanno, A.; Savvaris, A. Aircraft Control Using Fluidic Maneuver Effectors. In Proceedings of the 26th AIAA Applied Aerodynamics Conference, Honolulu, Hawaii, 18–21 August 2008. [CrossRef]
3. Mason, M.; Crowther, W. Fludic Thrust Vectoring for Low Observable Air Vehicles. In Proceedings of the 2nd AIAA Flow Control Conference, Portland, OR, USA, 28 June–1 July 2004. [CrossRef]
4. Flamm, J.; Deere, K.; Mason, M.; Berrier, B.; Johnson, S. Design enhancements of the Two-Dimensional, Dual Throat Fluidic Thrust Vectoring Nozzle Concept. In Proceedings of the 3rd AIAA Flow Control Conference, San Francisco, CA, USA, 5–8 June 2006. [CrossRef]
5. Orme, J.; Sims, R. Selected Performance Measurements of the F-15 ACTIVE Axisymmetric Thrust-Vectoring Nozzle. In Proceedings of the 4th ISABE Symposium, Florence, Italy, 5–10 September 1999.
6. Ferlauto, M.; Marsilio, R. Open and Closed-Loop Responses of a Dual-Throat Nozzle during Thrust Vectoring. In Proceedings of the 52nd AIAA/SAE/ASEE Joint Propulsion Conference, Salt Lake City, UT, USA, 25–27 July 2016. [CrossRef]
7. Ferlauto, M.; Marsilio, R. Numerical Investigation of the Dynamic Characteristics of a Dual-Throat Nozzle for Fluidic Thrust-Vectoring. *AIAA J.* **2017**, *55*, 86–98. [CrossRef]
8. Gu, R.; Xu, J. Dynamic experimental investigations of a bypass dual throat nozzle. *ASME J. Eng. Gas Turbines Power* **2015**, *137*, 084501. [CrossRef]
9. Warsop, C.; Crowther, W.J. Fluidic Flow Control Effectors for Flight Control. *AIAA J.* **2018**, *56*, 3808–3824. [CrossRef]

10. Warsop, C.; Crowther, W.; Forster, M. NATO AVT-239 Task Group: Supercritical Coanda based Circulation Control and Fluidic Thrust Vectoring. In Proceedings of the AIAA Scitech 2019 Forum, San Diego, CA, USA, 7–11 January 2019. [CrossRef]
11. Ferlauto, M.; Ferrero, A.; Marsicovetere, M.; Marsilio, R. Differential Throttling and Fluidic Thrust Vectoring in a Linear Aerospike. *Int. J. Turbomach. Propuls. Power* **2021**, *6*, 8. [CrossRef]
12. Marsilio, R.; Ferlauto, M.; Hadi Hamedi-Estakhrsar, M. Numerical simulation of a vectored axisymmetric nozzle. *AIP Conf. Proc.* **2020**, *2293*, 200018. [CrossRef]
13. Takahashi, H.; Munakata, T.; Sato, S. Thrust Augmentation by Airframe-Integrated Linear-Spike Nozzle Concept for High-Speed Aircraft. *Aerospace* **2018**, *5*, 19. [CrossRef]
14. Deere, K. Summary of Fluidic Thrust Vectoring Research Conducted at NASA Langley Research Center. In Proceedings of the 21st AIAA Applied Aerodynamics Conference, Orlando, FL, USA, 23–26 June 2003. [CrossRef]
15. Deere, K.; Flamm, J.; Berrier, B.; Johnson, S. Computational Study of an Axisymmetric Dual Throat Fluidic Thrust Vectoring Nozzle for a Supersonic Aircraft Application. In Proceedings of the 43rd AIAA/ASME/SAE/ASEE Joint Propulsion Conference & Exhibit, Cincinnati, OH, USA, 8–11 July 2007. [CrossRef]
16. Deng, R.; Kim, H. A study on the thrust vector control using a bypass flow passage. *Proc. IMechE Part G J. Aerosp. Eng.* **2015**, *5*, 1722–1729. [CrossRef]
17. Deng, R.; Setoguchi, T.; Dong Kim, H. Large eddy simulation of shock vector control using bypass flow passage. *Int. J. Heat Fluid Flow* **2016**, *62*, 474–481. [CrossRef]
18. Gu, R.; Xu, J.; Guo, S. Experimental and Numerical Investigations of a Bypass Dual Throat Nozzle. *ASME J. Eng. Gas Turbines Power* **2014**, *136*, 084501. [CrossRef]
19. Chouicha, R.; Sellam, M.; Bergheul, S. Effect of reacting gas on the fluidic thrust vectoring of an axisymmetric nozzle. *Propuls. Power Res.* **2020**. [CrossRef]
20. Zmijanovic, V.; Lago, V.; Sellam, M.; Chpoun, A. Thrust shock vector control of an axisymmetric conical supersonic nozzle via secondary transverse gas injection. *Shock Waves* **2014**, *24*, 97–111. [CrossRef]
21. Eilers, S.; Wilson, M.; Whitmore, S.; Peterson, Z. Side Force Amplification on an Aerodynamically Thrust Vectored Aerospike Nozzle. *J. Propuls. Power* **2012**, *28*, 811–819. [CrossRef]
22. Wu, K.; Kim, T.; Kim, H. Sensitivity Analysis of Counterflow Thrust Vector Control with a Three-Dimensional Rectangular Nozzle. *J. Aerosp. Eng.* **2021**, *34*, 04020107. [CrossRef]
23. Sieder, J.; Bach, C.; Propst, M.; Tajmar, M. Evaluation of the performance potential of aerodynamically thrust vectored aerospike nozzles. In Proceedings of the 67th International Astronautical Congress (IAC), Guadalajara, Mexico, 26–30 September 2016.
24. Cen, Z.; Smith, T.; Stewart, P.; Stewart, J. Integrated flight/thrust vectoring control for jet-powered unmanned aerial vehicles with ACHEON propulsion. *IMechE Part G J. Aerosp. Eng.* **2014**, *229*, 1057–1075. [CrossRef]
25. Capello, E.; Ferrero, A.; Ferlauto, M.; Marsilio, R. CFD-based Fluidic Thrust Vector model for fighter aircraft. In Proceedings of the 55th AIAA/SAE/ASEE Joint Propulsion Conference, Indianapolis, IN, USA, 19–22 August 2019. [CrossRef]
26. Ferlauto, M.; Marsilio, R. Numerical Simulation of the Unsteady Flowfield in Complete Propulsion Systems. *Adv. Aircr. Spacecr. Sci.* **2018**, *5*, 349–362. [CrossRef]
27. Carlson, T.; Lee, E. *Experimental and Analytical Investigation of Axisymmetric Supersonic Cruise Nozzle Geometry at Mach Numbers from 0.6 to 1.3*; NASA-TP-1953 L-14661; NASA: Washington, DC, USA, 1981.
28. Carlson, J.; Lee, E. *Computational Prediction of Isolated Performance of an Axisymmetric Nozzle at Mach Number 0.90*; Nasa Technical Memorandum NASA/TM-4506; NASA: Washington, DC, USA, 1994.
29. Batten, P.; Goldberg, U.; Chakravarthy, S.; Craft, T.; Leschziner, M. Afterbody Boattail- and Plume-Flow Modeling using Anisotropy-Resolving Turbulence Closures. In Proceedings of the 37th AIAA/ASME/SAE/ASEE Joint PropulsionConference and Exhibit, Salt Lake City, UT, USA, 8–11 July 2001. [CrossRef]
30. Star-CCM+ User Guide. Available online: https://docs.sw.siemens.com/documentation/external/PL20190509110447511/en-US/starccm/starccm/index.html#page/connect{%}2Fsplash.html (accessed on 1 April 2021).
31. Spalart, P.; Allmaras, S. A One-Equation Turbulence Model for Aerodynamic Flows. *Rech. Aerosp.* **1994**, *1*, 5–21.
32. Spalart, P.; Johnson, F.; Allmaras, S. Modifications and Clarifications for the Implementation of the Spalart-Allmaras Turbulence Model. In Proceedings of the 7th International Conference on Computational Fluid Dynamics (ICCFD7), Big Island, HI, USA, 9–13 July 2012.
33. Tian, C.; Lu, Y. Turbulence Models of Separated Flow in Shock Wave Thrust Vector Nozzle. *Eng. Appl. Comput. Fluid Mech.* **2013**, *7*, 182–192. [CrossRef]
34. Ferlauto, M.; Marsilio, R. Numerical Simulation of Fluidic Thrust Vectoring. *Aerotec. Missili Spaz. J. Aerosp. Sci. Technol. Syst.* **2016**, *95*, 53–62. [CrossRef]
35. Liou, M. A sequel to ausm: Ausm+. *J. Comput. Phys.* **1996**, *129*, 364–382. [CrossRef]
36. Waithe, K.; Deere, K. An Experimental and Computational Investigation of Multiple Injection Ports in a Convergent-Divergent Nozzle for Fluidic Thrust Vectoring. In Proceedings of the 21st AIAA Applied Aerodynamics Conference, Orlando, FL, USA, 23–26 June 2003. [CrossRef]
37. Ferlauto, M.; Marsilio, R. Influence of the External Flow Conditions to the Jet-Vectoring Performances of a SVC Nozzle. In Proceedings of the 55th AIAA/SAE/ASEE Joint Propulsion Conference, Indianapolis, IN, USA, 19–22 August 2019. [CrossRef]

38. Emelyanov, V.; Yakovchuk, M.; Volkov, K. Multiparameter Optimization of Thrust Vector Control with Transverse Injection of a Supersonic Underexpanded Gas Jet into a Convergent Divergent Nozzle. *Energies* **2021**, *14*, 4359. [CrossRef]
39. Ferlauto, M.; Marsilio, R. Computational Investigation of Injection Effects on Shock Vector Control Performance. In Proceedings of the 54nd AIAA/SAE/ASEE Joint Propulsion Conference, Cincinnati, OH, USA, 9–11 July 2018. [CrossRef]

fluids MDPI

Article

Schlieren Flow Visualization and Analysis of Synthetic Jets

John E. Pellessier [1], Heather E. Dillon [2,*] and Wyatt Stoltzfus [3]

1 Department of Mechanical Engineering, Texas A&M University, College Station, TX 77843, USA; jpellessier@gmail.com
2 Department of Mechanical Engineering, University of Washington Tacoma, Tacoma, WA 98402, USA
3 Department of Mechanical Engineering, University of Portland, Portland, OR 97203, USA; stoltzfu20@up.edu
* Correspondence: hedillon@uw.edu

Abstract: This work explores several low-cost methods for the visualization and analysis of pulsed synthetic jets for cooling applications. The visualization methods tested include smoke, Schlieren imaging, and thermography. The images were analyzed using Proper Orthogonal Decomposition (POD) and numerical methods for videos. The results indicated that for the specific nozzle studied, the optimal cooling occurred at a frequency of 80 Hz, which also corresponded to the highest energy in the POD analysis. The combination of Schlieren photography and POD is a unique contribution as a method for the optimization of synthetic jets.

Keywords: synthetic jet; Schlieren; smoke; proper orthogonal decomposition; heat transfer; fluids; cooling

Citation: Pellessier, J.E.; Dillon H.E.; Stoltzfus, W. Schlieren Flow Visualization and Analysis of Synthetic Jets. *Fluids* **2021**, *6*, 413. https://doi.org/10.3390/fluids6110413

Academic Editor: Olga A. Azarova

Received: 4 October 2021
Accepted: 8 November 2021
Published: 13 November 2021

1. Introduction

Synthetic jets have applications in cooling. The ability to tune the frequency and flow characteristics of a synthetic jet make cooling of electronic devices an important application. In this paper, we explore the complex heat transfer interactions of synthetic jets for cooling systems, to create tools for the optimization of the jets.

Visualization of flow in heat transfer systems has been used for decades to help understand and quantify the characteristics. In heat transfer, early flow visualization helped the research community develop Nusselt number correlations. In jet applications, the visualization of the fluid movement characterizes key features of the jet [1]. Advances in cameras and computational tools have increased the use of image analysis as a way to understand turbulent flows. The early work of Sirovich [2] pioneered the method of snapshots as a way to quantify flow structure in both time and space. The method relies on the proper orthogonal decomposition of a matrix image.

In heat transfer, image analysis methods have been used to study the frequency and turbulence behavior of many natural and forced convection systems [3,4]. The present work examines synthetic jets and compares several visualization methods for the analysis of the time and flow behaviors.

We compared low-cost smoke and Schlieren visualization for synthetic jet optimization applications. We tested each method and quantified the results using proper orthogonal decomposition. Both visualization tools represent an inexpensive and efficient way to optimize a synthetic jet application, but the Schlieren method we found to be more robust when coupled with the analysis methods.

2. Background

2.1. Synthetic Jets

Prior researchers have considered many aspects of synthetic jets, as shown in Table 1. Glezer and Amitay defined synthetic jets as the alternating momentary ejection and suction of a fluid across an orifice such that the net mass flux is zero. The range of frequencies possible for the jets makes them attractive for many applications in fluid control [5]. Kercher et al. [6] provided an overview of key features and characteristics of synthetic jets for cooling. Yi et al. considered the application of plate cooling using jets [7].

Table 1. Summary of experimental and computational work on jets.

Citation	Year	Frequency (Hz)	Jet Type	Study Type
Kercher et al. [6]	2003	1–1000	synthetic	
Travnicek and Tesar [8]	2003	106–692	synthetic	smoke
Smith and Swift [1]	2003		synthetic and continuous	Schlieren
Pavlova et al. [9]	2006	420–1200	synthetic	
Arik [10]	2008	4500	synthetic	
Chaudhari et al. [11]	2009	100–350	synthetic	
Bazdidi-Tehrani et al. [12]	2011	16–400	synthetic	
Biden et al. [13]	2012		unforced jet in crossflow	POD
Biden et al. [14]	2012		forced jet in crossflow	POD
Ghaffari et al. [15]	2016	20,000	ultrasonic microblower	
Ghadi et al. [16]	2016		pulsed impinging jet	smoke
Albright and Solovitz [17]	2016		variable-diameter synthetic jet	
Firdaus et al. [18]	2018	300–700	synthetic	
Solovitz et al. [19]	2018	350–2000	synthetic	PIV
Viggiano et al. [20]	2018		variable-density jet	POD and PIV
Kristo et al. [21]	2021		crossflow jets	POD and PIV
Present work	2021	20–100	synthetic	POD, smoke, Schlieren

Pavlova et al. [9] found that high-frequency jets removed heat more effectively, but that lower-frequency jets were better for larger separation distances. Many authors found that the optimal frequency for heat transfer was the resonance frequency for the device or the diaphragm [11,18,19]. Recent work has pushed devices to frequencies with lower acoustic impact [15] and explored the option of variable diameters [17]. In all these applications, there is a need for the optimization of the synthetic jet using low-cost visualization methods that can be applied in industry.

Other research teams have performed visualization studies of jets. Travnicek and Tesar [8] used smoke to identify vortical puffs and characterize the flow. Smith and Swift used Schlieren imaging to compare synthetic jets and continuous jets [1]. Solovitz et al. [19] used particle image velocimetry (PIV) to quantify the flow of synthetic jets at a range of frequencies. A summary of many of the relevant papers for this work is provided in Table 1. Visualization studies for related applications include a PIV study of thermal plumes [22].

We considered smoke and Schlieren methods with analysis methods for low-cost visualization. PIV and computational fluid methods are important tools for synthetic jet research, but expensive to apply in industrial applications. None of the prior research teams studied numerical methods such as POD using multiple visualization methods such as smoke and Schlieren.

2.2. *Turbulence Analysis Methods*

The visualization and analysis of turbulence has been the focus of many fluid research groups. Sirovich [2] pioneered the method of snapshots as a way to analyze turbulent flow structures. Proper Orthogonal Decomposition (POD) is a mathematical method based on the diagonalization of a matrix and the eigenvalues. POD has been used by prior researchers in the examination of thermal flows [3,4].

POD gives us a way to quantify the structures that we observe in the flow field of the jet. Using images of the jet, we can convert the image into a matrix and then use POD to understand important structures of the jet fluid behavior. This provides a method for quantifying the turbulent behavior of the jet during cooling operations.

One prior work analyzed synthetic jets using high-speed Schlieren photography [1], but they did not perform POD analysis of the images. Viggiano et al. [20] used POD to understand the flow of a variable-density jet using PIV, but did not consider synthetic jets. Bidan et al. used proper orthogonal decomposition to examine unforced and forced jets [13,14], but did not use Schlieren images nor consider synthetic jets.

3. Experimental Methods

3.1. *Design and Construction of the Nozzle*

A pulsed jet nozzle for this experiment was designed that would fit a Sony Xplod XS-GTF1027 Speaker, as shown in Figure 1 and outlined in Table 2. The nozzle design was based on the prior work of Albright and Solovitz [23]. This design redirects some of the primary flow inward to produce a tighter jet. To make the nozzle, four pieces were drawn in Solidworks to fit a four-inch speaker, then 3D printed. After assembly, the gaps in the nozzle were sealed with an adhesive to prevent the working fluid from leaking. A small hole was drilled into the lower side of the nozzle from the exterior to the inner chamber; an inch-long tube was then glued into the hole; this provided a way to fill the nozzle with smoke or low-density air for visualization tests. When not being used, the exit of the tube was sealed.

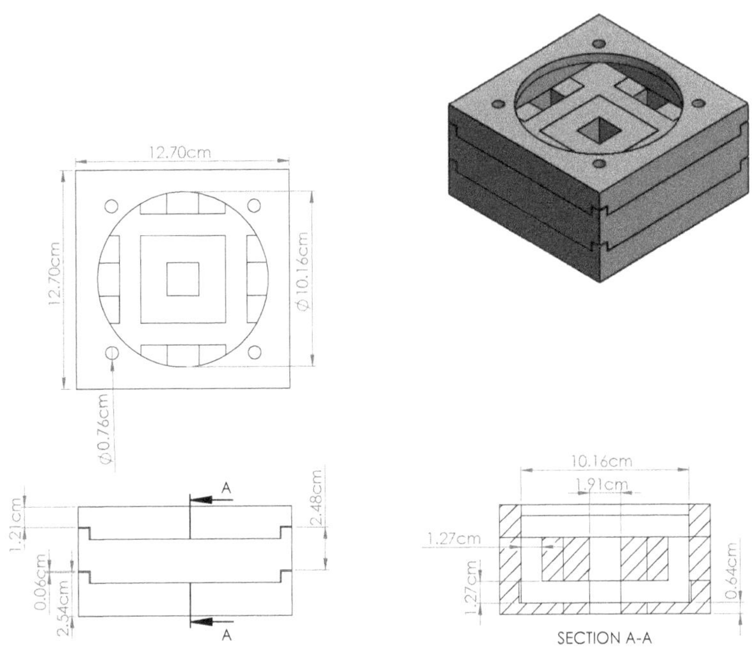

Figure 1. Schematic drawing of the assembled nozzle.

Table 2. Summary of the equipment used in the experiments.

Description	Model
Canon DSLR	EOS 7D Mark II (G)
Camera Lens	EF100/2.8 L MACRO IS USM
FLIR IR Camera	
PHANTOM Hi-Speed Camera	V711
Canon Point and Shoot	PowerShot ELPH 300 HS
Smoke Machine	
4″ Speaker	Sony Xplod XS-GTF1027
Amplifier	Adafruit Stereo 20 W Class D Amplifier
10 mm Blue LEDs	
Hot Plate	Thermolyne HP46515
Multimeter	RSR MAS830
Thermocouple	FLUKE 80TK
Thermocouple probe	FLUKE
Wane Anemometer	Fieldpiece EHdl1

3.2. Experimental Setup

The Schlieren system used in these experiments was a z-type, as described in prior work [24] and shown in Figure 2. A z-type Schlieren uses two concave mirrors to allow the visualization of density gradients in fluids. In a z-type Schlieren, a point light, positioned one focal length from the first mirror, creates a beam of collimated light, which passes through the test region. In the testing region, light that passes through a fluid with a different density than the medium is refracted. The second mirror focuses the light back to a point, one focal length from the mirror, where a razor edge was positioned to block half of the focused light, allowing visualization of the density gradient of the air.

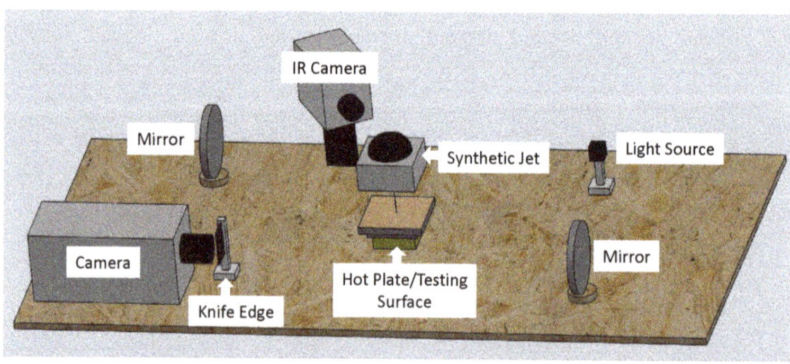

Figure 2. *Cont.*

Figure 2. Schematic drawing and photograph of the experimental Schlieren configuration.

3.3. *Schlieren Methods*

To calibrate POD, the jet was first photographed alone using the Schlieren in an upflow condition. This configuration is the simplest structure for POD to analyze over time:

1. The test was set up as seen in Figure 2 with the pulsed jet taking the center position in the Schlieren system, the nozzle exit facing upwards, using a 100mm micro lens on the camera;
2. To achieve the best-quality videos, the Schlieren system was calibrated using a candle, providing a high-density gradient. The light source and the knife edge were moved to optimize the image one focal length away from across the mirror direction. The camera was then placed behind the knife edge;
3. The synthetic jet was filled with low-density air. This was accomplished by sealing the tube on the side of the nozzle and gently pulling the trigger on the canned air;
4. The speaker in the jet was powered with the specified frequency between 20 Hz and 100 Hz and passed through an amplifier, set at about 75%;
5. The camera was set to record at least one full cycle of the jet. After the video was captured, the video was exported at either 1×, 2×, or 3× the playback speed.

For the second Schlieren experiment, the jet was photographed in a downflow condition impinging on a heated plate. The space between the jet exit and the heated plate was 12.7 cm, a distance selected based on guidance from the electronic cooling applications. The top of the plate was made up of different metal cylinders with uniform spacing to simulate the surface of an electronic device. The heated plate was allowed to reach the steady state, and then, the infrared camera was used to capture the temperature profile of the plate. The jet was activated at a 20–100 Hz frequency and recorded using the high-speed camera. This system is shown in Figure 3.

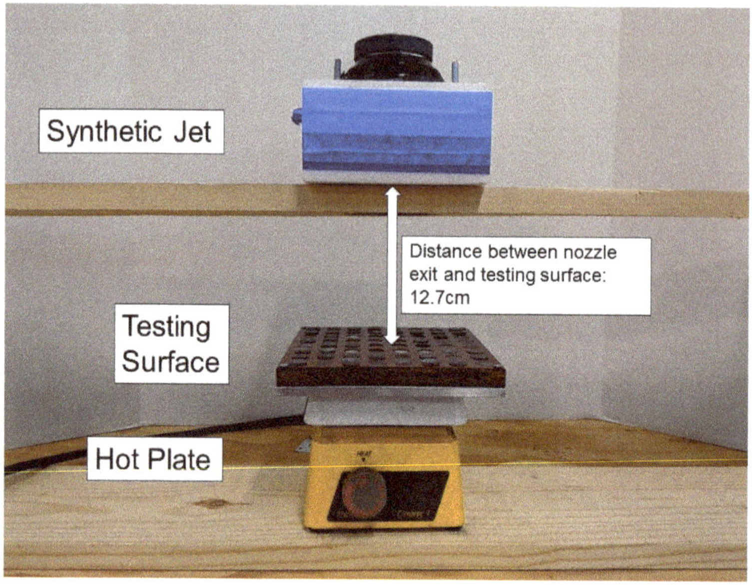

Figure 3. Photograph of the experimental setup for the heated plate test.

3.4. *Smoke Methods*

To capture specific characteristics of the flow from the jet alone, a smoke visualization was performed. The nozzle was placed facing up with a dark background. Using a 50 mm lens, the high-speed camera was used to capture the synthetic jet. Smoke was provided using a smoke machine with tubing connected to the side of the nozzle. This setup was also used to calibrate POD and estimate the jet velocity.

4. Computational Methods

4.1. *Frequency Analysis*

To determine velocity and confirm the frequency, f, of the synthetic jet, the diameter of the smoke plume passing two fixed vertical points was analyzed. The video was converted to black and white in MATLAB. The smoke plume was primarily white, making it easy to identify relative to the dark background. The next step in the analysis was to collect all the color values of pixels in both fixed row vectors of each frame. The color pixels were on a scale of 0 to 255 with black equaling 0 and white equaling 255. A tolerance of pixel color values larger than 150 was set in order to only count the white pixels of the smoke plume. The threshold of 150 was determined from the calculation of pixel differences in the image. This scaled the vector to only the pixels that made up the smoke plume. Calculating the length of this vector produced the number of pixels that made up the smoke plume. The diameter was then found in units of centimeters using the set pixel conversion factor determined from measuring a known width in the video.

This analysis was completed for each frame in the video. The frame vector was converted to units of seconds using the frame rate conversion factor. The smoke plume diameter was then plotted for each respective row vector over the calculated time span of the video. The maximum diameter change over time was used to estimate the jet velocity near the exit.

4.2. *POD Analysis*

Proper Orthogonal Decomposition (POD) is based on the diagonalization of a matrix. The mathematical procedure linearly transforms the number of possibly correlated variables into a smaller number of uncorrelated variables. The first component contains as much of the variation in the system as possible. In this case, the matrix X is an $m \times n$ matrix composed of multiple observation frames from a video in time.

For the POD analysis, the data were centered by the mean of each row. Then, the covariance matrix Cx was calculated. The covariance matrix is a square, symmetric $m \times m$ matrix, whose diagonal represents the variance of particular measurements.

$$C_x = \frac{1}{n-1} X X^T \tag{1}$$

Singular Value Decomposition (SVD) was used to diagonalize the matrix. The SVD diagonalization is shown in Equation (2), where $U \in \mathbb{C}^{m \times m}$ is unitary, $V \in \mathbb{C}^{n \times n}$ is unitary, and $\Sigma \in \mathbb{R}^{m \times n}$ is diagonal.

$$X = U \Sigma V^* \tag{2}$$

POD gives us a way to quantify the structures that we observe in the flow field of the jet. Each mode of POD is a characteristic of the flow field. Proper orthogonal decomposition was applied to the matrix from the videos, and the energy associated with each of the eigenvalues and eigenvectors was calculated.

The image analysis of the synthetic jets followed a standard approach. First, the video file was imported into MATLAB as a series of images. Each image file was converted to grayscale, and each pixel value became a number in a matrix for each image. The matrix for each frame of the video was reshaped to become one long row of data. The image matrices were then combined to form one large matrix that represents both time and space dimensions, X.

5. Results

5.1. Experimental Results

Figure 4 shows one cycle of the synthetic jet at 20 Hz using the smoke visualization. Figure 5 shows the same frequency for the jet using the Schlieren visualization. The first three images of each set show the major jet, which was a result of the membrane moving towards the exit. In the fourth image, some of the working fluid is being pulled back into the nozzle as the membrane moves away from the exit. An unexpected observation was the formation of a secondary minor jet, the fifth image in each set. This minor jet was small, but not well defined, and only existed for around 3–4 in vertically (7.62–10.16 cm).

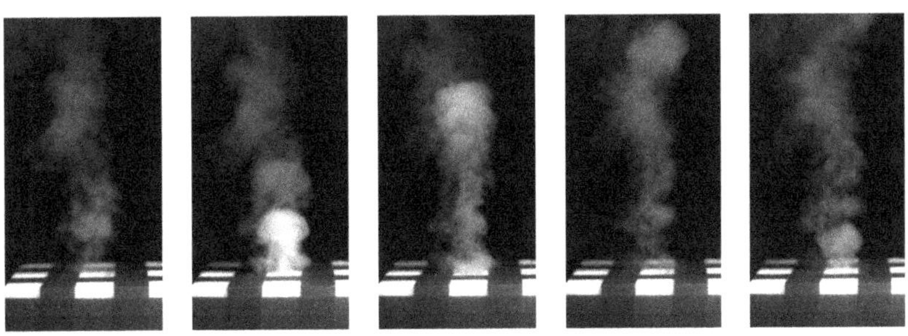

Figure 4. Smoke machine, 20 Hz, 3500 fps. Exported at 2× speed.

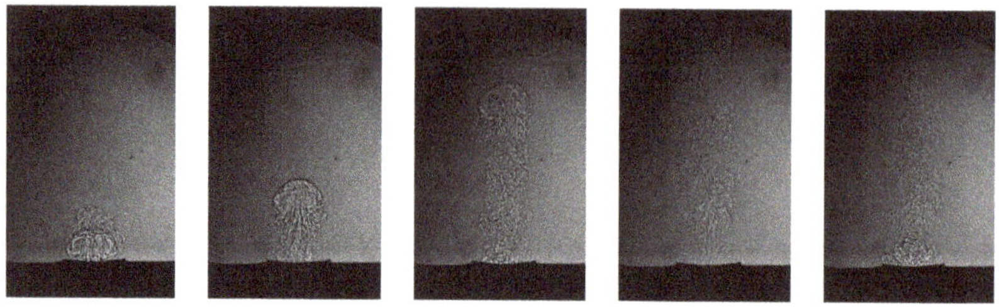

Figure 5. Schlieren, 20 Hz, 3500 fps. Exported at 2× speed.

For the experiments using the heated plate, the plume created by the jet was tight and easily visualized with the density gradient of the Schlieren. It was experimentally determined with trial and error that the optimal distance to place the jet away from the testing surface was 5 in (12.7 cm), or a distance of $5 \cdot r$, with r defined as the side length of the square exit on the nozzle.

At frequencies ranging from 20–100 Hz, there was significant fluid movement as a result of the synthetic jet. For the jet at 20 Hz, the Schlieren images and infrared images are shown in Figures 6 and 7. The steady-state figure for 20Hz was labeled to show the region of laminar natural convection below the plate where the jet was not interacting significantly. Above the heated plate, the complex mixing of the jet with the air created a turbulent region, cooling the plate. Figure 7 shows that the center of the plate was initially warmer due to the metal cylinders in the center of the plate. After the jet was activated, the temperature of the plate became more uniform as the jet distributed the heat and cooled the surface.

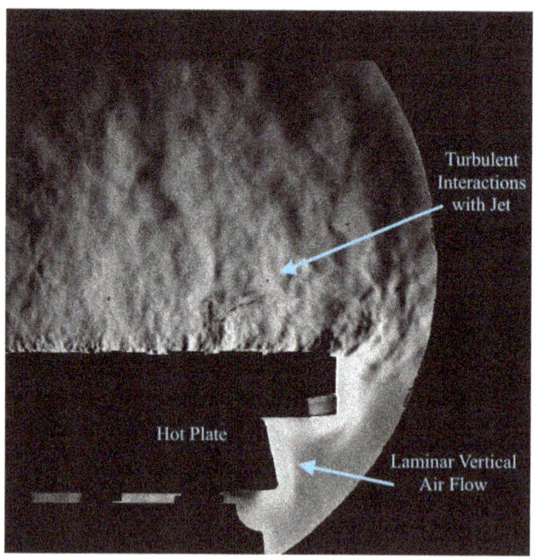

Figure 6. Schlieren image showing the steady-state interaction of air from the synthetic jet with a hot testing surface, 20 Hz.

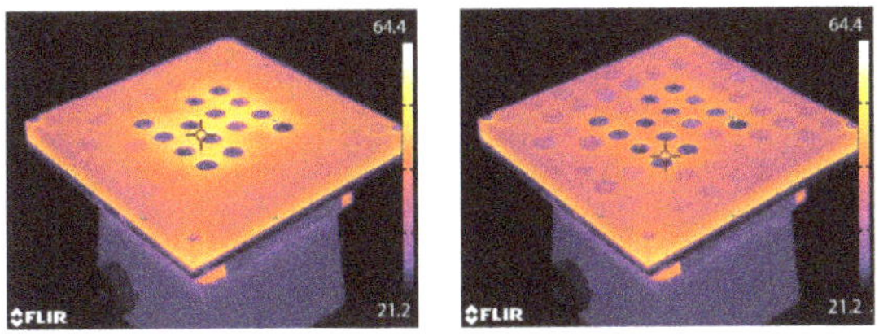

Figure 7. Infrared photo of the hot testing surface at time 0 s and then 60 s with a jet at 20 Hz. The temperature scale is °C.

The greatest fluid movement (250 ft/min (11.94 m/s)) and cooling effect (approximately 50 °F (10 °C) in 60 s) was found at 80Hz, as shown in Figures 8 and 9. The infrared images show significant cooling when compared to the same time period for the 20 Hz jet (Figure 7). Above 100 Hz, fluid movement was insignificant, and therefore, no tests were run above that frequency. A more formal study of this relationship is planned for future work.

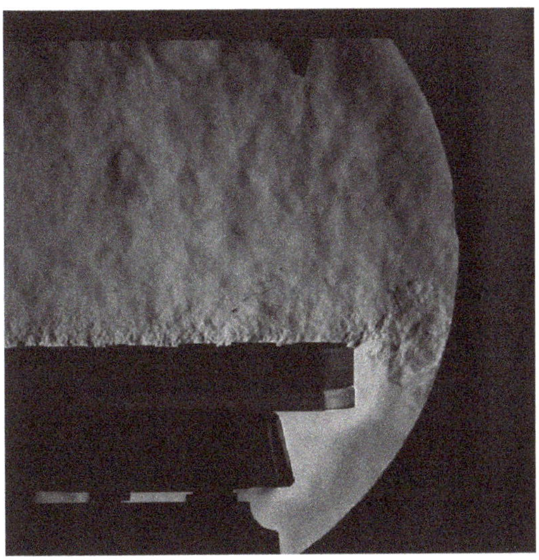

Figure 8. Schlieren image showing the steady-state interaction of air from the synthetic jet with the hot testing surface, 80 Hz.

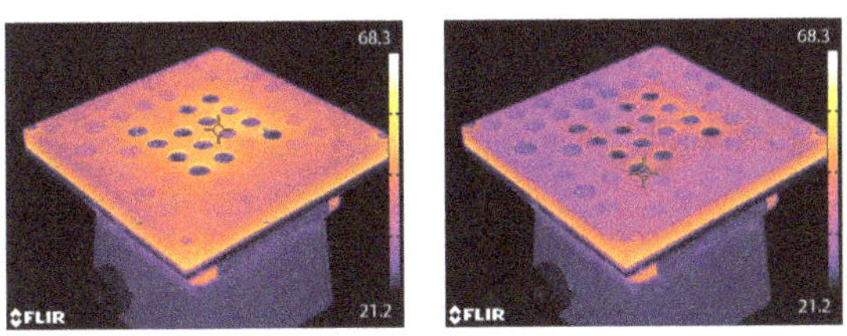

Figure 9. Infrared photo of the hot testing surface at time 0 s and then 60 s with a jet at 80 Hz. The temperature scale is °C.

5.2. Frequency Analysis Results

The frequency plots created by implementing the MATLAB algorithm represent the true frequency of 20 Hz, as shown in Figure 10. The peaks show where the smoke plume diameter was the largest and passed by the designated pixel row in the video. Each peak was indexed within the MATLAB code. As seen in the plots, the primary peaks for both measurement locations (1 cm and 2 cm) were separated by 0.05 s, or a frequency of approximately 20 Hz. Smaller peaks show the negative flow of the jet as the smoke returns back toward the surface.

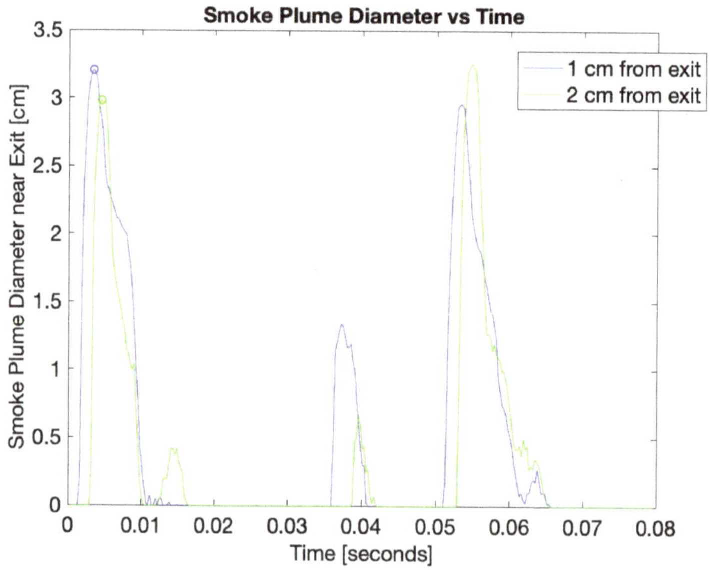

Figure 10. Diameter of the jet over time for two vertical locations. The spacing of the largest peaks (both 1 cm and 2 cm from the exit) are both approximately 0.05 s, near 20 Hz.

The velocity of the smoke plume was estimated using the same method. The calculation was repeated at a distance of 1 cm, and the time for the maximum diameter to travel was calculated. Near the exit of the jet, the velocity estimate was approximately 9 m/s.

5.3. POD Analysis Results

One smoke video was processed using proper orthogonal decomposition for the full length of the video (approximately three cycles of the jet). Figure 11 compares the first mode, which contains the basic structural elements of the flow with a sample frame from the video. Figure 12 shows the first 10 modes of the POD analysis. In each mode, important flow structures are shown. For example, Modes 5, 6, and 7 each show vortex structures, and Mode 6 shows the repeated patterns associated with the synthetic flow.

When using POD, many systems are well characterized mathematically when the modes capture 90% of the energy in the system [3]. For most systems, this is less than 10 modes; however, for the synthetic jet at 80 Hz, 27 modes were required to reach the 90% energy threshold. This indicates that the flow in the system is complex and requires additional modes to characterize it well mathematically. This was expected for the transient version of a synthetic jet. The first five modes and the energy are shown in Table 3 for the smoke jet at 80 Hz.

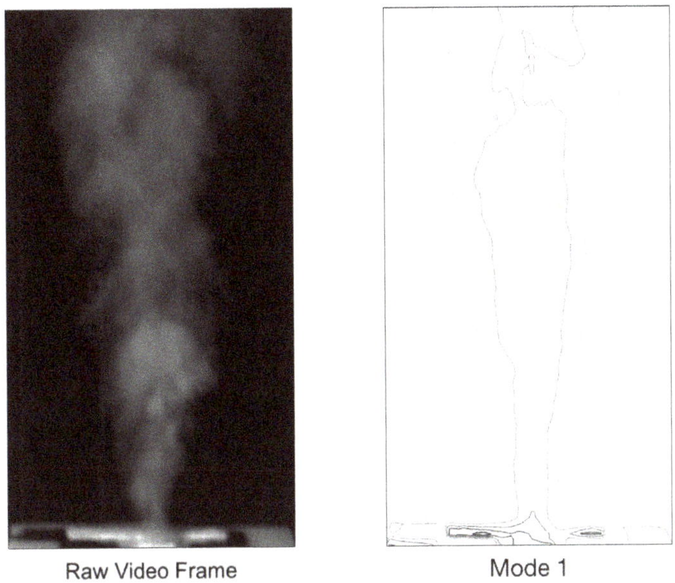

Figure 11. The first mode (**right**) of the POD of the jet operating at 80 Hz captures key characteristics of the flow when compared with the raw video frames (**left**).

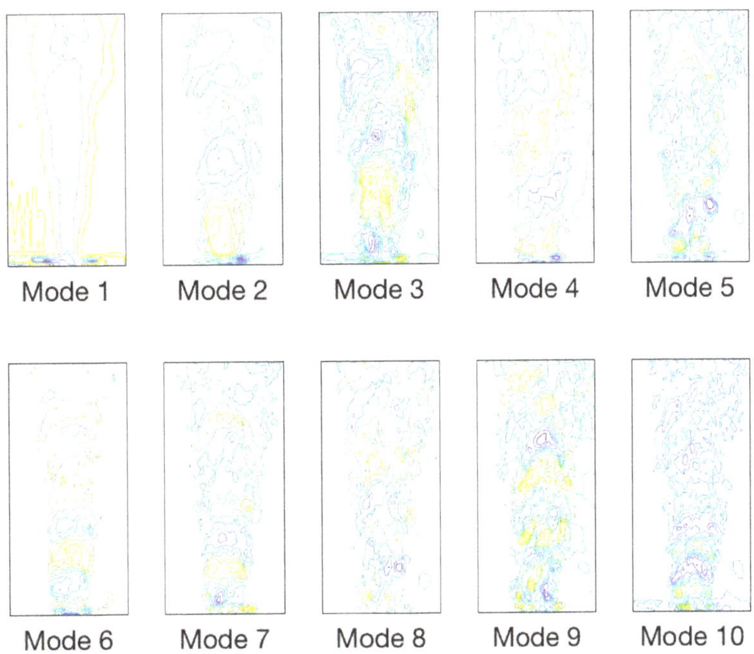

Figure 12. The first 10 modes of the POD of the jet operating at 80 Hz.

Table 3. Energy associated with the modes developed from the POD at 80 Hz for the synthetic jet in smoke.

Mode	Eigenvalue λ	Total Energy %
1	114.64	44.46
2	18.23	7.07
3	15.92	6.17
4	14.48	5.62
5	8.31	3.22
27	0.96	0.37

The same technique was then applied to the more complex flow where the jets were impinging on the heated plate. In this case, the modes required were significantly higher, with 200–239 modes required to reach the same 90% threshold. These values confirmed that the flow was erratic and high in movement, which enhanced the cooling. The results are shown in Tables 4 and 5. The highest number of modes required was for the 80 Hz frequency, also found to represent the highest heat transfer for the system. This confirmed the trend expected for a more turbulent flow.

Table 4. Energy associated with the modes developed from the POD at 20 Hz for the jet on the heated plate with the Schlieren.

Mode	Eigenvalue λ	Total Energy %
1	480.74	15.76
2	99.83	3.27
3	80.12	2.62
4	73.41	2.41
5	64.07	2.10
200	2.82	0.09

Table 5. Energy associated with the modes developed from the POD at 80 Hz for the jet on the heated plate with the Schlieren.

Mode	Eigenvalue λ	Total Energy %
1	433.65	14.27
2	98.70	3.25
3	82.12	2.70
4	67.73	2.23
5	52.27	1.72
239	2.85	0.09

6. Discussion

The research team designed a synthetic jet for cooling and then optimized the thermal performance of the synthetic jet. We then tested the best low-cost visualization methods to quantify the flow behaviors. Smoke visualization works well, but has limitations in capturing the complex interaction of the heat transfer and the fluid. The smoke visualization method does offer more opportunities for the calculation of the frequency and velocity in a single jet. The Schlieren visualization was determined to be the best way to visualize the synthetic jet during cooling, since the density gradient visualization captures the influence of the heat transfer. The method also captures more subtle elements of the flow structure, as shown in Figure 5. Schlieren methods were also used to successfully capture the characteristic behavior in a single jet.

The frequency calculation (Figure 10) confirmed peaks spaced out by 0.05 s in both pixel rows of the video where data were taken. This matches the 20 Hz frequency that the modeled cooling jet was pulsed at through the testing iterations. It can be concluded that using this form of video analysis to determine or confirm the frequency is very effective. A limitation of this method is that it requires the pixel row data to be taken a small distance away from where the jet exits the nozzle to avoid visual interference. This allows some time for the fluid to interact with its surroundings, which may have added viscous or drag forces. This could introduce error in the timing of when the smoke plume passes the row where data are collected as compared to the actual frequency at which the jet is pulsating. In general, this type of photography is a low-cost way to determine the jet performance in some applications.

7. Conclusions

At 80 Hz, the plate experienced optimal cooling by the synthetic jet. This frequency of the system represents a more complex optimal than often considered in synthetic jets since the scale of the device and cooling object influences the behavior of the system rather than just the device frequency. This makes characterizing the fluid behavior using POD an important part of the study.

Proper orthogonal decomposition was used to analyze the jet behavior in free air and during cooling applications. POD provides a way to quantify the complexity of the synthetic jet flow for electronic cooling applications. A high number of POD modes to quantify the system indicates a more complex flow. The optimal cooling of the plate occurred at 80 Hz, also the conditions that required the highest number of modes to reach the 90% threshold.

We considered smoke and Schlieren methods with analysis methods for low-cost visualization. We tested each method and quantified the results using proper orthogonal decomposition. Both visualization tools represent a low-cost and efficient way to optimize a synthetic jet application, but the Schlieren method we found to be more robust coupled with the analysis methods.

The results indicated that POD may be used to quantify the relative complexity of a synthetic jet. In future work, this could be used to more fully optimize the frequency of the synthetic jet for cooling and the placement of the synthetic jet. We also confirmed that Schlieren methods would be preferred for the study of jets since including the density information is helpful to optimize the system.

Author Contributions: Experimental work by J.E.P. and W.S. Design of the nozzle, writing, data analysis, and writing—draft preparation by J.E.P. Frequency analysis, writing, and draft preparation by W.S. Conceptualization, methodology, POD analysis, writing, review, editing, supervision, project administration, and funding acquisition by H.E.D. All authors have read and agreed to the published version of the manuscript.

Funding: This research was funded in part by W.M. Keck Foundation.

Institutional Review Board Statement: Not applicable.

Informed Consent Statement: Not applicable.

Data Availability Statement: Videos available from authors upon request.

Acknowledgments: Special thanks to Jared Reese and Jacob Amos for technical support. Thanks to Steve Solovitz for the discussion and suggestions.

Conflicts of Interest: The authors declare no conflict of interest.

Abbreviations

The following abbreviations are used in this manuscript:

C_x	The covariance matrix
f	Frequency
POD	Proper Orthogonal Decomposition
r	Side length of the square exit on the nozzle
SVD	Singular-Value Decomposition
t	Time
v	Velocity
X	The matrix of spatial and temporal data

References

1. Smith, B.L.; Swift, G.W. A comparison between synthetic jets and continuous jets. *Exp. Fluids* **2003**, *34*, 467–472. [CrossRef]
2. Sirovich, L. Turbulence and the dynamics of choerent structures. *Q. Appl. Math.* **1987**, *45*, 561–590. [CrossRef]
3. Dillon, H.E.; Emery, A.F.; Mescher, A.M.; Sprenger, O.; Edwards, S.R. Chaotic Natural Convection in an Annular cavity with non-isothermal walls. *Front. Heat Mass Transf.* **2011**, *2*. [CrossRef]
4. Dillon, H.E.; Emery, A.F.; Mescher, A.M. Analysis of Chaotic Natural Convection in a Tall Rectangular Cavity with Non-Isothermal Walls. *Front. Heat Mass Transf. (FHMT)* **2013**, *4*. [CrossRef]
5. Glezer, A.; Amitay, M. Synthetic Jets. *Annu. Rev. Fluid Mech.* **2002**, *34*, 503–529. [CrossRef]
6. Kercher, D.S.; Lee, J.B.; Brand, O.; Allen, M.G.; Glezer, A. Microjet cooling devices for thermal management of electronics. *IEEE Trans. Components Packag. Technol.* **2003**, *26*, 359–366. [CrossRef]
7. Yi, S.J.; Kim, M.; Kim, D.; Kim, H.D.; Kim, K.C. Transient temperature field and heat transfer measurement of oblique jet impingement by thermographic phosphor. *Int. J. Heat Mass Transf.* **2016**, *102*, 691–702. [CrossRef]
8. Travnicek, Z.; Tesar, V. Annular synthetic jet used for impinging flow mass-transfer. *Int. J. Heat Mass Transf.* **2003**, *46*, 3291–3297. [CrossRef]
9. Pavlova, A.; Amitay, M. Electronic Cooling Using Synthetic Jet Impingement. *J. Heat Transf.* **2006**, *128*, 897. [CrossRef]
10. Arik, M. Local Heat Transfer Coefficients of a High-Frequency Synthetic Jet during Impingement Cooling over Flat Surfaces. *Heat Transf. Eng.* **2008**, *29*, 763–773. [CrossRef]
11. Chaudhari, M.; Puranik, B.; Agrawal, A. Heat transfer characteristics of synthetic jet impingement cooling. *Int. J. Heat Mass Transf.* **2010**, *53*, 1057–1069.10.1016/J.IJHEATMASSTRANSFER.2009.11.005. [CrossRef]
12. Bazdidi-Tehrani, F.; Karami, M.; Jahromi, M. Unsteady flow and heat transfer analysis of an impinging synthetic jet. *Heat Mass Transf.* **2011**, *47*, 1363–1373. [CrossRef]
13. Bidan, G.; Vézier, C.; Nikitopoulos, D.E. Study of Unforced and Modulated Film-Cooling Jets Using Proper Orthogonal Decomposition—Part I: Unforced Jets. *J. Turbomach.* **2012**, *135*, 021037. [CrossRef]
14. Bidan, G.; Vézier, C.; Nikitopoulos, D.E. Study of Unforced and Modulated Film-Cooling Jets Using Proper Orthogonal Decomposition—Part II: Forced Jets. *J. Turbomach.* **2012**, *135*, 021038. [CrossRef]
15. Ghaffari, O.; Solovitz, S.A.; Ikhlaq, M.; Arik, M. An investigation into flow and heat transfer of an ultrasonic micro-blower device for electronics cooling applications. *Appl. Therm. Eng.* **2016**, *106*, 881–889. [CrossRef]
16. Ghadi, S.; Esmailpour, K.; Hosseinalipour, S.; Mujumdar, A. Experimental study of formation and development of coherent vortical structures in pulsed turbulent impinging jet. *Exp. Therm. Fluid Sci.* **2016**, *74*, 382–389.10.1016/J.EXPTHERMFLUSCI.2015.12.007. [CrossRef]
17. Albright, S.O.; Solovitz, S.A. Examination of a Variable-Diameter Synthetic Jet. *J. Fluids Eng.* **2016**, *138*, 121103. [CrossRef]
18. Firdaus, S.M.; Abdullah, M.Z.; Abdullah, M.K.; Mazlan, A.Z.A.; Ripin, Z.M.; Amri, W.M.; Yusuf, H. Synthetic Jet Study on Resonance Driving Frequency for Electronic Cooling. In *Regional Conference on Science, Technology and Social Sciences (RCSTSS 2016)*; Springer: Singapore, 2018; pp. 435–443. [CrossRef]
19. Solovitz, S.A.; Ghaffari, O.; Arik, M. FREQUENCY-DEPENDENT FLOW RESPONSE OF A HIGH-SPEED RECTANGULAR SYNTHETIC JET. *J. Flow Vis. Image Process.* **2016**, *23*, 93–116. [CrossRef]
20. Viggiano, B.; Dib, T.; Ali, N.; Mastin, L.G.; Cal, R.B.; Solovitz, S.A. Turbulence, entrainment and low-order description of a transitional variable-density jet. *J. Fluid Mech.* **2018**, *836*, 1009–1049. [CrossRef]
21. Kristo, P.J.; Kimber, M.L.; Girimaji, S.S. Towards Reconstruction of Complex Flow Fields Using Unit Flows. *Fluids* **2021**, *6*, 255. [CrossRef]
22. Li J, Liu, J., Pei, J., Mohanarangam K., Yang, W. Experimental study of human thermal plumes in a small space via large-scale TR PIV system.*Int. J. Heat Mass Transf.* **2018**, *127*, 970–980. [CrossRef]
23. Albright, S.O.; Solovitz, S.A. Development of a Variable Diameter Synthetic Jet Actuator. In *Advances in Aerospace Technology*; ASME: New York, NY, USA, 2014; Volume 1, p. V001T01A002,
24. Kaessinger, J.C.; Kors, K.C.; Lum, J.S.; Dillon, H.E.; Mayer, S.K. Utilizing Schlieren Imaging to Visualize Heat Transfer Studies. In Proceedings of the American Society of Mechanical Engineers 2014 International Mechanical Engineering Conference, Montreal, QC, Canada, 14–20 November 2014; pp. 2014–38329.

fluids

MDPI

Article

A Study of the Dependence of the Mach Stem Height on the Trailing Edge Height

Chen-Yuan Bai [1,†] and Zi-Niu Wu [2,*,†]

1 Ministry of Education Key Laboratory of Fluid Mechanics, School of Aeronautic Science and Engineering, Beihang University, Beijing 100191, China; baicy@buaa.edu.cn
2 Department of Engineering Mechanics, Tsinghua University, Beijing 100084, China
* Correspondence: ziniuwu@tsinghua.edu.cn
† These authors contributed equally to this work.

Abstract: The Mach stem height is an important parameter in the Mach reflection of steady supersonic flow. Various experimental, numerical, and theoretical works have been conducted to study this parameter in the past. However, much of the established work focuses around a single set of trailing edge heights. Here, we perform a study to show the dependence of Mach stem height on the trailing edge height for a wider range of geometry. Through numerical simulation for a set of trailing edge heights, we found that the normalized Mach stem height is almost linear with respect to the normalized wedge trailing edge height. The parameter used for normalization can be either the inlet height or the length of the lower wedge surface. The observation of this linear trend is justified through a simplified analysis, which leads to an expression of the Mach stem height that linearly depends on the trailing edge height. The present study extends our knowledge about how the geometry affects the Mach stem height, and provides a basis for future work to elaborate analytical models for Mach stem height.

Keywords: gas dynamics; shock waves; shock reflection

Citation: Bai, C.-Y.; Wu, Z.-N. A Study of the Dependence of the Mach Stem Height on the Trailing Edge Height. *Fluids* **2021**, *6*, 313. https://doi.org/10.3390/fluids6090313

Academic Editor: Olga A. Azarova

Received: 2 August 2021
Accepted: 25 August 2021
Published: 2 September 2021

1. Introduction

Shock reflection is an important phenomenon in high-speed flow [1]. Both regular reflection and Mach reflection are possible, and the conditions to have regular reflection and Mach reflection and their transition have been well studied [1–11]. For instance, the regions of various forms of reflection were defined for both air and nitrogen by Bazhenova, Fokeev and Gvozdeva [3].

Figure 1 is a schematic configuration of a typical Mach reflection with some necessary details. The incident shock wave (i), produced by the wedge (with wedge angle θ_w) in supersonic flow (with Mach number M_0), reflects over the reflecting surface to produce a reflected shock wave (r), a Mach stem (m) and a slipline (s). These four discontinuities are connected by a point (T), known as the triple point. The flow region behind the Mach stem, bounded by the slipline and the reflecting surface, forms a flow duct that is initially convergent since the slipline deflects towards the reflecting surface near point T.

The size of the Mach stem in the case of Mach reflection has received great interest. The mechanism by which the size of a Mach reflection can be estimated is an issue raised long ago by Courant and Friedrichs [12] and Liepmann and Roshko [13]. This issue was considered to be unsolved prior to the 1990s [14–16]. Chow and Chang [17] proposed an integral approach to estimate the Mach stem height for a slightly different problem where Mach reflection lies in an over-expanded jet flow outside of a nozzle. Hornung and Robinson [18] performed an experimental study for the Mach stem height, and proposed a mechanism by which the size of the Mach stem is determined. They pointed out that the pressure decreasing information from the wedge trailing edge expansion fan is carried out to the quasi-one-dimensional flow duct, which then moves upstream through the subsonic pocket to adjust the position of the triple point, or the Mach stem height. They pointed out a functional form of the Mach stem height and proposed that the normalized Mach stem height depends on M_0, θ_w, γ (ratio of specific heats) and g/w (trailing edge height g normalized by the wedge length w). If viscosity is accounted for, a boundary layer is attached to the wedge surface. Schmisseur and Gaitonde [19] used numerical simulation to show that this boundary layer increases the Mach stem height, due to the increased displacement effect of the wedge boundary-layer to increase the effective shock angle.

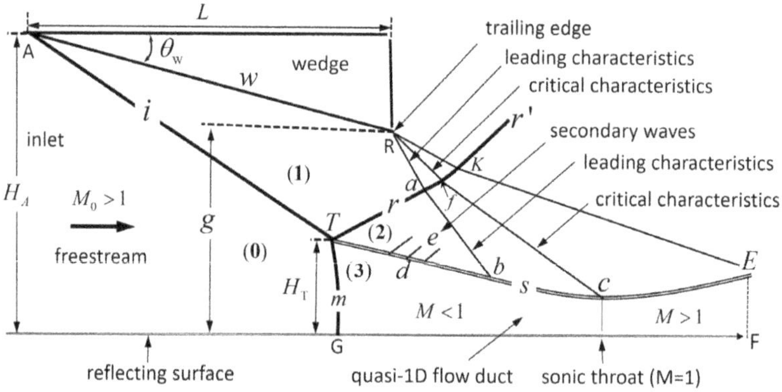

Figure 1. Mach reflection configuration. The inflow Mach number M_0, the wedge angle θ_w, the inlet height H_A and the lower wedge surface length w or the trailing edge height g are given conditions.

Since the 1990s, various attempts have been made to derive simplified models for predicting the Mach stem height [16,20,21]. A key issue in these models is how the pressure variation inside the expansion fan of the trailing edge (R) is carried out to the slipline, particularly to the sonic throat, where the height of the duct is minimal and related to the Mach stem height through the quasi-one-dimensional area Mach number relation plus the isentropic flow relation for pressure. Azevedo and Liu [20] assumed the sonic throat to be at point b, where the leading characteristics of the transmitted expansion waves intersects with the slipline. Li and Ben-Dor [16] allowed the sonic throat to be determined by the transmitted expansion waves, using the assumption that the flow just above the sonic throat is parallel to the free stream flow. This assumption was later adopted by Mouton and Hornung [21], who improved the model of Azevedo and Liu [20] and determined the Mach stem height using an unsteady approach. In the work of Azevedo and Liu [20], Li and Ben-Dor [16] and Mouton and Hornung [21], the slipline ahead of point b is treated as a straight line. Gao and Wu [22] and Bai and Wu [23] considered secondary generated expansion waves on the initial part of the slipline and demonstrated that the Mach stem height is sensitive to these pressure waves. Including the influence of these expansion waves appears to greatly increase the accuracy of the modeling. Recently, these works were extended to asymmetric shock reflection [24,25].

For symmetrical Mach reflection, past studies have shown how the Mach stem height depends on the inflow Mach number and the wedge angle. The normalized Mach stem height is a decreasing function of the inflow Mach number (Gao and Wu [22], Figure14a,b) and is an increasing function of the wedge angle or shock angle (Gao and Wu [22], Figure 14c,d, as can also be seen in Hornung and Robinson [18]). However, much of the established literature studying Mach stem height focused art approximately $g/w = 0.4$ [16,18,20–23], and the influence of a geometric setup on shock reflection was considered in the transition study [6,16]. It is thus desirable to study the Mach stem height for a wide range of geometry.

In this paper, we will use numerical simulation by computational fluid dynamics (CFD) to show that the normalized Mach stem height is almost a linear function of the normalized wedge height (Section 2). We then use some assumptions to derive a simplified expression for the Mach stem height, to show that this linearity can be predicted even with a simplified analysis (Section 3). Finally, we make our conclusions.

In the following, M is the Mach number, p is the pressure, θ is the flow deflection angle with respect to the free-stream direction which is positive when deflected to towards the reflecting surface, β is the shock angle, and γ is the ratio of specific heats.

2. Numerical Simulation for Dependence of Mach Stem Height on the Trailing Edge Height

Numerical results for Mach stem height are obtained through solving the full set of nonlinear Euler equations in gas dynamics, using the second-order Roe scheme based on finite difference approximation and second-order upwinding for the flux [26]. The grid number we used is 400 × 600, which is two times denser than the grid used by Gao and Wu [22] for similar purposes.

To ensure the accuracy of CFD computation, we performed calculations using various grid densities or accuracy for $\frac{g}{H_A} = 0.6$ or $\frac{g}{w} = 0.634$ or $\frac{H_A}{w} = 1.05$, when $M_0 = 4$, $\theta_w = 25°$. The Mach contours with four different density of grids are displayed in Figure 2.

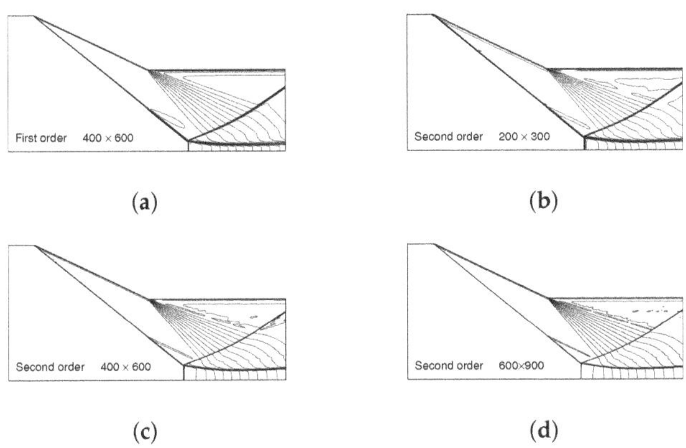

Figure 2. Mach number contours for $M_0 = 4$, $\theta_w = 25°$ and $\frac{g}{H_A} = 0.6$: (**a**) first order with a 400 × 600 mesh; (**b**) second order with a 200 × 300 mesh; (**c**) second order with a 400 × 600 mesh; and (**d**) second order with a 600 × 900 mesh.

It can be seen that the global flow structures are similar, but the Mach stem height with first-order accurate method with a grid 400 × 600 and second order method with a coarse grid 200 × 300 yield a Mach stem height much lower than that with the second order accurate method with grids 400 × 600 and 600 × 900. Moreover, the second order method with grids 400 × 600 and 600 × 900 results in Mach stem heights that are marginally close. Thus, we will use a second-order method with a grid 400 × 600 for simulations, since it needs less computational time than with the grid 600 × 900.

Now we display in Figure 3 the numerical results of normalized Mach stem height as a function of the normalized wedge trailing height for several different values of the inflow Mach number and wedge angle.

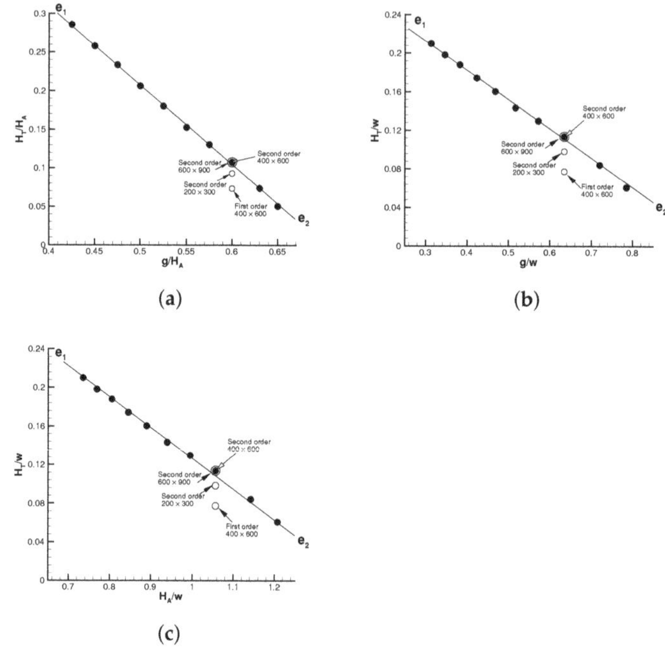

Figure 3. Numerical results for normalized Mach stem heights as a function of the normalized wedge trailing edge height for $M_0 = 4$ and $\theta_w = 25°$: (**a**) $\frac{H_T}{H_A}$ versus $\frac{g}{H_A}$; (**b**) $\frac{H_T}{w}$ versus $\frac{g}{w}$; (**c**) $\frac{H_T}{w}$ versus $\frac{H_A}{w}$. The line e_1e_2 is a straight line passing through the CFD data. Open circles are CFD data obtained using grids different to 400 × 600 points or a first-order accurate method.

Figure 3a displays the variation of $\frac{H_T}{H_A}$ with respect to $\frac{g}{H_A}$, for $M_0 = 4$, $\theta_w = 25°$ and a set of $\frac{g}{H_A}$ varying from 0.425 to 0.65. Figure 3b is the result for $\frac{H_T}{w}$ for various $\frac{g}{w}$, with $M_0 = 4$, $\theta_w = 25°$ and a set of $\frac{g}{w}$ varying from 0.3124 to 0.7849. Figure 3c displays the variation of $\frac{H_T}{w}$ versus $\frac{H_A}{w}$ for $M_0 = 4$, $\theta_w = 25°$ and a set of $\frac{H_A}{w}$ varying from 0.735 to 1.2075. The additional marks in Figure 3 show that numerical results with a grid of 400 × 600 points and with second order of accuracy are acceptable since further refining the grid does not change the Mach stem height. This supports the previous claim that the use of a second-order method with the grid 400 × 600 is accurate enough.

We observe that, for the conditions tested, the normalized Mach stem height $\frac{H_T}{H_A}$ is almost linear with respect to $\frac{g}{H_A}$, the normalized Mach stem height $\frac{H_T}{w}$ is almost linear with respect to $\frac{g}{w}$, and the normalized Mach stem height $\frac{H_T}{w}$ is almost linear with respect to $\frac{H_A}{w}$. In Figure 3, a straight line e_1e_2 is marked across the numerical data and this straight line is given by

$$\begin{cases} \frac{H_T}{H_A} = -1.0314\frac{g}{H_A} + 0.7227, R^2 = 0.9994 \\ \frac{H_T}{w} = -0.3087\frac{g}{w} + 0.3054, R^2 = 0.9985 \\ \frac{H_T}{w} = -0.3087\frac{H_A}{w} + 0.4359, R^2 = 0.9985 \end{cases}$$

where R^2 stands for the linear correlation coefficient.

The Mach contours for $M_0 = 4$, $\theta_w = 25°$ and $\frac{g}{H_A} = 0.425, 0.475, 0.525, 0.575$ are displayed in Figure 4a–d. From the Mach contours, we can see how the Mach stem height decreases with increasing $\frac{g}{H_A}$. It is noted that when $\frac{g}{H_A}$ increases, the triple point moves in the downstream direction and the flow duct below the slipline is narrowed.

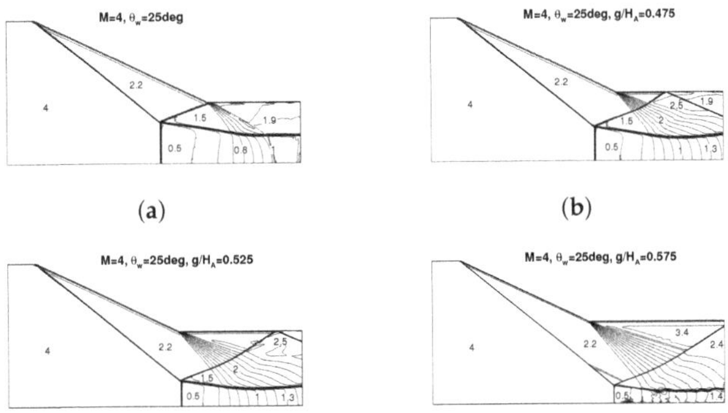

(a) (b)

(c) (d)

Figure 4. Numerical results for Mach contours with $M_0 = 4$ and $\theta_w = 25°$: (**a**) $\frac{g}{H_A} = 0.425$; (**b**) $\frac{g}{H_A} = 0.475$; (**c**) $\frac{g}{H_A} = 0.525$; and (**d**) $\frac{g}{H_A} = 0.575$.

The linearity is also observed for another set of M_0 and θ_w. Figure 5a displays the variation of $\frac{H_T}{H_A}$ with respect to $\frac{g}{H_A}$ for $M_0 = 4$, $\theta_w = 30°$ and a set of $\frac{g}{H_A}$ varying from 0.55 to 0.62. Figure 5b displays the variation of $\frac{H_T}{H_A}$ with respect to $\frac{g}{H_A}$ for $M_0 = 3$, $\theta_w = 25°$ and a set of $\frac{g}{H_A}$ varying from 0.57 to 0.70. In Figure 5, a straight line e_1e_2 is marked across the numerical data and this straight line is given by

$$\begin{cases} \frac{H_T}{H_A} = -2.3695\frac{g}{H_A} + 1.6993, R^2 = 0.9996 \\ \frac{H_T}{H_A} = -1.8537\frac{g}{H_A} + 1.4308, R^2 = 0.9990 \end{cases}$$

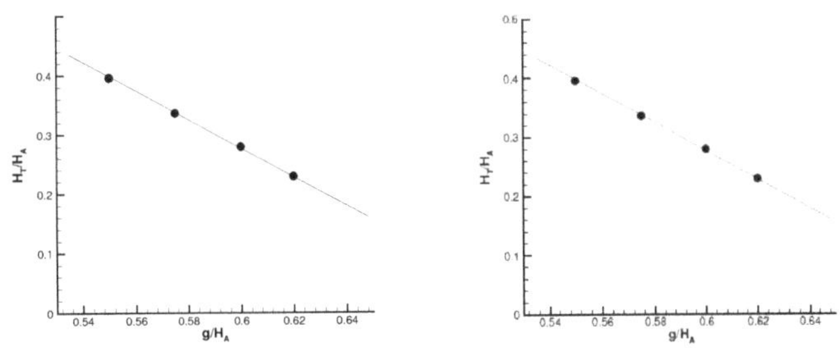

(a) (b)

Figure 5. Numerical results for normalized Mach stem heights as a function of the normalized wedge trailing edge height: (**a**) $\frac{H_T}{H_A}$ versus $\frac{g}{H_A}$ for $M_0 = 4$ and $\theta_w = 30°$; (**b**) $\frac{H_T}{H_A}$ versus $\frac{g}{H_A}$ for $M_0 = 3$ and $\theta_w = 25°$.

The numerical trend that the normalized Mach stem height decreases with increasing normalized trailing edge height seems to be counter-intuitive (and this trend has been demonstrated by Vuillon, Zeitoun and Ben-Dor [6] for a particular set of conditions), since apparently the Mach stem height should be proportional to the inlet height and one would expect the Mach stem height increase with increasing wedge trailing edge height. However, according to a simplified theoretical analysis given in Section 3, this trend can be justified.

3. A Simplified Analysis Showing Linearity of the Mach Stem Height with Geometry

Analytical models of various degrees of accuracy or complexity have been proposed in the past [16,20–23], none of which have been put into a linear form and have been used to predict the dependence of the Mach stem height on the wedge trailing length. Here, we derive an expression for the Mach stem height, which can be indeed put into a linear form. This expression is obtained by relating the sonic throat point c to the trailing edge R and to the triple point T, as done in previous works [16,20–23]. This analysis requires the slopes of the shock waves and of the slipline in the vicinity of the triple point as shown in Figure 6, as well as the slopes of the characteristic line Rfc as marked in Figure 1. The method to estimate these slopes is given in Appendix A.

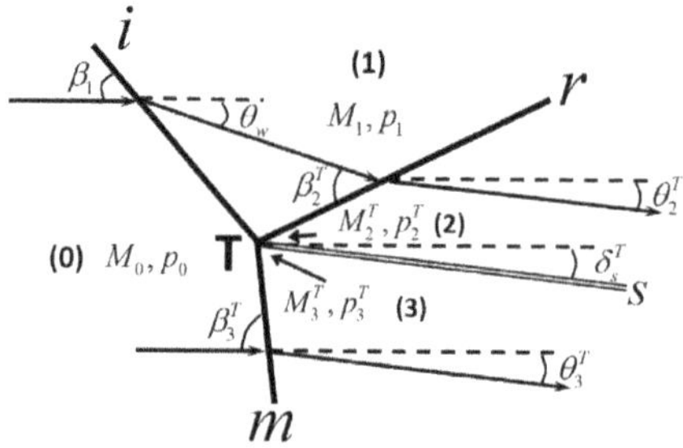

Figure 6. Triple point solution.

3.1. *Preliminary Geometric Relations*

Here, we establish geometric relations that relate the sonic throat position c to the trailing edge R through the characteristic line Rfc (called the critical characteristic line) and to the triple point T through the slipline Tbc. Notations can be seen in Figure 1. The critical characteristic line brings the required pressure to the sonic throat c on the slipline to balance the critical pressure p^* in the quasi-one-dimensional flow duct below the slipline.

The critical characteristic line Rfc intersects with the reflected shock wave at point f and is composed of two segments Rf and fc, both of which are assumed to be straight as by Bai and Wu [23]. Point f lies on the reflecting shock wave, which is generally curved according to Li and Ben-Dor [16]. Bai and Wu [23] gave a differential relation for the curved shape of this reflected shock. Here, for the purpose of evaluating the position of f within the present context of a simplified analysis, we omit the curvature of the shock segment Tf and relate the position of f to the triple point as

$$y_f - y_T = (x_f - x_T)\tan(\beta_2^T - \theta_w) \tag{1}$$

where β_2^T is the shock angle of the reflected shock wave as shown in Figure 6.

The position of point c at the sonic throat is related to the positions of the intersection point f and the trailing edge R by

$$\begin{cases} y_f - y_R = -(x_f - x_R)S_{Rf} \\ y_c - y_f = -(x_c - x_f)S_{fc} \end{cases} \tag{2}$$

where S_{Rf} is the slope of the critical characteristic line Rf upstream of the reflected shock wave and is determined by

$$S_{Rf} = \tan(\theta_f^* + \mu_f^*) \tag{3}$$

and S_{fc} is the slope of the transmitted critical characteristic line fc which is given by

$$S_{fc} = \tan(\theta_r^* + \mu_r^*). \tag{4}$$

The parameters θ_f^* and μ_f^* are the flow deflection angle and Mach angle along the critical characteristic line Rf. The parameters θ_r^* and μ_r^* are the flow deflection angle and Mach angle along the critical characteristic line fc. The method to evaluate these parameters is given in Appendix A.

Solving (1) and (2) gives the following geometric relation that relates the sonic throat point c to the trailing edge R via the critical characteristic line Rfc :

$$L_1 y_c + L_2 x_c = O_1 y_R + O_2 x_R + N_1 y_T + N_2 x_T, \tag{5}$$

Here:

$$\begin{cases} L_1 = 1 + \frac{S_{fc}}{\Psi} + \frac{\tan(\beta_2^T - \theta_w)}{\Psi} \\ L_2 = S_{fc} + \frac{S_{fc}^2}{\Psi} + \frac{S_{fc}\tan(\beta_2^T - \theta_w)}{\Psi} \\ O_1 = \frac{S_{fc}}{\Psi} + \frac{\tan(\beta_2^T - \theta_w)}{\Psi} \\ O_2 = \frac{S_{Rf} S_{fc}}{\Psi} + \frac{S_{Rf}\tan(\beta_2^T - \theta_w)}{\Psi} \\ N_1 = 1 \\ N_2 = -\tan(\beta_2^T - \theta_w) \end{cases} \tag{6}$$

and:

$$\Psi = \tan(\mu_f^* + \theta_f^*) - \tan(\mu_r^* + \theta_r^*) \tag{7}$$

Now, we establish a geometric relation that relates the sonic throat c to the triple point T through the slipline. This slipline is assumed to be composed of two segments Tb and bc, where b is the intersection of the slipline and the leading characteristics Rab.

Since the intersection point b also lies on the leading characteristics Rab, we have the following geometric relations for point b:

$$\begin{cases} y_a - y_T = (x_a - x_T)\tan(\beta_2^T - \theta_w) \\ y_b - y_a = -(x_b - x_a)\tan(\mu_2^T + \theta_2^T) \\ y_a - y_R = -(x_a - x_R)\tan(\mu_1 + \theta_w) \end{cases} \tag{8}$$

where $\mu_1 = \arcsin\frac{1}{M_1}$ and $\mu_2^T = \arcsin\frac{1}{M_2^T}$ are the Mach angles. Solving (8) yields:

$$l_1 y_b + l_2 x_b = m_1 y_R + m_2 x_R + n_1 y_T + n_2 x_T \tag{9}$$

where:

$$\begin{cases} l_1 = \frac{1}{\Phi_2 \tan(\mu_2^T + \theta_2^T)} + \frac{\tan(\beta_2^T - \theta_w)}{\Phi_1} \\ l_2 = \frac{1}{\Phi_2} + \frac{\tan(\beta_2^T - \theta_w)\tan(\mu_2^T + \theta_2^T)}{\Phi_1} \\ m_1 = \frac{1}{\Phi_2 \tan(\mu_1 + \theta_w)} + \frac{\tan(\beta_2^T - \theta_w)}{\Phi_1} \\ m_2 = \frac{1}{\Phi_2} + \frac{\tan(\mu_1 + \theta_w)\tan(\beta_2^T - \theta_w)}{\Phi_1} \\ n_1 = -1 \\ n_2 = \tan(\beta_2^T - \theta_w) \end{cases} \tag{10}$$

with:

$$\begin{cases} \Phi_1 = \tan(\mu_2^T + \theta_2^T) - \tan(\mu_1 + \theta_w) \\ \Phi_2 = \frac{1}{\tan(\mu_1 + \theta_w)} - \frac{1}{\tan(\mu_2^T + \theta_2^T)} \end{cases} \tag{11}$$

Gao and Wu (2010) found that the segment Tb has some curvature due to secondary expansion waves which serve to balance the pressure change in the quasi-one-dimensional flow duct. Here, we omit this curvature in order to have an explicit relation between c and T. Notations can be seen in Figure 1.

Since b lies on Tb, which is treated to be a straight line, we have:

$$y_T - y_b = (x_b - x_T)\tan\delta_s^T \tag{12}$$

Meanwhile, the slipline bc is a curve that has a vanishing slope at point c (throat). We approximate bc by a second order curve:

$$y - y_b = -(x - x_b)\tan\delta_s^T + \tau(x - x_b)^2 \tag{13}$$

Here, τ is a parameter ensuring that the second order curve has a vanishing slope at point c (throat), i.e., $\left.\frac{dy}{dx}\right|_{x=x_c} = 0$, which, when (13) is used, gives $-\tan\delta_s^T + 2\tau(x_c - x_b) = 0$, or:

$$\tau = \frac{\tan\delta_s^T}{2(x_c - x_b)}$$

Using this value of τ, (13) takes the following simple form:

$$y_c - y_b = -\frac{1}{2}(x_c - x_b)\tan\delta_s^T \tag{14}$$

Solving (12) and (14) gives:

$$\begin{cases} x_b = (y_T - y_c)\frac{2}{\tan\delta_s^T} + 2x_T - x_c \\ y_b = 2y_c - y_T - x_T\tan\delta_s^T + x_c\tan\delta_s^T \end{cases} \tag{15}$$

Now, putting (15) into (9) gives the geometric relation that relates the sonic throat to the triple point via the slipline:

$$X_1 y_c + X_2 x_c = Y_1 y_R + Y_2 x_R + Z_1 y_T + Z_2 x_T \tag{16}$$

where:

$$\begin{cases} X_1 = 2l_1 - \frac{2l_2}{\tan\delta_s^T} \\ X_2 = l_1\tan\delta_s^T - l_2 \\ Y_1 = m_1 \\ Y_2 = m_2 \\ Z_1 = l_1 - \frac{2l_2}{\tan\delta_s^T} + n_1 \\ Z_2 = l_1\tan\delta_s^T - 2l_2 + n_2 \end{cases} \tag{17}$$

In summary, the geometric relation (5) connects the sonic throat c to the trailing edge R following the critical characteristic line, and the relation (16) connects the sonic throat c to the triple point T following the slipline. They will be used to derive the expression for the Mach stem height below.

3.2. *Mach Stem Height Expression Showing Linearity*

In the following, we will use:

$$y_T = H_T, y_C = H^* = \frac{H_T}{\psi(M_m)}, y_R = g, x_R - x_A = L = \frac{H_A - g}{\tan\theta_w}, x_A = 0, y_A = H_A \tag{18}$$

Here, H^* is the height of the sonic throat which can be related to the Mach stem height H_T by the quasi-one-dimensional area Mach number relation $\frac{H_T}{H^*} = \psi(M_m)$ where:

$$\psi(M_m) = \frac{1}{M_m}\left(\frac{2}{\gamma+1}\left(1 + \frac{\gamma-1}{2}M_m^2\right)\right)^{\frac{\gamma+1}{2(\gamma-1)}} \tag{19}$$

The triple point is on the incident shock wave, so that $y_A - y_T = (x_T - x_A)\tan\beta_1$, meaning that:

$$x_T = x_A + \frac{y_A - y_T}{\tan\beta_1} \tag{20}$$

For $x_A = 0$, the expression (20) simplifies to:

$$x_T = \frac{H_A - H_T}{\tan\beta_1} \tag{21}$$

Solving (5) and (16) gives:

$$Qy_c = C_1 y_R + C_2 x_R + D_1 y_T + D_2 x_T \tag{22}$$

where:

$$\begin{cases} Q = X_2 \frac{L_1}{L_2} - X_1 \\ C_1 = X_2 \frac{O_1}{L_2} - Y_1 \\ C_2 = X_2 \frac{O_2}{L_2} - Y_2 \\ D_1 = X_2 \frac{N_1}{L_2} - Z_1 \\ D_2 = X_2 \frac{N_2}{L_2} - Z_2 \end{cases} \tag{23}$$

Using (21) to replace x_T in (22), we obtain:

$$Qy_c = C_1 y_R + C_2 x_R + \left(D_1 - \frac{D_2}{\tan\beta_1}\right) y_T + D_2 x_A + \frac{D_2}{\tan\beta_1} y_A \tag{24}$$

Putting (18) into (24) yields:

$$\begin{cases} \frac{H_T}{H_A} = A(M_0, \theta_w) \frac{g}{H_A} + B(M_0, \theta_w) \\ A(M_0, \theta_w) = \frac{C_1 - \frac{C_2}{\tan\theta_w}}{\frac{Q}{\psi(M_m)} - D_1 + \frac{D_2}{\tan\beta_1}} \\ B(M_0, \theta_w) = \frac{\frac{C_2}{\tan\theta_w} + \frac{D_2}{\tan\beta_1}}{\frac{Q}{\psi(M_m)} - D_1 + \frac{D_2}{\tan\beta_1}} \end{cases} \tag{25}$$

The expression (25) indeed shows the linearity of $\frac{H_T}{H_A}$ with respect to $\frac{g}{H_A}$, for fixed M_0, θ_w. This linearity has been observed in numerical simulation, as shown in Figure 3a.

If we introduce the obvious geometric relation $H_A = g + w\sin\theta_w$ into (25), we obtain an equivalent form:

$$\begin{cases} \frac{H_T}{w} = D^{(k)}(M_0, \theta_w) \frac{g}{w} + E^{(k)}(M_0, \theta_w) \\ D^{(k)} = A^{(k)}(M_0, \theta_w) + B^{(k)}(M_0, \theta_w) \\ E^{(k)} = B^{(k)}(M_0, \theta_w) \sin\theta_w \end{cases} \tag{26}$$

which shows the linearity of $\frac{H_T}{w}$ with respect to $\frac{g}{w}$, for fixed M_0, θ_w. This linearity was observed in numerical simulation, as shown in Figure 3b.

If we put $g = H_A - w\sin\theta_w$ into (26), we obtain another equivalent form:

$$\frac{H_T}{w} = D^{(k)}(M_0, \theta_w) \frac{H_A}{w} + E^{(k)}(M_0, \theta_w) \tag{27}$$

where $E^{(k)} = B^{(k)}(M_0, \theta_w)\sin\theta_w$. Thus, if normalized inlet height $\frac{H_A}{w}$ is used as an input parameter representing geometry, $\frac{H_T}{w}$ is still a linear function of $\frac{H_A}{w}$, for fixed M_0 and θ_w. This linearity has been observed in numerical simulation, as shown in Figure 3c. Note that Li and Ben-Dor [16] considered a situation where the wedge length w is fixed and the height g is increased, and showed that the foot of the Mach stem follows a linear trajectory. This observation, which they pointed out needs to be explained, may be associated with the linearity pointed out by the present work.

3.3. *Assessment of the Accuracy of the Mach Stem Height Expression*

The Mach stem height expressions (25)–(27) were obtained under some simplifications as stated in Appendix A. It is thus interesting to see whether these are also accurate enough for quantitative prediction. Here, we will assess their accuracy by comparing them with previous results and the present CFD data.

The experimental data of Hornung and Robinson [18] are usually used for comparison. Here, we consider their case with $M_0 = 3.98$ and $g/w = 0.4$, with the varying incident shock angle (β_1). Figure 7 displays the comparison of the Mach stem height expression (25) with some various previous works. In Figure 7, the experimental data of Hornung and Robinson [18] and Mouton and Hornung [27], and the CFD data of Mouton and Hornung [21], and Vuillon et al.[6] are displayed. It can be seen that the expression (25) provides a curve lying between the curves of Bai and Wu [23] and Gao and Wu [22].

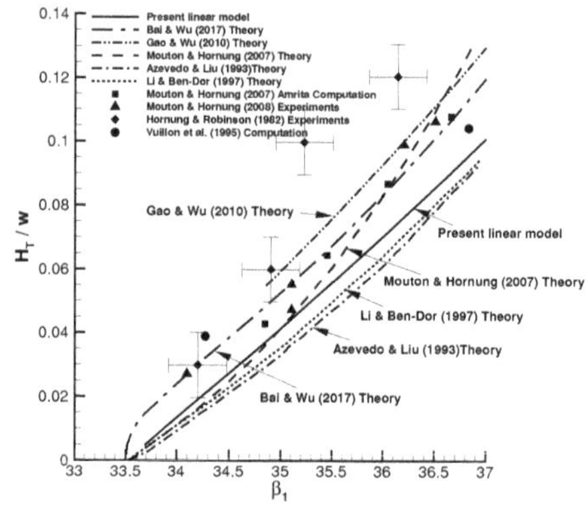

Figure 7. Comparison of the linear model (25) against previous authors for $M_0 = 3.98$ and $g/w = 0.4$.

The comparison of linear expression (25) with the present numerical solutions (already shown in Figure 3a–c and Figure 5a,b of Section 2) is given in Figure 8a–e. It is seen that, though the expression (25) displays linearity as CFD simulation, the slopes of the linear curves signficantly differ from the CFD results.

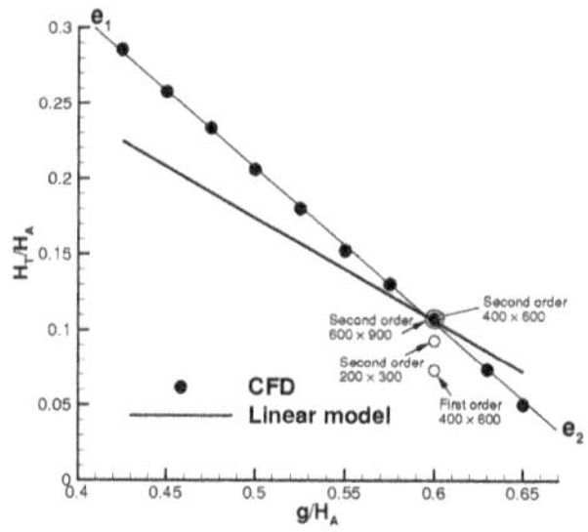

(a)

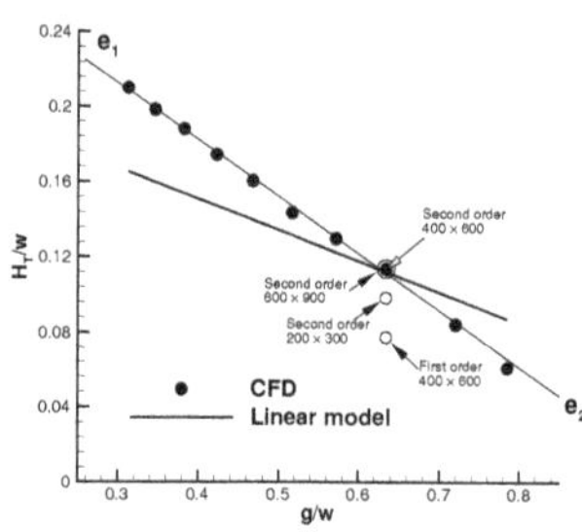

(b)

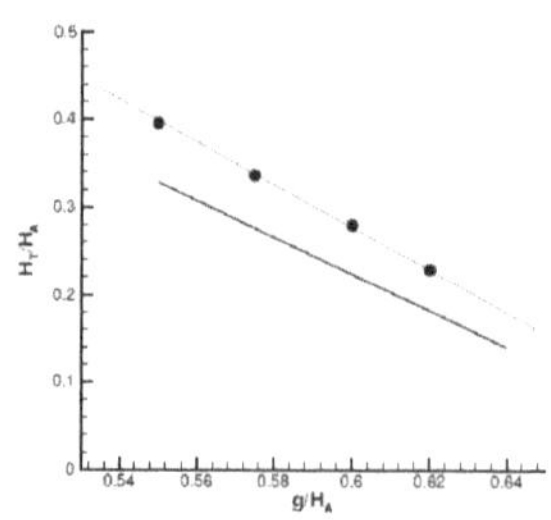

(c)

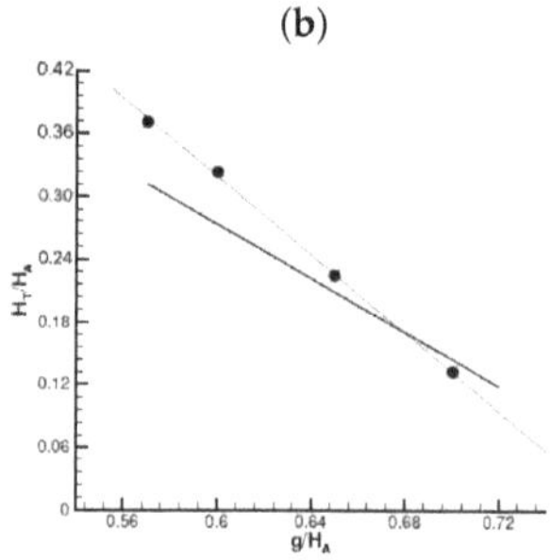

(d)

Figure 8. Cont.

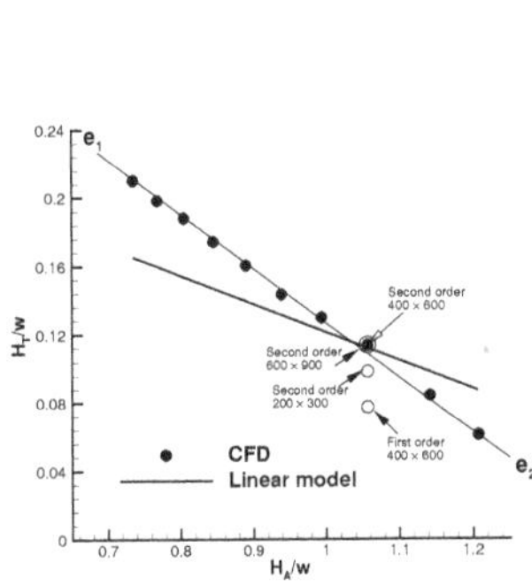

(e)

Figure 8. Comparison between the linear model (25) and CFD data: (**a**) $\frac{H_T}{H_A}$ versus $\frac{g}{H_A}$ for $M_0 = 4, \theta_w = 25°$; (**b**) $\frac{H_T}{w}$ versus $\frac{g}{w}$ for $M_0 = 4, \theta_w = 25°$; (**c**) $\frac{H_T}{H_A}$ versus $\frac{g}{H_A}$ for $M_0 = 4, \theta_w = 30°$; (**d**) $\frac{H_T}{H_A}$ versus $\frac{g}{H_A}$ for $M_0 = 3, \theta_w = 25°$; and (**e**) $\frac{H_T}{w}$ versus $\frac{H_A}{w}$ for $M_0 = 4, \theta_w = 25°$.

3.4. Summary and Significance of the Linear Analysis

The linearity predicted in the present simplified analysis appears to suggest that the linearity observed in the CFD simulation for a finite number of input parameters has some generality.

A comparison of the linear model (25) obtained from the simplified analysis of CFD data suggests that the slope, say $\frac{A(M_0,\theta_w)}{B(M_0,\theta_w)}$, in the linear curve does not yet have the required accuracy to be comparable with CFD data. The factors A and B in the expression (25) thus need elaboration before the expression (25) can be used for accurate prediction.

Despite the difference between the slope in the model (25) and the slope in CFD results, the agreement of linearity is meaningful in the future study of model elaboration since in the suggested linear model, the influence of wedge height and the influence of the inflow Mach number M_0 and wedge angle θ_w are separate. For model elaboration, one can focus on working out more accurate A and B, which only depend on the inflow Mach number M_0 and wedge angle θ_w. Moreover, even without knowing the exact values of the linear coefficients A and B, the conclusion that the normalized Mach stem height is linear with respect to the normalized wedge height is already useful in specific application to consider the precise influence of geometry, since one can just perform numerical or experimental work for the two sets of wedge height to fit the values of the slope in the linear model and then apply this linear model to predict the Mach stem height for other wedge height. This could greatly reduce the cost.

4. Conclusions

In this paper, we considered the dependence of the Mach stem height on the geometry when the geometry parameter (such as the trailing edge height) has a wide range. Such a study complements the past studies since many of the previous studies have focused on a narrow range of geometry.

Numerical simulation showed that the normalized Mach stem height is almost linear with respect to the normalized trailing edge height, independently of how they are normalized. When the trailing edge height is increased, keeping the inflow Mach number and the wedge angle fixed, the triple point moves in the downstream direction and the flow region between the slipline and the reflected surface is narrowed.

A simplified analysis showed that the linearity observed in CFD could be explained. This analysis leads to an expression of the normalized Mach stem height with respect to the normalized trailing edge height, which has the linear form $\frac{H_T}{H_A} = A\frac{g}{H_A} + B$ or $\frac{H_T}{w} = A\frac{g}{w} + B$. The coefficients A and B only depend on the inflow Mach number and the wedge angle.

The present work suggested that a Mach stem height model can be expressed as a linear function of the geometry. Further work could be done by working out more accurate coefficients A and B for purpose of quantitative prediction. In a specific application to consider the precise influence of geometry, one can also perform numerical or experimental work for two sets of wedge height to fit the values of the slope in the linear model and then apply this linear model to predict the Mach stem height for other wedge height.

Author Contributions: C.-Y.B. and Z.-N.W. equally contributed to the development of the present paper, including the conceptualization, methodology, data acquisition and analysis, interpretation of the results, and manuscript preparation. All authors have read and agreed to the published version of the manuscript.

Funding: The authors are thankful to the editors and referees who provided valuable comments to improve the manuscript. This work was partly supported by the National Key Project (Grant No. GJXM92579), by the National Science and Technology Major Project 2017-II-003-0015, and by the National Natural Science Foundation of China (Grants No. 11721202).

Institutional Review Board Statement: Not applicable.

Informed Consent Statement: Not applicable.

Data Availability Statement: Not applicable.

Conflicts of Interest: The authors declare no conflict of interest.

Appendix A. Method to Evaluate the Slopes of Shock Waves and Critical Characteristic Line

In this Appendix A, we provide the estimation of the slopes of various waves required in the Mach stem height model presented in Section 3.

Appendix A.1. Slopes of Waves in the Vicinity of Triple Point

The triple point theory, due to von Neumann[28], provides solutions in the vicinity of the triple point (T) as shown in Figure 6. The solutions in regions (1), (2) and (3), in the vicinity of the triple point, follow from the oblique shock wave relations for shock i, r and m:

$$\begin{cases} \tan\theta_w = f_\theta(M_0,\beta_1), M_1^2 = f_M(M_0,\beta_1), p_1 = f_p(M_0,\beta_1)\ (i) \\ \tan(\theta_w-\theta_2^T) = f_\theta(M_1,\beta_2^T), (M_2^T)^2 = f_M(M_1,\beta_2^T), p_2^T = f_p(M_1,\beta_2^T)\ (r) \\ \tan\theta_3^T = f_\theta(M_0,\beta_3^T), (M_3^T)^2 = f_M(M_0,\beta_3^T), p_3^T = f_p(M_0,\beta_3^T)\ (m) \end{cases} \quad \text{(A1)}$$

Here, the flow parameters in the vicinity of the triple point are denoted with superscript T, and $f_\theta(M,\beta), f_M(M,\beta), f_p(M,\beta)$ are functions for oblique shock waves defined by

$$\begin{cases} f_\theta(M,\beta) = 2\cot\beta\frac{M^2\sin^2\beta-1}{M^2(\gamma+\cos 2\beta)+2} \\ f_M(M,\beta) = \frac{M^2+\frac{2}{\gamma-1}}{\frac{2\gamma}{\gamma-1}M^2\sin^2\beta-1} + \frac{M^2\cos^2\beta}{\frac{\gamma-1}{2}M^2\sin^2\beta+1} \\ f_p(M,\beta) = \frac{2\gamma}{\gamma+1}M^2\sin^2\beta - \frac{\gamma-1}{\gamma+1} \end{cases} \quad \text{(A2)}$$

For shock i and r, the oblique shock wave relations are solved for weak solutions (i.e., for the smaller value of shock angle β). For shock m, the oblique shock wave relations are solved for strong solution (i.e., for the larger value of shock angle β).

Across the slipline, the flow streams are parallel, i.e.,

$$\theta_3^T = \theta_2^T = \delta_s^T \quad \text{(A3)}$$

and the pressures are balanced, meaning that:

$$p_2^T = p_3^T \quad \text{(A4)}$$

Giving M_0 and θ_w, Equations (A1)–(A4) form a closed set to determine the flow parameters near the triple point, i.e., the Mach numbers M_2^T, M_3^T, the pressures $p_2^T = p_3^T$, the initial shock angle β_2^T of the reflected shock wave and the initial angle of the slipline δ_s^T. The shock angle β_2^T and the slipline angle δ_s^T were used in Equations (6) and (17) to establish the required geometrical relations for the Mach stem height model.

Appendix A.2. Slopes of the Critical Characteristic Line

In order to compute the slopes S_{Rf} and S_{fc} of the critical characteristic line Rfc by Equations (3) and (4), we need to evaluate θ_f^*, μ_f^*, θ_r^* and μ_r^*. Here, θ_f^* is the local flow deflection angle inside the trailing edge expansion fan, at which the pressure p_f^*, when amplified through the reflected shock wave to become p_r^*, will be carried out through the critical characteristic line fc to the sonic throat to balance the critical pressure p_s^* (pressure at the sonic throat in the quasi-one-dimensional flow duct), meaning that:

$$p_r^* = p_s^* \tag{A5}$$

The critical pressure p_s^* can be determined by using an isentropic flow assumption for pressure. This gives a relation between the pressure p_s^* at the sonic point ($M_s^* = 1$) and the pressure p_m behind the Mach stem:

$$\frac{p_s^*}{p_m} = \left(\frac{1+\frac{\gamma-1}{2}M_m^2}{1+\frac{\gamma-1}{2}}\right)^{\frac{\gamma}{\gamma-1}}. \tag{A6}$$

where M_m is the Mach number behind the Mach stem. The parameters p_m and M_m can be computed as $p_m = \frac{1}{2}(p_3^T + p_n)$, $M_m = \frac{1}{2}(M_3^T + M_n)$, where p_n and M_n are normal shock solutions computed as $p_n = f_p(M_0, \frac{\pi}{2})$, $M_n^2 = f_M(M_0, \frac{\pi}{2})$.

The amplified pressure p_r^*, the pressure along the critical characteristic line fc, depends on the interaction between the expansion fan and the reflected shock wave. Bai and Wu [23] gave a differential form to predict this. Here, within the present context of simplicity, we assume that $p_r^* = \frac{p_2^T}{p_1}p_f^*$, i.e., the pressure at point f, is amplified by the reflected shock wave by the same factor as through the initial segment Ta of the reflected shock wave.

Using (A5), $p_r^* = \frac{p_2^T}{p_1}p_f^*$, and (A6), we obtain:

$$p_f^* = \left(\frac{1+\frac{\gamma-1}{2}M_m^2}{1+\frac{\gamma-1}{2}}\right)^{\frac{\gamma}{\gamma-1}} \frac{p_1 p_m}{p_2^T} \tag{A7}$$

which is the pressure at $\theta = \theta_f^*$. This pressure is further related to the pressure p_1 in region (1) by the Prandtl–Meyer relation:

$$\begin{cases} p_f^* = \left(\frac{1+\frac{\gamma-1}{2}M_1^2}{1+\frac{\gamma-1}{2}M_f^{*2}}\right)^{\frac{\gamma}{\gamma-1}} p_1 \\ \nu(M_f^*) - \nu(M_1) = \theta_w - \theta_f^* \end{cases}. \tag{A8}$$

where $\nu(M) = \sqrt{\frac{\gamma+1}{\gamma-1}}\arctan\sqrt{\frac{\gamma-1}{\gamma+1}(M^2-1)} - \arctan\sqrt{M^2-1}$ is the Prandtl–Meyer function. Once (A7) is used to obtain p_f^*, the first expression in (A8) is used to obtain the Mach number M_f^* and the last expression in (A8) is used to obtain θ_f^*. The Mach angle μ_f^* can then by computed through $\sin\mu_f^* = 1/M_f^*$. These provide values of θ_f^* and μ_f^*, needed in (3), to compute the slope of the segment Rf of the critical characteristic line.

By (A5) and by (A6), we obtain:

$$p_r^* = \left(\frac{1+\frac{\gamma-1}{2}M_m^2}{1+\frac{\gamma-1}{2}}\right)^{\frac{\gamma}{\gamma-1}} p_m$$

We then consider the oblique shock wave relation across the reflected shock wave at point f. After solving the pressure relation $p_r^* = f_p\left(M_f^*, \beta_r^*\right)$ to obtain the local shock angle β_r^*, we then solve $\tan(\theta_r^* - \theta_f^*) = f_\theta\left(M_f^*, \beta_r^*\right)$ to obtain the local flow deflection angle θ_r^*, and then solve $M_r^{*2} = f_M\left(M_f^*, \beta_r^*\right)$ to obtain the Mach number M_r^*, before finally using $\sin\mu_r^* = 1/M_r^*$ for the Mach angle. These provide values of θ_r^* and μ_r^*, needed in (4), to compute the slope of the segment fc of the critical characteristic line.

References

1. Ben-Dor, G. *Shock Wave Reflection Phenomena*; Springer: Berlin/Heidelberg, Germany, 2007.
2. Henderson, L.F.; Lozzi, A. Experiments on transition of Mach reflection. *J. Fluid Mech.* **1975**, *68*, 139–155. [CrossRef]
3. Bazhenova, T.V.; Fokeev, V.P.; Gvozdeva, L.G. Regions of various forms of Mach reflection and its transition to regular reflection. *Acta Astronaut.* **1976**, *3*, 131–140. [CrossRef]
4. Hornung, H.G.; Oertel, H.; Sandeman, R.J. Transition to Mach reflection of shock waves in steady and pseudo-steady flows with and without relaxation. *J. Fluid Mech.* **1979**, *90*, 541–560. [CrossRef]
5. Teshukov, V.M. On stability of RR of shock waves. *Prikl. Mekh. Technol. Fiz* **1989**, *2*, 26–33.
6. Vuillon, J.; Zeitoun, D.; Ben-Dor, G. Reconstruction of oblique shock wave reflection in steady flows. Part 2. Numerical investigation. *J. Fluid Mech.* **1995**, *301*, 37–50 . [CrossRef]
7. Chpoun, A.; Passerel, D.; Li, H.; Ben-Dor, G. Reconsideration of the oblique shock wave reflection in steady flows. Part 1. Experimental investigation. *J. Fluid Mech.* **1995**, *301*, 19–35. [CrossRef]
8. Li, H.; Ben-Dor, G. Application of the principle of minimum entropy production to shock wave reflections. I. Steady flows. *J. Appl. Phys.* **1996**, *80*, 2027–2037. [CrossRef]
9. Ivanov, M.S.; Ben-Dor, G.; Elperin, T.; Kudryavtsev, A.N.; Khotyanovsky, D.V. Flow–Mach-number-variation-induced hysteresis in steady flow shock wave reflections. *AIAA J.* **2001**, *39*, 972–974. [CrossRef]
10. Ben-Dor, G.; Ivanov, M.; Vasilev, E.I.; Elperin, T. Hysteresis processes in the regular reflection to Mach reflection transition in steady flows. *Prog. Aerosp. Sci.* **2002**, *38*, 347–387. [CrossRef]
11. Hornung, H. Mach reflection in steady flow. I. Mikhail Ivanovs contributions, II. Caltech stability experiments. *AIP Conf. Proc. AIP Publ.* **2014**, *1628*, 1384–1393.
12. Courant, R.; Friedrichs, K.O. *Supersonic Flow and Shock Waves*; Wiley-Interscience: Hoboken, NJ, USA, 1948.
13. Liepmann, H.; Roshko, A. *Elements of Gasdynamics*; John Wiley and Sons: Hoboken, NJ, USA, 1957.
14. Emanuel, G. *Gasdynamics: Theory and Applications*; American Institute of Aeronautics and Astronautics: New York, NY, USA, 1986.
15. Ben-Dor, G.; Takayama, K. The phenomena of shock wave reflection-a review of unsolved problems and future research needs. *Shock Waves* **1992**, *2*, 211–223. [CrossRef]
16. Li, H.; Ben-Dor, G. A parametric study of Mach reflection in steady flows. *J. Fluid Mech.* **1997**, *341*, 101–125. [CrossRef]
17. Chow, W.L.; Chang, I.S. Mach reflection associated with over-expanded nozzle free jet flows. *AIAA J.* **1975**, *13*, 762–766. [CrossRef]
18. Hornung, H.G.; Robinson, M. Transition from regular to Mach reflection of shock waves Part 2. The steady-flow criterion. *J. Fluid Mech.* **1982**, *123*, 155–164. [CrossRef]
19. Schmisseur, J.; Gaitonde, D. Numerical simulation of Mach reflection in steady flows. *Shock Waves* **2011**, *21*, 499–509. [CrossRef]
20. Azevedo, D.J.; Liu, C.S. Engineering approach to the prediction of shock patterns in bounded high-speed flows. *AIAA J.* **1993**, *31*, 83–90. [CrossRef]
21. Mouton, C.A.; Hornung, H. Mach stem height and growth rate predictions. *AIAA J.* **2007**, *45*, 1977–1987. [CrossRef]
22. Gao, B.; Wu, Z.N. A study of the flow structure for Mach reflection in steady supersonic flow. *J. Fluid Mech.* **2010**, *656*, 29–50. [CrossRef]
23. Bai, C.Y.; Wu, Z.N. Size and shape of shock waves and slipline for Mach reflection in steady flow. *J. Fluid Mech.* **2017**, *818*, 116–140. [CrossRef]
24. Roy, S.; Gopalapillai, R. An analytical model for asymmetric Mach reflection configuration in steady flows. *J. Fluid Mech.* **2019**, *863*, 242–268. [CrossRef]
25. Lin, J.; Bai, C.Y.; Wu, Z.N. Study of asymmetrical whock wave reflection in steady supersonic flow. *J. Fluid Mech.* **2019**, *864*, 848–875. [CrossRef]
26. Roe, P. Characteristic based schemes for the Euler equations. *Annu. Rev. Fluid Mech.* **1986**, *18*, 337–365. [CrossRef]
27. Mouton, C.A.; Hornung, H. Experiments on the mechanism of inducing transition between regular and Mach reflection Mach stem height and growth rate predictions. *Phys. Fluids* **2008**, *20*, D541–D572. [CrossRef]
28. Von Neumann, J. *Oblique Reflection of Shock*; Explos. Res. Rep. 12 Navy Dept.; Bureau of Ordinance: Washington, DC, USA, 1943.

fluids

MDPI

Article

Experiments in Shock-Vortex Interactions

Beric Skews

School of Mechanical, Industrial and Aeronautical Engineering, University of the Witwatersrand, Johannesburg 2000, South Africa; beric.skews@wits.ac.za

Abstract: Studies of shock-vortex interactions in the past have predominantly been numerical, with a number of idealizations such as assuming an isolated vortex and a plane shock wave. In the present case the vortex is generated from flow separation at a corner. A shear layer results which wraps up into a spiral vortex. The flow is impulsively initiated by the diffraction of a shock wave over the edge. The strength of the shock determines the nature of the flow at the corner and that induced behind the diffracted wave. A wide variety of cases are considered using different experimental arrangements such as having two independent shock waves arriving at the corner at different times, to reflecting the diffracting wave off different surfaces back into the vortex, and to examining the flow around bends where the reflection off the far wall reflects back onto the vortex. The majority of studies have shown that the vortex normally retains its integrity after shock transit. Some studies with curved shock waves and numerous traverses have shown evidence of vortex breakup and the development of turbulent patches in the flow, as well as significant vortex stretching. Depending on the direction of approach of the shock wave it refracts through the shear layer thereby changing the strength and direction of both. Of particular note is that the two diffracted waves which emerge from the vortex as the incident wave passes through interact with each other resulting in a pressure spike of considerable magnitude. An additional spike is also identified.

Keywords: spiral vortex; shock diffraction; unsteady flow

PACS: 4740Nm

Citation: Skews, B. Experiments in Shock-Vortex Interactions. *Fluids* **2021**, 6, 303. https://doi.org/10.3390/fluids6090303

Academic Editor: Olga A. Azarova

Received: 19 July 2021
Accepted: 20 August 2021
Published: 26 August 2021

1. Introduction

Shock-vortex interactions have been the subject of extensive computational, theoretical and experimental studies over many years. One of the earliest experiments was a schlieren study of the interaction [1]. They observed the generation of acoustic waves which became a major area of subsequent research. The response to this finding led to an early investigation [2], followed by a more extensive treatment some years later [3]. The theory developed showed good agreement with pressure measurements of the emerging wave [4] done by Dosanjh and Weeks. Following this work important numerical results on the nature of the acoustic waves of different strengths was done by Ellzey et al. [5] who commented on the development of the complex regular and Mach reflection wave systems that evolve between the incident and transmitted waves, and the quadrupolar nature of the acoustic emission. In a later paper [6] more details of the interaction and the reasons for the the development of the acoustic wave were given, as being due to distortion of the shock wave and the associated vortex compression. A later detailed numerical study of the early development in the interaction placed the emphasis on the development of the reflected shock waves [7]. A further detailed numerical study by Zhang et al. [8] between strong shock waves interacting with a strong vortex identified a multistage feature with the development of shocklets and multiple sound waves. The multiple wave feature was examined further by Chatterjee and Vijayaraj [9] using a large computational domain and high-order simulation which showed more waves than previously identified which were produced by the deformed rotating vortex. Various cases of the interaction depending on relative strengths of the shock wave and vortex have been given by Chang et al. [10] using numerical shadowgraphs in order to discuss the interaction process.

It is evident from the above discussion that the development in the study of shock-vortex interactions has predominantly been numerical, almost entirely so for the past 30 years, and has concentrated on the mechanism for the development of acoustic waves. The predominant model used is for an isolated vortex and a plane shock wave. The pattern of wave interaction as described in [5] is shown in Figure 1 and is that dealt with in many other works.

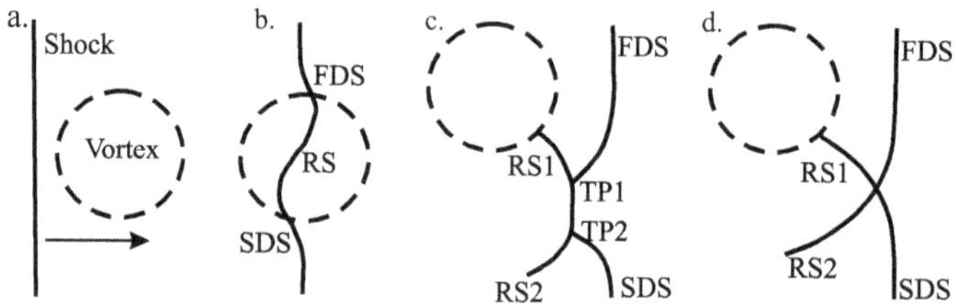

Figure 1. The complex shock wave-vortex interaction. (**a**) Prior to interaction, (**b**) early stages, (**c**) stage for a strong shock and strong vortex, (**d**) stage for a weak shock and strong vortex.

As the shock enters the vortex it is distorted due the the vortex velocity distribution which results in refraction of the shock profile, RS, as shown in Figure 1b. This results in the fast diffracted shock, FDS, as the shock is pushed forward and a slow diffracted shock, SDS, where it is retarded. For a strong shock interacting with a strong vortex the waves then interact producing a pair of Mach reflections with associated reflected waves, RS1 and RS2, and triple points, TP1 and TP2. For the interaction of a weak shock and strong vortex the interaction between the diffracted shocks results in a regular reflection.

The experimental work reviewed in this paper was initiated based on the study of shock wave propagation around bends [11,12]. In that case and others discussed subsequently the vortex produced results from the diffraction of a shock wave around a corner and is significantly different from the ideal isolated vortex case. This shock wave diffraction is treated in [13]. The vortex in that realistic case is spiral in nature with typical cases given in Figure 2. For a shock Mach number greater than 2.07 the flow behind the shock becomes supersonic resulting in a markedly change in the downstream flow.

The row of sketches show the main features for different incident shock Mach numbers. The main items of interest are the shear layer which wraps around the shed vortex, the vortex structure, and the appearance of embedded shock waves. As the incident shock Mach number is about 1.6 the flow immediately above the shear layer becomes supersonic with the appearance of weak shocks in a lambda format. A further shock may appear further downstream as the flow accelerates around the vortex. Additional shocks can appear on the opposite side as the flow must decelerate as it approaches the underside of the shear layer. The vortex is also surrounded by a contact surface which is the dividing surface between the gas that was there before the shock arrived at the corner and has moved downstream and the gas that has been influenced by the diffracting shock. The first two frames in the bottom row shows typical experimental results of the full flow field and the remaining frames an unusually large scale experiment showing the evolution of the shear layer and vortex. Both the lambda shock and embedded shock are evident but in addition small vortices develop along the length of the shear layer as well as it becoming increasingly turbulent.

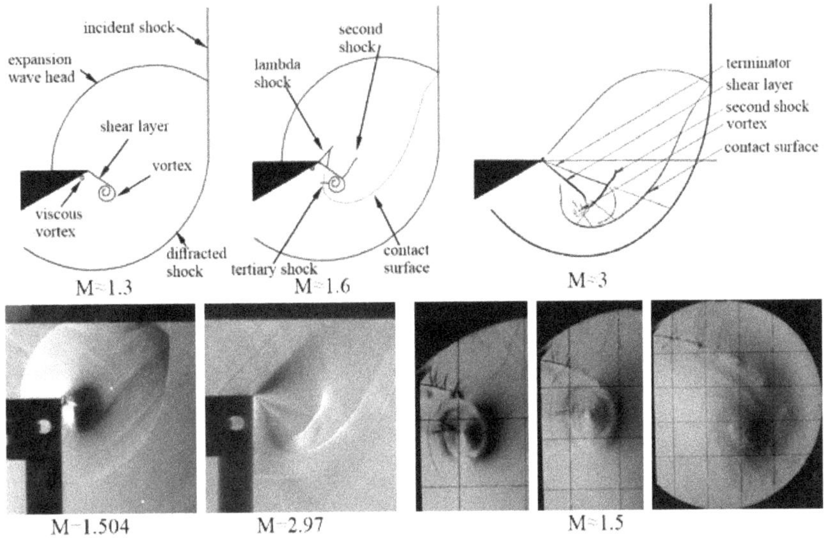

Figure 2. Flow development due to shock wave diffraction. Sketches show the main features noting the differences between subsonic and supersonic post-shock flow. The first two images of the bottom row show flow features, and the remaining three images a large scale experiment showing the nature of the shear layer and vortex. The reference lines are 50 mm apart.

The flow around a bend is particularly interesting since it gives evidence of some vortex disruption and regions of turbulence, which contrasts to the general view of a vortex being very stable. Examples for propagation around a number of bends was given in [12]. A detailed interferometric study for comparison with some results of the above experimental study was undertaken by Timofeev et al. [14] for the case of a Mach 2 wave around a 90° bend. A sketch of two cases, as the flow progresses, are given in Figure 3 together with corresponding experimental results and the comparative interferograms. In both cases a sketch of the main features is given followed by a schlieren image and an interferogram taken at similar times. The first set covers the case after the first transit of the vortex. The shock, D, is part of the initial diffracted wave on the corner which has reflected off the right-hand surface as shock, 1, which passes through the vortex and has just started to reflect off the left-hand surface as wave 2. Wave 1 propagates all the way up to the upper surface of the bend. The center of the vortex is well defined in the schlieren image, with the contact surface below it, but the upper region of vortex flow is influenced by the expansion wave and shear layer arising from the bend corner and their interaction with shock 1 giving rise to a reflected wave, R. The wide region behind wave 1 at the top of the bend is due to the development of a lambda shock on the tests section window as the wave propagates into the oncoming flow.

As wave 2 propagates through the region disturbed by wave 1 it merges with it near the top surface and interesting effects occur in the region of the vortex center at C. The upper part of wave 2 is accelerated due to the vortex motion whilst the lower part is slowed down. These then meet in a regular reflection resulting in a reflected wave which appears to fan out in the region of the vortex. This unusual feature is examined later. An important feature of the overall interaction is that the flow adjacent to the left-hand surface becomes turbulent.

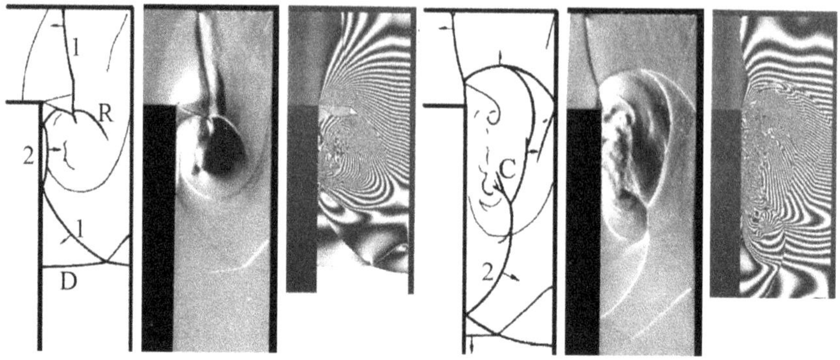

Figure 3. Sketch, schlieren and interferograms. The first three are after first shock transit and the second three after second transit.

2. Materials and Methods

The major facilities for generating shock waves are shock tubes in a variety of arrangements. In the current study two formats are used: In the first; a simple tube with two chambers initially at different pressures separated by a frangible diaphragm. A shock wave propagates down one section on diaphragm removal. The arrangement in the test section is shown in Figure 4. Part of the incident shock is cut out to pass over the diffracting edge which can have different profiles. The diffracting shock wave is then reflected off a surface situated downstream in order to generate a reflected wave to impact the vortex.

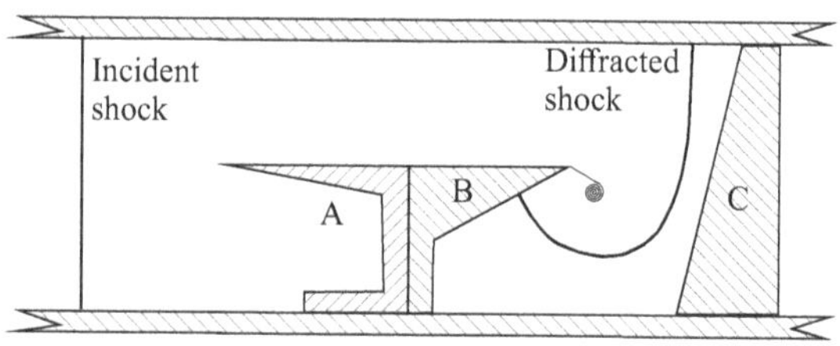

Figure 4. Testpiece arrangemnt. Part A is the 'cookie cutter', part B the diffracting wedge and part C the reflecting surface to produce the transmitting wave into the vortex.

In the second format, shown in Figure 5, the shock wave from the shock tube is spit into two waves which then interact with each other in the test section with arrival times determine by the lengths of the rolled tubes. The tube heights are 44 mm. In all cases the vortex is generated from the diffraction of the early arriving shock around the edge of a test piece which is then impacted by the late arriving wave.

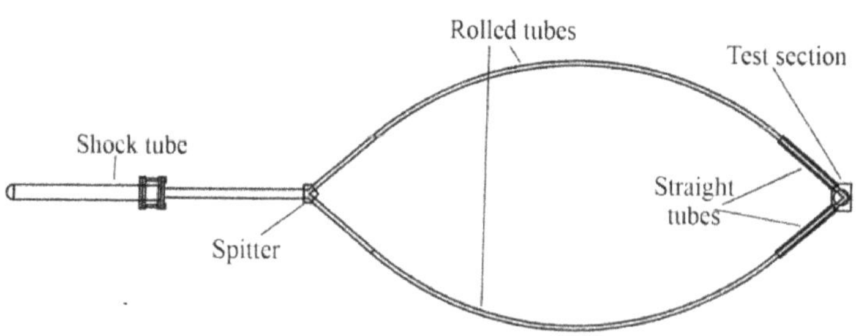

Figure 5. Bifurcated shock tube for generating two independent shock waves.

Contact laser shadowgraphs were employed in many cases. This was done by placing a 4 inch by 5 inch holographic film against the far-side test-section window.A pulsed ruby laser with an expanded beam was used, with a pulse duration of 30 ns, for illumination. Thus a high resolution image was obtained with a magnification of unity and minimal optical aberration.

For standard shadowgraphs a schlieren visualization system is used based on the standard z-format shown in Figure 6 and is used in either schlieren or shadowgraph mode. Either single shot images are taken with a short duration light source or movies with a continuous source and a high-speed camera.

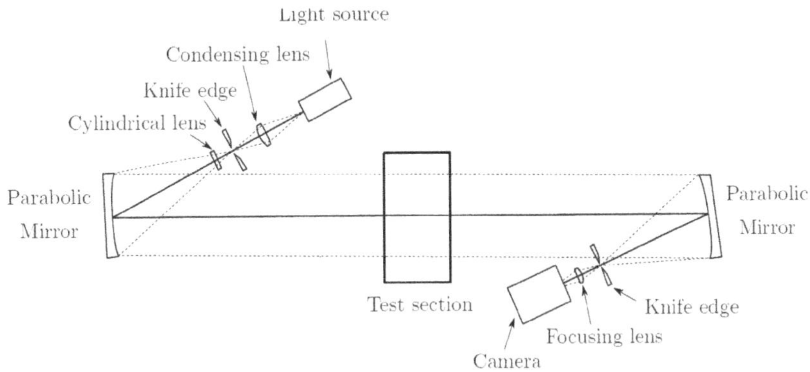

Figure 6. Z-format schlieren optical arrangement.

Numerical simulations were done with an in-house code using a finite element Taylor-Galerkin scheme [15] which has been validated through a number of studies as well as in this work, as discussed below. A Flux-Corrected Transport algorithm together with an adaptive refinement algorithm was used to obtain high resolution of discontinuities. In order to facilitate comparison with experiment and to enhance the visibility of the flow structure, numerical schlieren images were computed. These were formed by calculating the first derivative of the density matrix. In order to compute numerical colour schlieren images the density gradient matrices in the x and y directions were first calculated. The components of the density gradients perpendicular to three virtual cut-off colour filters (knife-edges) inclined at 120° to each other, (red along the x-axis, green inclined at 120° to the x-axis and blue inclined at 240° to the x-axis), were then obtained using the x and y density gradient matrices. The three resulting matrices were first thresholded by a factor of 10 and then combined to form the red, green and blue components of a true-colour RGB image. The resulting colour schlieren image is more sensitive to gradual changes in density than shadowgraphs and thus shows additional features such as expansion waves.

3. Results

3.1. Single Shock Passage

Preliminary experimental results using the facility of Figure 5 were reported in [16]. A number of magnified images using a laser light source are given in Figure 7. The test piece at the exit of the two legs of the rig is a 80° wedge. The shock wave from the top leg of the rig arrives early entering from the upper right, shedding a vortex, and that from the lower leg, bottom right, transverse over the vortex. The difference between the arrival times at the corner, δt, are given in the figure. The left and right images, with similar delay times, are taken at different times thereby showing the evolution of the interaction in time. The first image shows the development of the double Mach reflection with the shear layers propagating to the vortex with the two triple points corresponding to the pattern in Figure 1c. The centre image is at an earlier stage of development because of the much larger delay time between the two waves arriving at the corner. The fast diffracted shock, FDS, and and associated reflected shock, RS2, appear to be doubled into separate waves. Similar effects are evident for the reflected shock in previous work [4,5]although the geometry of the experiment is somewhat different. There is no clear Mach stem and the reflection appears to be regular but with a clear shear layer. Whether this is due to different shock strengths on either side which would produce a shear layer, or a small Mach stem which is not resolved is not clear. There is an additional feature of a shock between the vortex and the wall which does not appear in the two cases with a lower synchronization delay. There are also two vortices shed from the corner. The details of this flow is that the vortex induced flow is up the surface towards the corner and separates there with the shock acting to decelerate the flow to meet the shear layer.

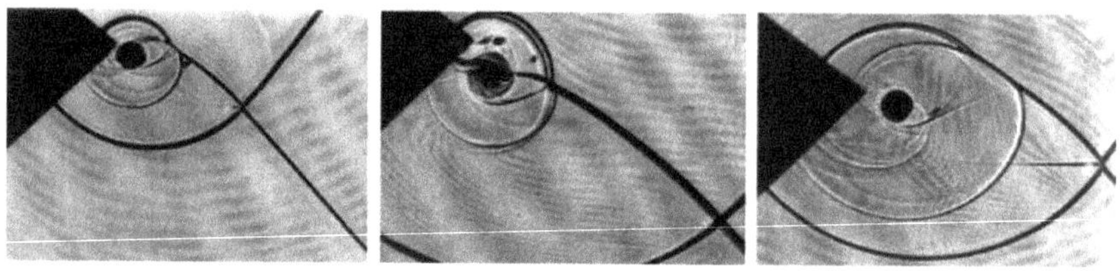

Figure 7. Contact shadowgraphs, M = 1.33. left: $\delta t = 20$ µs; centre: $\delta t = 52$ µs; right $\delta t = 22$ µs.

Additional work was presented in [17]. Figure 8 shows more images from a series of tests at the same nominal Mach number and same delay time between arrival of the two shock waves at the corner. The vortex appears as dark blob because of the high density gradients and corresponding refractive index gradients resulting in total internal reflection of the laser beam. There is some correspondence with previous numerical predictions for the isolated vortex case but noticeable differences due to the presence of the shear layer and with the vortex being close to the wall. Both the last frame in the top row and first frame in the bottom row show small vortices being shed off the corner and a shock between the vortex and the wall as also noted in Figure 7. These slowly dissipate as the vortex moves away from the wall. The shock moves away from the corner in the opposite direction to the clockwise upward flow of the vortex. The last frame shows a complex interaction as the reflected wave from the upper triple point enters the influence of the vortex where it becomes distorted.

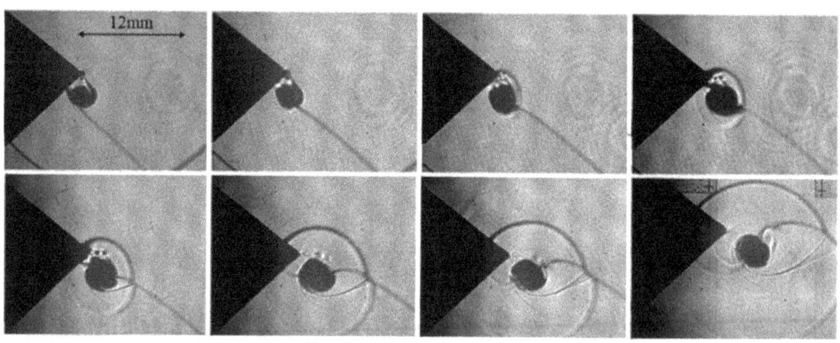

Figure 8. Contact shadowgraphs, M = 1.33. synchronization delay time $\delta t \approx 40$ µs.

Detailed numerical simulations of this case were undertaken. Figure 9 shows a series of colour schlieren images for a single Mach number and constant difference in time interval between arrival time of the two shocks. This technique clearly defines the vortex centre where the six colours: red, magenta, blue, cyan, green, and yellow meet at a point. At 45 µs the diffraction pattern of the first wave and associated vortex are well developed. This is the fast diffracted shock, FDS, defined earlier. The later arriving second wave is about halfway into the vortex. There is no slow diffracted shock near the wall but rather a jet like flow as described when dealing with Figure 8. As the wave propagates through the diffracted part emerges and meets the blue incident wave in a regular reflection, as shown at 45 μs, with the development of associated reflected waves. The highly curved diffracted wave expands and moves up the upper surface of the wedge. The angle between the incident and diffracted wave becomes larger resulting in the reflection becoming a Mach reflection. The slipstreams that develop from the triple points are wound up into the vortex, with the end of the reflected wave from the upper triple point also merging with the vortex. More detail, at much larger scale, is given in the numerical shadowgraph of Figure 10. The physical area covered in these images is about 7 mm^2 with the vortex being about 0.4 mm in diameter. As the wave enters the vortex field it is distorted more and more, part becoming almost parallel to the shear layer which results from the the earlier shock diffraction. It then develops a cusp with the forming of a regular reflection after having passed through the shear layer. These images also identify the development of the shock between the vortex and the wall, which propagates away from the corner. It is associated with the diffracted wave reflecting off the corner. This interaction, with the jetting effect up the surface, also results in significant distortion of the shear layer.

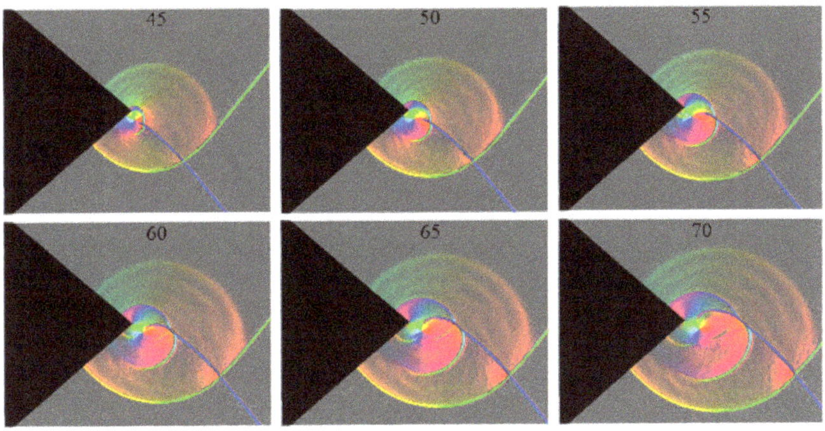

Figure 9. Colour schlieren, M=1.33. synchronization delay time $\delta t = 40$ μs. Numbers refer to time, in μs, since the first shock reached the corner.

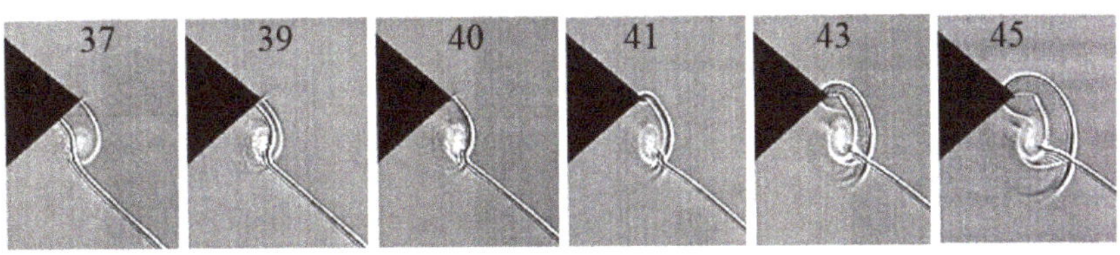

Figure 10. Detail of passage of shock wave through the vortex field.

Previous work dealing with shock interaction with an ideal isolated vortex paid particular attention to the pressure field. Figure 11 shows numerical results for the present case with the spiral vortex from shock diffraction, corresponding to the test case above. The slightly variable yellow region correspond to the gas which has experienced passage by both shocks being modified slightly by the changing strength of the diffraction of the early arriving wave, which also results in a slight curvature of the late arriving wave. This plateau of pressure changes significantly in the region of the vortex, shown in blue. A high pressure region shown in red, grows in space and time behind the reflected wave which results from the appearance of the regular and Mach reflections at the cusp between the incident and diffracted waves. This pressure spike is very close in position to that of the vortex, resulting in extremely high pressure gradients. As the flow progresses the pressure at the intersection between the two incident waves moving into undisturbed gas starts to rise as noted in the figure and can exceed that of the pressure spike developed at the cusp but is not part of the shock-vortex interaction itself. There is also a pressure increase at the wedge corner when the diffracted shock reflects off it.

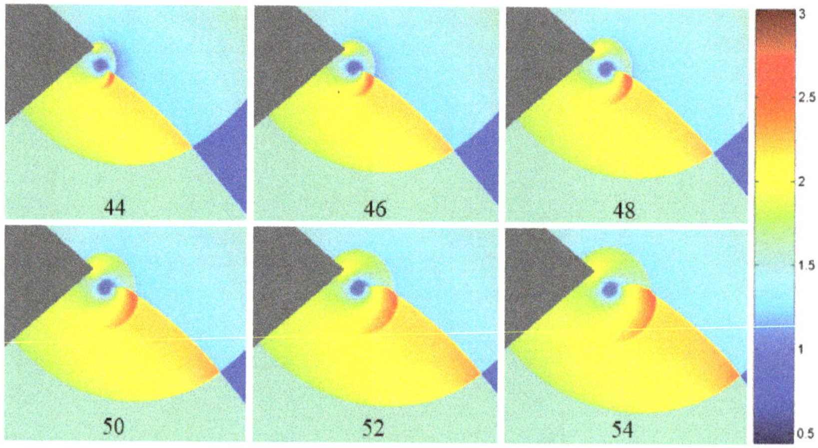

Figure 11. Pressure flood plots, M = 1.33. Synchronization delay time $\delta t = 40$ µs. Image numbers refer to time, in µs, since the first shock reached the corner. Pressure values are factors of the initial pressure of 83 kPa.

Animations of the pressure distributions have been done with Figure 12 showing the pressure surface at 43 µs viewed from different angles and Figure 13 showing evolution in time. The gray wedge surface is set at an arbitrary pressure level. The green area is the uniform pressure behind the plane shock waves. Both pressure spikes are evident.

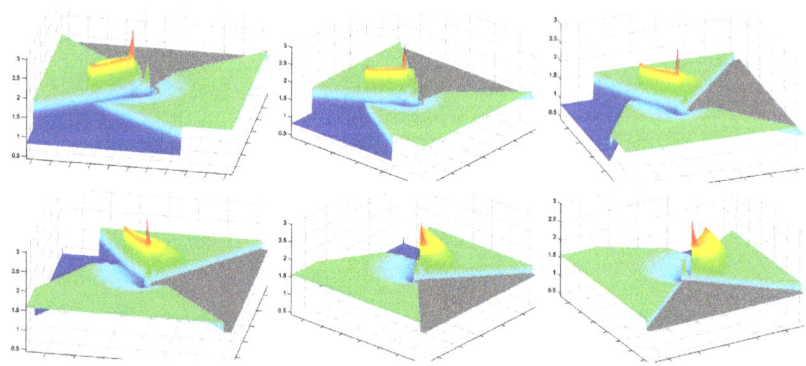

Figure 12. Rotated images of pressure surface at 43 µs. Ambient pressure coloured dark blue and wedge coloured gray.

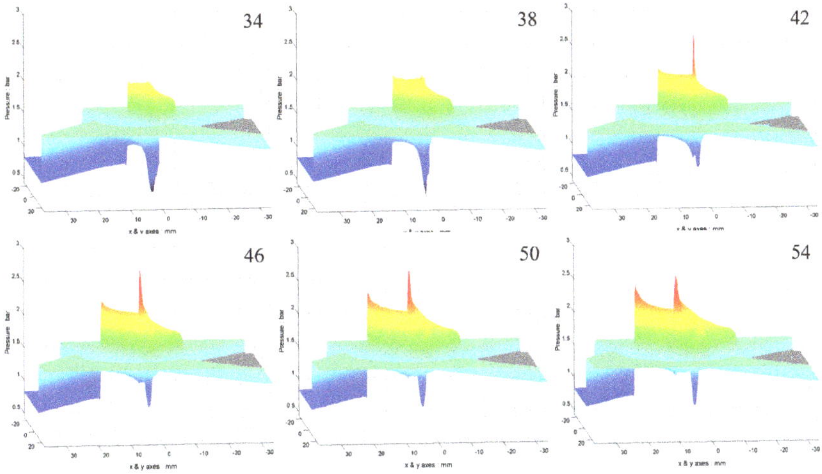

Figure 13. Images of pressure surface variation with time, Ambient pressure coloured dark blue and wedge coloured gray.

Whilst the numerical results provide a comprehensive description of the flow experiments are needed in order to assess their validity. All the main features found in the experiment, shown in Figure 8, correspond to those in the simulation shown in Figure 9. The pressure spike is extremely narrow, about 0.5 mm in diameter at mid-height and of short duration, thus making it difficult to measure experimentally. The track of the position of the peak pressure is obtained from the numerical data and small transducers are positioned in the region of the track. Three transducers each having an outside diameter of 3.8 mm and a sensor element radius of 0.4 mm are arranged in rosette pattern as shown in the inset of Figure 14, and situated in a plate replacing a window in the rig. The position relative to that of both the numerical and experimental spike positions is indicated in the figure. The virtual transducer is the size of the sensing element. Transducer 1 was placed where the pressure spike is a maximum and transducer 2 in the region of the spike trajectory.

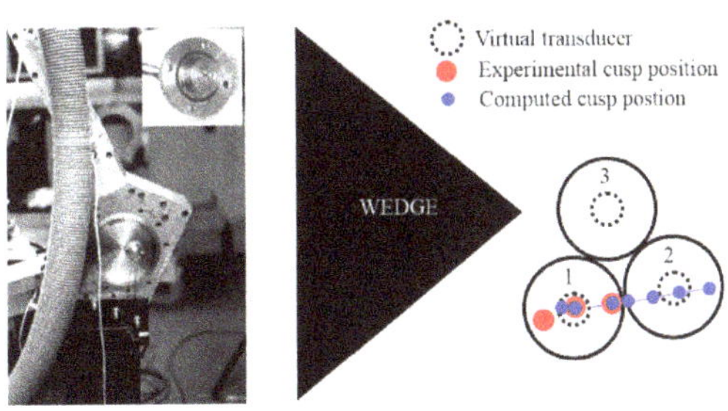

Figure 14. Positioning of the transducers and spike trajectories. Black circles represent transducer physical diameter.

The CFD essentially gives a point measurement of the peak pressure but the transducer is exposed to pressure across its sensing face. To account for this further simulations were done by averaging the pressure over the sensor face area to represent what would be measured by a virtual transducer of 0.4 mm radius. The effect of this averaging comparing the point pressure to the averaged value for comparison to experiment is given in Figure 15. The point pressure traces from the numerical simulation at the positions of the three transducers is given together with the effect of simulations for a transducer of 0.4 mm radius. The point measurement for transducer 2 shows that the pressure spike is about 230% above that behind the diffracted shock wave, at 2.75 bar. Transducer 2 still records the spike but is lower because it is not directly on the track of the spike.

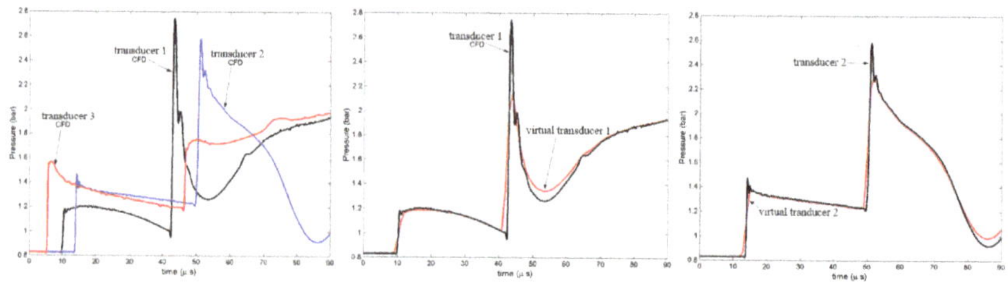

Figure 15. Transducer pressure traces, M = 1.33 δ = 40 µs. (**Left**) CFD for all transducers, (**Right**) Computed virtual transducer traces for transducers 1 and 2 compared to CFD point values.

It is evident that although the virtual transducer, having a finite size, reduces the peak pressure it is sufficient to measure and identify the spike reasonably well. Figure 16 compares the virtual transducer output with that obtained experimentally. There is considerable agreement considering the complexity of the experiment. The spike width and arrival time are well predicted, the former within 5% and the latter within 2 µs.

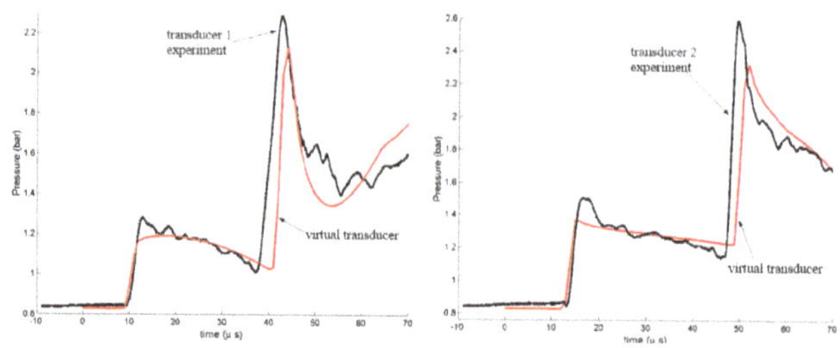

Figure 16. Comparison between experiment and simulation.

Limited items from the above study using the bifurcated tube at Mach 1.33 and $\delta t = 40$ µs have been published in [18]. Preliminary work at higher Mach numbers, particularly when the flow behind the incident waves is supersonic, which occurs with a Mach number graeater than 2.07, raises practical issues since the tube needs to be evacuated. A few results in this region have been done [19]. Results for a variety of Mach numbers are given in Figure 17. The much lower test section density results in some loss of definition in the vortex. In the first image in which the flow behind the shocks is subsonic the overall pattern is similar to the Mach 1.33 case discussed above. However, as the Mach number increases the overall pattern becomes extended in the flow direction. The shear layers from the two triple points merge as they get caught up in the vortex flow. For the higher Mach number cases where the post-shock flow is supersonic the diffraction pattern on a corner is different as given in Figure 2.

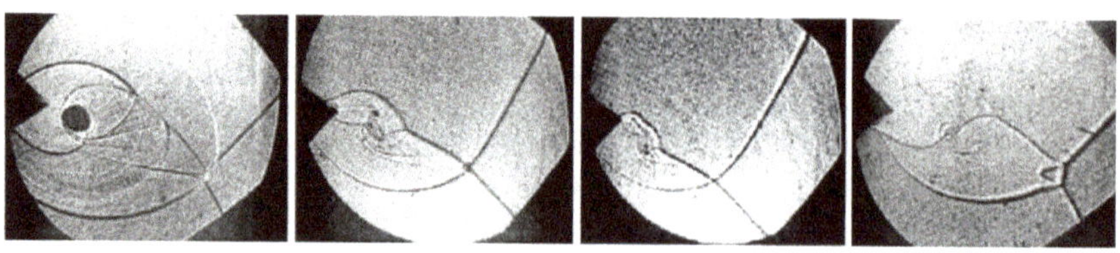

Figure 17. Shadowgraph images. Shock Mach numbers: 1.48, 2.24, 2.3, and 2.5.

At Mach 2.24 only one triple point is clearly evident although there is a second slipstream emanating from the region where the incident shocks intersect. No second reflected shock is evident. The reflected shock propagating on the upper surface is inclined as an oblique shock in a supersonic stream. At the slightly higher Mach number, which was taken at a lower difference between arrival times at the corner, the reflected wave becomes distorted as it meets the oncoming supersonic flow. At Mach 2.5 the incident shock intersection is a strong Mach reflection with shear layers that terminate into a pair of adjacent vortices which curl up rather than moving downstream. This is characteristic of strong Mach reflection from a surface. Since they are not of equal strength a trailing slipstream emanates from where they meet and is caught up in the vortex. In all the supersonic cases the vortex pattern itself is not well defined. This is a case which has not been explored previously and additional tests are required together with numerical simulation in order to establish the details.

3.2. Multiple Shock Passage

The reason for examining the interaction where a vortex is traversed by a shock wave more than once is that it shows possible vortex disruption and occurs in a number of practical cases.

A very interesting experimental case of a complex shock-vortex interaction with multiple transits and a spiral vortex is partly contained in a video 'Waves in Fluids' [20]. It deals with shock wave propagation over a fence. This case did not deal with the shock/vortex interactions at all but was an illustration of complex flows resulting from shock interaction with a simple boundary. A number of frames have been extracted from the video and annotated in Figure 18. The incident wave diffracts over the fence with the curved portion reflecting off the bottom surface and passing through the vortex in the typical S-shaped format, giving rise to the diffracted wave D, emerging from the opposite side just below the shear layer.There are embedded shocks in the vortex on either side of the vortex core. These indicate acceleration to supersonic velocity due to the vortex action followed by deceleration due to the downstream boundary conditions. In the second frame the diffracted wave then passes trough the shear layer and moves towards the top surface. It is circular in shape and reflects off the bottom surface as well as Dr. The reflected wave R is distorted downwards as it encounters the clockwise flow of the vortex. In the region between the vortex and the fence a shock wave V develops, as discussed earlier, moving down the surface in the opposite direction to the vortex flow and reducing the induced supersonic flow to subsonic. In the third frame the wave F is distorted as it passes though the shear layer and then enters the vortex. It passes through and then reflects off the bottom surface as Fr. There are thus multiple transits, first the reflected wave R, then the reflected wave F, followed by the further reflections of waves D and F. There is insufficient detail in the vortex region itself to track the internal flow.

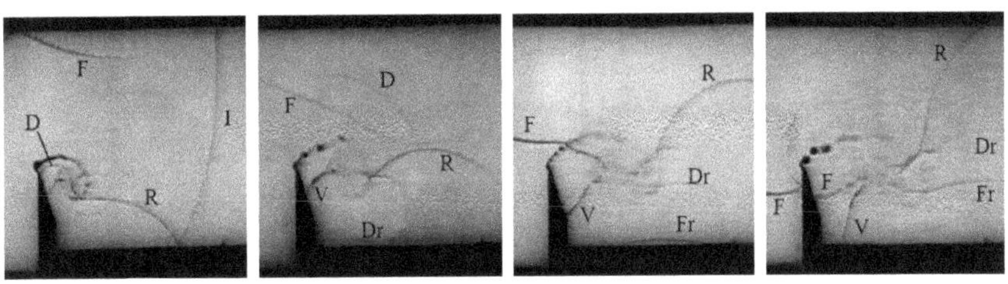

Figure 18. Annotated frames from video of shock wave propagation over a fence [20].

A simulation of this experiment has been done [21]. Figure 19 shows gray-scale images of Mach number and a numerical shadowgraph. These are at a time beyond the end of the video where both waves F and V terminate on the bottom surface. The instability in the shear layer is evident, as it is in the experiment, but most striking is the extreme distortion of the vortex core.

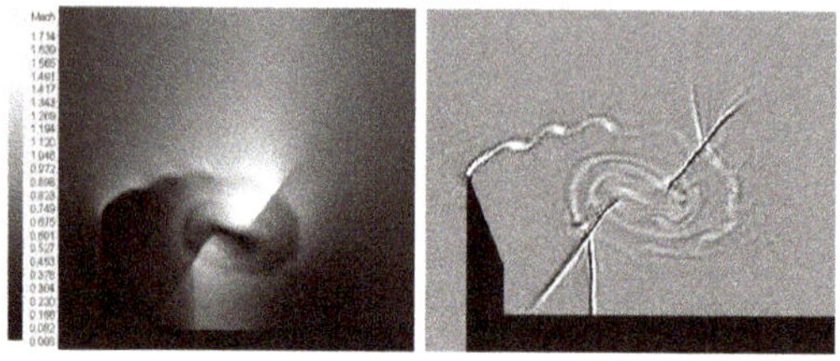

Figure 19. Numerical late stage Mach number flood plot and shadowgraph image.

A more complete set from the simulation is given in Figure 20 showing both density and Mach number flood plots at the corresponding times to the frames from the video. The density plots clearly show the movement of the stronger shock waves, particularly the wave, F, between the vortex and the fence which moves in an opposite direction to the adjacent vortex flow. The Mach number plot emphasizes the extreme distortion of the vortex core. There are distinct supersonic patches on either side of this narrow elongated core. The flow underneath the shear layer also becomes supersonic as it is accelerated by the vortex motion. There are clearly many components of this complex flow deserving future attention.

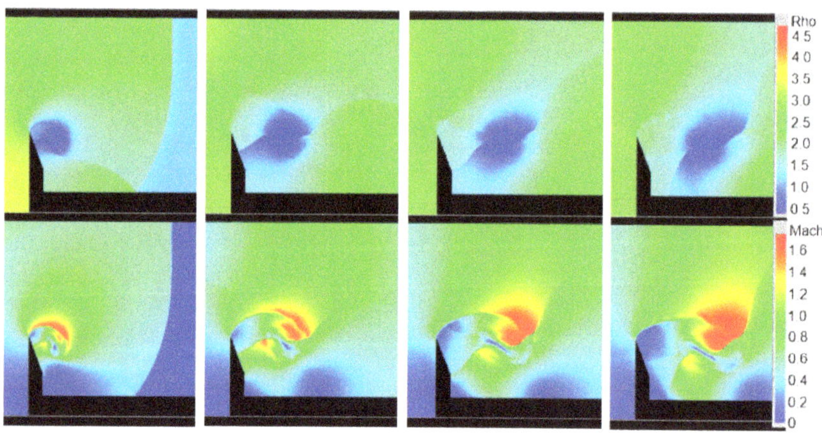

Figure 20. Density and Mach number flood plots corresponding to the images of Figure 18.

Anther practical study showing multiple shock traverses is that of transient flow around a bend, referred to in the introduction. At the stage when that was done the interest was in the behavior of the transmitted wave rather than the flow in the region of the shed vortex. Figure 21 shows the evolution of the interaction for propagation over a 90° bend for incident shocks of Mach 1.5 and 2.0. The first frame, for Mach 1.5, shows a conventional diffraction over the corner with its associated vortex and with the wave reflecting back off the far wall. This wave then passes through the vortex and in the third frame starts reflecting off the near wall as the start of the second transit through the vortex. A typical shock-vortex interaction then occurs with the upper part of the wave accelerated and the lower part slowed down, producing the wave system external to the vortex as discussed before. Similar results occur for Mach 2 with the first frame showing the flow after the first shock transit and the following three frames as it passes through the vortex field. In the final frame a curious shock structure is noted where a shock arising from the reflected shock pattern discussed earlier appears to fan out in space. The spiral vortex region degenerates into an elongated turbulent patch after the double shock transit and remains adjacent to the near wall.

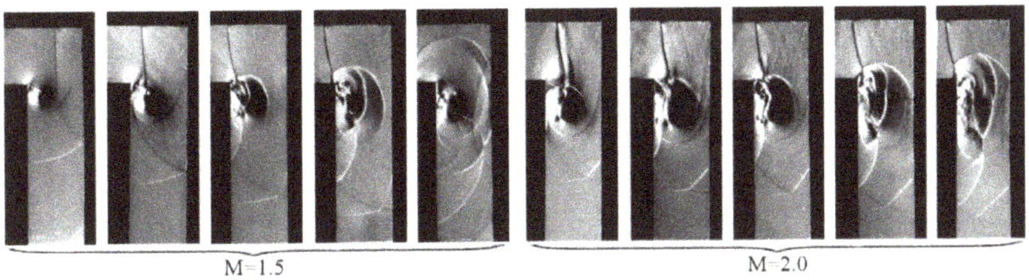

Figure 21. Evolution of the shock-vortex interaction resulting from shock diffraction over a 90° bend.

More details of these features are given in Figure 22 [22]. The fan-shaped structure is circled and is most evident in the experimental and shadowgraph images with the latter suggesting it consists of a shock with a series of weak waves above it resulting from its curvature. The simulations were done using the SAS model in conjunction with the $\kappa - \omega$ SST RANS model. This was found to more closely correspond to the experiment than other models. A vortex center is not well defined and the turbulent patch covers a significant part of the near wall surface below the shear layer emanating from the corner. The turbulent patch elongation does not happen for a Mach 1.5 incident shock wave and is a separated region from the surrounding flow. The fan-shaped flow slows the flow just outside the turbulent patch to allow the flow to follow the highly rotating flow in the turbulent patch. A jetting flow occurs near the wall where the flow is directed upwards towards the corner at supersonic speeds. The shock-fanning was found to be more marked for a Mach 2.5 case. For a Mach 1.5 incident shock, there is no evidence of shock-fanning and the turbulent patch is not elongated along the near wall.

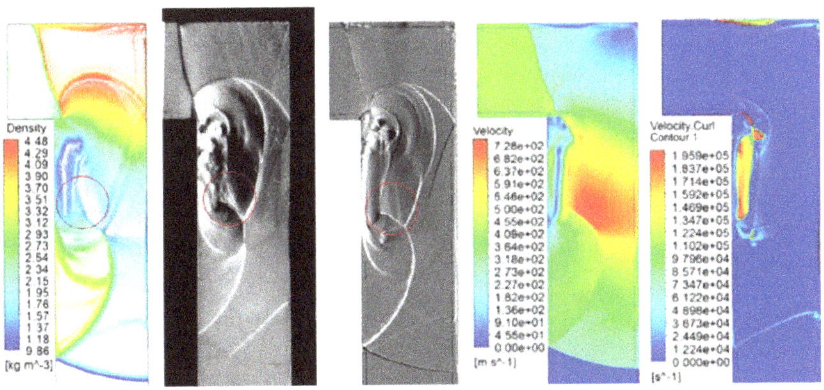

Figure 22. Mach 2 shock propagating around a 90 degree bend. The images in order are: density contour; experiment; numerical schlieren (2 million nodes); velocity; and vorticity.

Less detailed studies have been executed on the effect of bend angle, although experimental results have been obtained. It is important to note that for an incident Mach number of 2.5 the flow following the incident shock is supersonic, for the Mach 2 case is close to being sonic and for Mach 1.5 it is subsonic. This affects the ability for waves to propagate up into the section of the bend before the corner. Figure 23 is for a Mach 2.5 shock around a 120° bend. There is a Prandtl-Meyer expansion wave at the corner which accelerates the oncoming flow limiting passage of a wave upstream. There is also a complex wave reflection at the acute corner at the upper right. This will eventually interact with the vortex flow. The first frame shows the situation after first passage of the reflected shock from the far wall. It has just reached the near wall and is then forced significantly downstream as it passes through the shear layer into the high-speed flow coming from the upper leg of the bend. It turns almost at a right angle with influence on the upper portion of the vortex as particularly noted in the numerical schlieren image.The flow then becomes more complex as shown in the last two plots as well as the images in Figure 24 with increasing turbulence becoming evident.

Figure 23. Mach 2.5 shock propagation around a 120° bend. The images in order are: experiment; density contours; numerical schlieren (1.5 million nodes); later density contour plots.

At Mach 2 the features are similar although the upstream flow is now close to being sonic. For this and the Mach 2.5 case there is an almost normal wave passing down the bend which arises from that in the vortex. During the early stage at Mach 1.5 the reflected wave off the far wall passes through the shear layer and is influenced by the difference in velocity on either side of it as well as that of the vortex flow. The flow then becomes more complex due to multiple reflections from the walls.

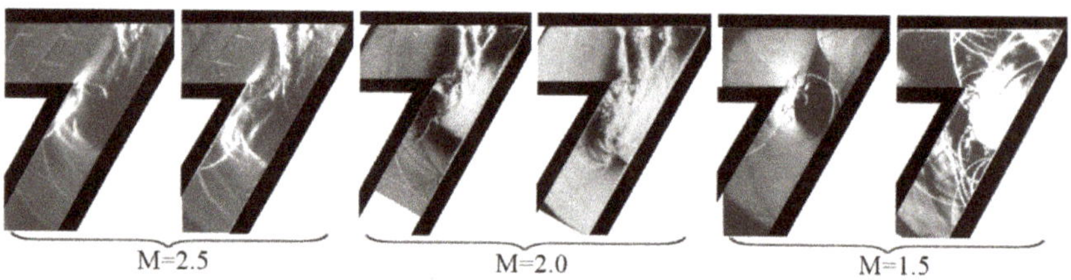

Figure 24. Shock propagation around a 120° bend.

The pattern of behavior for a sixty degree bend is similar to that for a ninety degree bend as shown in Figure 25 for three Mach numbers. The fanning effect, with the second pass of the shock through the vortex, becomes more noticeable at Mach 2.5 and 2.0. It is not evident at Mach 1.5. The turbulent path on the near wall is less elongated for the weaker shocks.

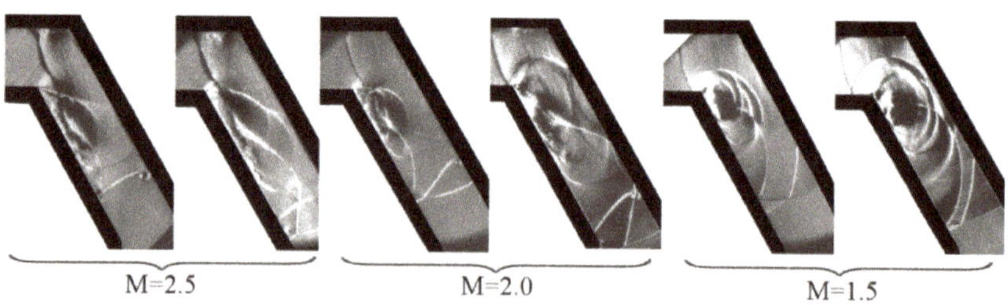

Figure 25. Shock propagation around a 60° bend.

As indicated above the nature of a shock-vortex interaction is influenced by the curvature of the transiting shock wave, the direction from which the shock approaches, and whether there is more than one transit. An experiment, done by some visiting students from Holland, explored some of these issues, particularly that of shock curvature and transit direction. A plane Mach 1.36 shock wave was diffracted over a 170° convex corner, in the rig shown in Figure 4, thereby shedding a spiral vortex. The curved diffracted wave is then reflected off plane reflection surfaces at angles of 60°, 90° and 120° back over the vortex as well as from a curved reflecting surface [23].

Images for the 90° case are given in Figure 26. The reflecting surface is 23 mm from the tip of the wedge and visible to the right in the images. The interaction of the shock with the shear layer results in both undergoing significant bending as well as the formation of a reflected wave. As the shock passes into the vortex it travels against the vortex motion as a slow diffracted shock (SDS) whereas that passing through is accelerated and is pulled around the vortex, as a fast diffracted shock (FDS). These two waves meet in a cusp in the second frame with a reflected wave which would result in a pressure peak as discussed earlier. The FDS then propagates upwards striking the bottom of the slipstream and the wedge surface. It reflects from the bottom surface and diffracts around the tip where it develops into a Mach reflection with its shear layer which is pulled into the vortex. Small vortices develop on the shear layer. The one reflected shock formed at the cusp propagates away and the other, together to the one formed at the cusp, move into the vortex and terminate there. They propagate together downward against the flow direction of the vortex.

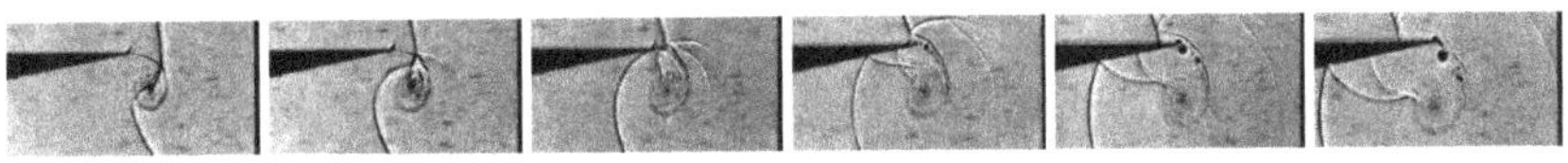

Figure 26. Shock-vortex interaction due to a reflected shock diffraction from a 90° surface.

Results for the 60° and 120° reflecting walls are given in Figures 27 and 28. In the case of all plane reflecting surfaces the vortex becomes more elongated as the surface angle increases, with the major axis tilting in the direction of the wall angle. The 60° case has the same features and development as the 90° case with the development of a cusp and associated reflected waves and refraction at the shear layer. This refraction does not occur for the 120° case since the reflected wave has a similar inclination as the shear layer, as shown in the first frame. A Mach reflection develops on the lower surface of the wedge with a slipstream winding into the vortex.

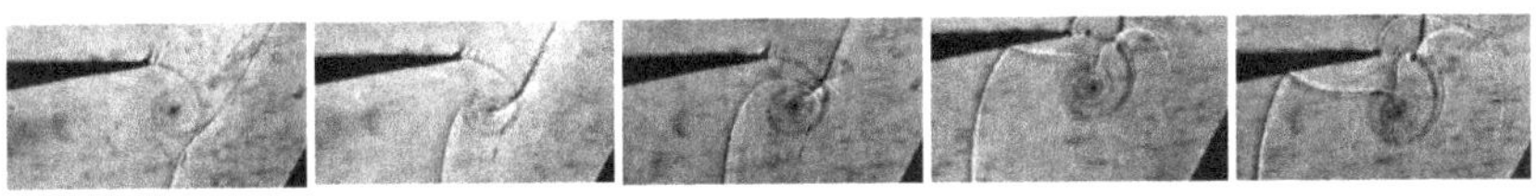

Figure 27. Shock-vortex interaction due to a reflected shock diffraction from a 60° surface.

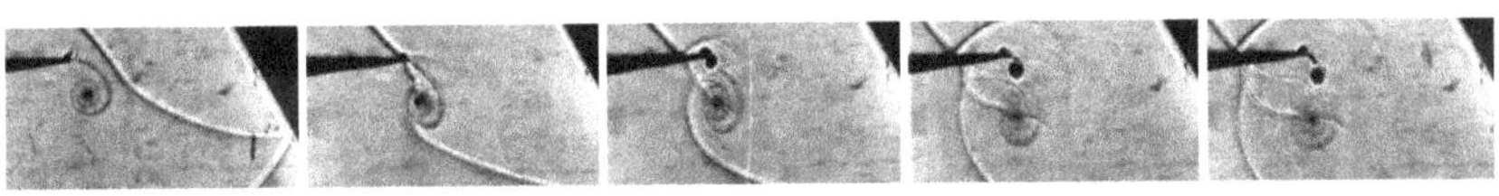

Figure 28. Shock-vortex interaction due to a reflected shock diffraction from a 120° surface.

In order to get some insight into the effect of an imploding shock on the vortex the arrangement shown in the first frame of Figure 29 was used. The shape of the converging reflected diffracted wave off the test piece is shown in the first image. It is refracted at the shear layer which is already wrapping around the vortex. On the opposite side the wave from the test piece also engages the vortex with the development of a reflected wave and a cusp which suggests the development of a pressure peak. There are other cusps resulting from the interaction of the fast diffracted shock with the slow diffracted shock as dealt with previously. The reflected wave from the underside of the wedge then enters the vortex where it dissipates.

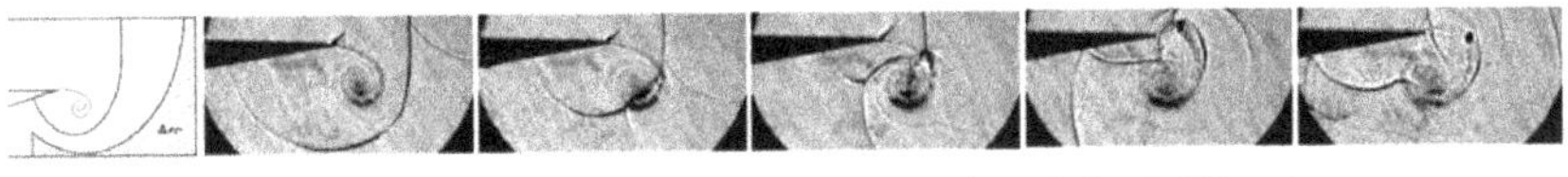

Figure 29. Shock-vortex interaction from a curved surface reflected shock diffraction.

Additional tests were conducted with the idea of using a parabolic reflector to generate a stronger and much more complex incoming wave [21]. This was with a Mach 1.42 incident shock. The arrangement is given in the first frame of Figure 30. Due to cutting off part of the incident shock generated in the shock tube the wave generated in the reflection off the curved cavity surface consists of a strong Mach stem between associated reflected waves and shear layers as shown in the second frame. Due to the higher Mach number the shear layer from the corner is also more complex with shocklets and small embedded vortices. The reflected wave from the top triple point curves slightly as it encounters the shocklets and then bends as it passes through the shear layer. As the triple point passes into the vortex some wave distortion occurs with the fast diffracted shock then emerging and joining the slow diffracted shock in a cusp. Thereafter there is clear evidence of vortex disruption and a complex resulting flow.

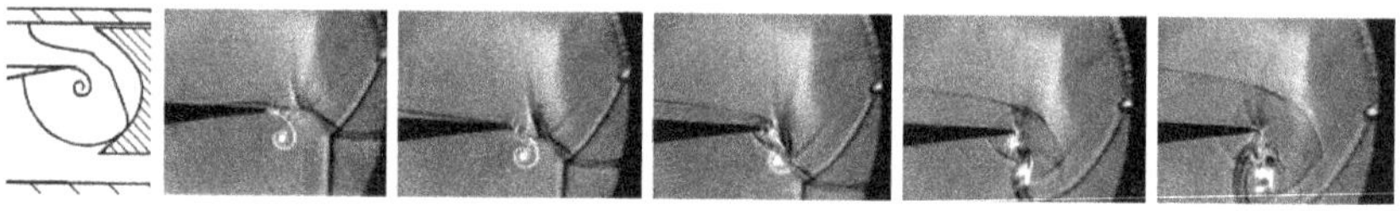

Figure 30. Shock-vortex interaction from a focusing reflected shock.

4. Discussion

The spiral vortex structure of the experiments reported represents a common and realistic flow pattern as is found in practice. It contrasts to previous studies most commonly treated in the literature using an idealized isolated vortex, mainly handled through numerical simulation. The generation of a spiral vortex arises due to the separation of a flow at a corner and is characterized by a shear layer wrapping around into a spiral. Although there are many similarities with previous studies dealing with an isolated idealized vortex there are many significant differences. The incident shock wave is refracted as it passes through the shear layer, changing in both strength and inclination and resulting in shear layer deflection as well. Vortex proximity to a surface has a significant influence inducing jet type flows and additional shocks between it and the surface with the shock propagating in a direction opposite tot he vortex flow.

A cusp is generated as the diffracted waves emerge from either side of the vortex as the incident wave passes through it due to the development of a regular reflection between the waves. This results in a pressure spike, determined experimentally for the first time, and confirmed numerically. The magnitude of the spike can be nearly three times that of the pressure in its surroundings due to the pressue behind the incident shocks, and occupies a very small spatial area. A physical explanation of the shock wave focusing process that generates the spike is elucidated as well as the existence and development of a second spike.

Experiments show that the vortex is extremely stable even when transited more than once by a shock and in many cases it can become highly distorted, often becoming more and more elliptical and extended depending on the approach direction of the shock. There is an indication of disruption when struck by a very complex shock wave pattern. In some cases significant areas of turbulent motion result.

Experimental cases when the incident shock has a Mach number sufficiently high to cause the flow behind it to be supersonic show the interaction to become increasingly complex. This is an area needing further investigation.

The indications are that transit direction and wave curvature are not primary factors in potential vortex disruption. Nevertheless there are clearly consequences of both of these variables on the nature of the interaction, which could then have an effect at later times, or if the vortex is impacted a second or more times.

Funding: This review covers a number of projects primarily done by students over a period of some 25 years, and was partially funded by the South African National Research Foundation through a series of annual grants for material and equipment averaging about $10,000 per year under a scheme 'Competitive Program for Rated Researchers'.

Acknowledgments: This work is a compilation of research over a number of years conducted in the Flow Research Unit of the University of the Witwatersrand. The contributions of the following students are acknowledged: L. Felthun, F Barbosa, G. du Sauytoy, C. Erasmus, U. Kaka and two visiting students from Holland: M Heemskerk and J Kooijmans. The significant input from the School's workshop staff is gratefully acknowledged for the manufacture of the experimental rigs.

Conflicts of Interest: The author declares no conflict of interest.

References

1. Hollingsworth, M.A.; Richards, E.J. *A Schlieren Study of the Interaction between a Vortex and a Shock Wave in a Shock Tube*; Technical Report FM 2323; Aeronautical Research Council, Fluid Motion Subcommittee 17985: Cranfield, UK, 1955.
2. Ribner, H.S. *The Sound Generated by Interaction of a Single Vortex with a Shock Wave*; Technical Report 61, UTIA Reports; UTIA: Springfield, MA, USA, 1959.
3. Ribner, H.S. Cylindrical sound wave generated by shock-vortex interaction. *AIAA J.* **1985**, *23*, 1708–1715. [CrossRef]
4. Dosanjh, D.S.; Weeks, T.M. Interaction of a starting vortex as well as a vortex street with a traveling shock wave. *AIAA J.* **1965**, *3*, 216–223. [CrossRef]
5. Ellzey, J.L.; Henneke, M.R.; Picone, J.M.; Oran, E.S. The interaction of a shock with a vortex: Shock distortion and the production of acoustic waves. *Phys. Fluids* **1995**, *7*, 172–184. [CrossRef]
6. Ellzey, J.L.; Henneke, M.R. The shock vortex interaction: The origins of the acoustic wave. *Fluid Dyn. Res.* **1997**, *21*, 171. [CrossRef]
7. Inoue, O.; Hattori, Y. Sound generation by shock-vortex interactions. *J. Fluid Mech.* **1999**, *380*, 81–116. [CrossRef]
8. Zhang, S.; Zhang, Y.T.; Shu, C.W. Multistage interaction of a shock wave and a strong vortex. *Phys. Fluids* **2005**, *17*, 116101. [CrossRef]
9. Chatterjee, A.; Vijayaraj, S. Multiple sound generation in interaction of shock wave with strong vortex. *AIAA J.* **2008**, *46*, 2558–2567. [CrossRef]
10. Chang, K.S.; Barik, H.; Chang, S.M. The shock-vortex interaction patterns affected by vortex flow regime and vortex models. *Shock Waves* **2009**, *19*, 349–360. [CrossRef]
11. Skews, B.W. An experimental study of the interaction of shock waves with bends in a duct. In *Symposium on Internal Flows*; Salford University: Manchester, UK, 1971; pp. D41–D45.
12. Skews, B.W. Shock/vortex interaction in shock wave propagation around sharp-angled bends. In *2nd International Workshop on Shock/Vortex Interaction*; Takayama, K., Jiang, Z., Eds.; Shock Wave Research Center, Tohoku University: Sendai, Japan, 1998; pp. 140–148.
13. Skews, B.W. The perturbed region behind a diffracting shock wave. *J. Fluid Mech.* **1967**, *29*, 705–719. [CrossRef]
14. Timofeev, E.V.; Voinovich, P.A.; Takayama, K. On shock-vortex interaction in a rectangular channel bend. In *3rd International Workshop on Shock-Vortex Interaction*; Higashino, F.; Takayama, K.; Timofeev, E., Eds.; Shock Wave Research Center, Tohoku University: Kanagawa, Japan, 1999; pp. 87–96.
15. Felthun, L. Finite Element Analysis of Compressible Flows. Master's Thesis, Mechanical Engineering, University of the Witwatersrand, Johannesburg, South Africa, 1995.
16. Barbosa, F.; Skews, B.W.; Felthun, L. Reflection of plane shock waves and vortex-shock wave interaction studies using holographic interferometry and high speed videography. In *Proceedings of the 21st International Symposium on Shock Waves*; Houwing, A.F.P., Ed.; University of Queensland, Department of Mechanical Engineering: Great Keppel Island, Australia, 1997; Number 5288.
17. Barbosa, F.; Skews, B.W. Experimental investigations of shock wave-vortex interactions. In *2nd International Workshop on Shock-Vortex Interactions*; Takayama, K., Ed.; Shock Wave Research Center, Tohoku University: Mount Zao, Japan, 1998; pp. 161–172.
18. Barbosa, F.J.; Skews, B.W. Shock wave interaction with a spiral vortex. *Phys. Fluids* **2001**, *13*, 3049–3060. [CrossRef]
19. Skews, B.; du Sautoy, G. Strong shock/vortex interactions in a bifurcated shock tube. In *4th International Workshop on Shock Wave/Vortex Interaction*; Jiang, Z.L., Ed.; Institute of Mechanics, Chinese Academy of Sciences, China: HuangShan, China, 2001; pp. 19–26.
20. Bryson, A.E. *Waves in Fluids*; National Committee on Fluid Mechanics Films; NSF: Chcago, IL, USA, 1964.
21. Skews, B.W.; Erasmus, C. Interaction of complex shock geometry with a spiral vortex. In *5th international Workshop on Shock Wave/Vortex Interactions*; Liang, S.M., Ed.; National Cheng Kung University: Kaohsiung, Taiwan, 2004; pp. 1–6.
22. Kaka, U. *Shock Wave Propagation around Bends*; Final year project report; University of the Witwatersrand: Johannesburg, South Africa, 2019.
23. Skews, B.W. Curved shock wave interaction with a spiral vortex. In *23rd International Symposium on Shock Waves*; Lu, F., Ed.; University of Texas at Arlington: Fort Worth, TX, USA, 2002; pp. 1317–1322.

fluids

MDPI

Article

Impact of High Inertia Particles on the Shock Layer and Heat Transfer in a Heterogeneous Supersonic Flow around a Blunt Body

Andrey Sposobin and Dmitry Reviznikov *

Moscow Aviation Institute, National Research University, Volokolamskoye sh. 4, 125993 Moscow, Russia; spise@inbox.ru
* Correspondence: reviznikov@mai.ru

Abstract: One of the most important and complex effects associated with the presence of particles in the flow is the gas-dynamic interaction of particles with the shock layer. Of particular interest is the intensification of heat transfer by high inertia particles rebounding from the surface or by the products of erosion destruction, which reach the front of the bow shock wave and violate the gas-dynamic structure of the flow. In this case, according to experimental data, the increase in heat fluxes is much greater than it could be predicted based on the combined action of the kinetic energy of particles and a high-speed flow. The problem is related to the destruction of the flow structure. In this paper, the problem is studied with numerical simulation. We show that the key role in the intensification of heat transfer is played by the formation of an impact jet flowing onto the surface. An area of increased pressure and heat flux is formed in the zone of action of the impact jet. This effect is maintained over time by the successive action of particles.

Keywords: heterogenous supersonic flow; gasdynamical particle—shock wave interaction; numerical simulation

Citation: Sposobin, A.; Reviznikov, D. Impact of High Inertia Particles on the Shock Layer and Heat Transfer in a Heterogeneous Supersonic Flow around a Blunt Body. *Fluids* **2021**, *6*, 406. https://doi.org/10.3390/fluids6110406

Academic Editor: Olga A. Azarova

Received: 8 October 2021
Accepted: 4 November 2021
Published: 9 November 2021

1. Introduction

Flows with suspended solid or liquid particles (heterogeneous flows) are present in many applications. Heterogenous flows are used in some technologies and structures for transport and energy purposes, such as the pneumatic conveying of bulk materials, the jet-abrasive treatment of machine parts, and for rocket engines running on metalized fuel. In many devices, particles enter the fluid flow against the will of the developers. For example, we can cite steam and gas turbine installations and gas-dynamic research stands, where it is problematic to get rid of particles completely despite the use of powerful filtration systems. In the atmosphere of the Earth and other planets, there is dust or clouds containing liquid (rain) or solid (snow, hail) particles. Therefore, one of the important problems of high-speed flight in the lower atmosphere is overcoming areas with different dispersion formations.

Various aspects of heterogeneous flows and their interaction with bodies are considered in a number of monographs and reviews [1–6]. From the point of view of the effect of a heterogeneous flow on an obstacle, the following main mechanisms can be distinguished: shock action, leading to additional heating and, possibly, erosion of the surface, enhancement of the convective heat transfer, and the radiative heat transfer between a dispersed phase and the body surface. It should be noted that these mechanisms are interrelated. The intensity of the erosion depends on the temperature of the material, which is formed under the action of the convective and radiative heat transfer. On the other hand, the release of the erosion products into the flow and changes in the body's shape due to erosion affect the heat transfer processes [2,7].

In this work, we focus on the issues of convective heat transfer enhancement. A number of new effects complicate the convective heat transfer in a heterogeneous supersonic flow around bodies compared to the traditional dust-free gas flow. Among them are the intensification of heat transfer due to an interphase energy exchange in the boundary layer, the acceleration of the laminar-turbulent transition, and the intensification of heat transfer due to the formation of craters on the exposed surface. The issues of heat transfer intensification in heterogeneous flows around bodies are considered in [8–13] for various inertial properties of particles and modes of particle deposition on the surface. Heat transfer enhancement caused by craters on the surface is discussed in [14,15].

Of particular interest is the intensification of heat transfer by high inertia particles rebounding from the surface or by the products of erosion destruction, which reach the front of the bow shock wave and violate the gas-dynamic structure of the flow. In this case, according to the experimental data [16–20], the increase in heat fluxes is much greater than it could be predicted based on the combined action of the kinetic energy flux of particles and the high-velocity flow. Thus, it is impossible to explain the increase in the model heating by the complete transition of the kinetic energy of particles into thermal energy. The problem is related to the destruction of the flow structure. This case is challenging for the numerical simulation due to the significant difference in scales between the body and particles.

In our previous work, we developed algorithms for the numerical simulation of a supersonic flow around bodies, taking into account the gas-dynamic interaction of the shock layer with a high inertia particle [21–24]. Distinctive features of our technique are the use of high-resolution adaptive sliding Cartesian grids, the immersed boundary ghost cell method for boundary conditions treatment, and the parallelization of computations on GPUs.

We used the developed computer model to carry out a series of computational experiments aimed at identifying characteristic shockwave and vortex structures formed when a single particle reflected from the surface passes through the bow shock wave [21,22]. Variants of flow around a cylinder with spherical bluntness and a flat end have been investigated. Detailed spatio-temporal pictures of the gas-dynamic interaction of the disturbed region in the vicinity of the particle with the macroscopic flow in the shock layer and the bow shock were obtained. It is shown that the shock wave and vortex flow structures are to a certain extent similar to those observed in the flow around spiked bodies [25,26]. Furthermore, a study of the oscillatory flow and heat transfer regimes induced by the gas-dynamic interaction of a high inertia particle with the shock layer was carried out. The shockwave structures and oscillation frequencies obtained through numerical simulation [23] agree well with experimental data [17].

The numerical simulation [24] showed significant growth of the heat flux even under the gas-dynamic action of a single particle. However, it lasts for a relatively short period and does not lead to overall heat transfer intensification. The next question is whether several particles' subsequent actions can keep the high heat flux level over time. This effect is the subject of the present article.

2. Model and Methods

In our previous work [21–24], we considered the gas-dynamic interaction of a single particle with a shock layer in the flow around an axisymmetric body. In this case, the particle moved strictly along the axis of symmetry, which made it possible to solve the problem in a two-dimensional formulation. This is of principal importance since modeling the gas-dynamic interaction requires a high grid resolution near a moving particle. Even in a two-dimensional version, solving the problem requires enormous computational costs. The deviation of a particle from the axis of symmetry violates the axisymmetric structure of the flow and requires three-dimensional modeling. Considering that in this work we studied the collective action of a group of particles on the shock layer, we were forced to simplify the problem by considering the process in a two-dimensional formulation (plane flow). Of course, it was difficult to talk about the quantitative agreement between the results and experimental data with this approach. However, it was possible to trace the qualitative features of the group effect of particles on the flow structure and heat transfer.

Thus, a supersonic flow around a flat, blunt body was considered. The flow was assumed to be laminar. Particles successively left the body surface towards the flow. Figure 1 represents the computational domain.

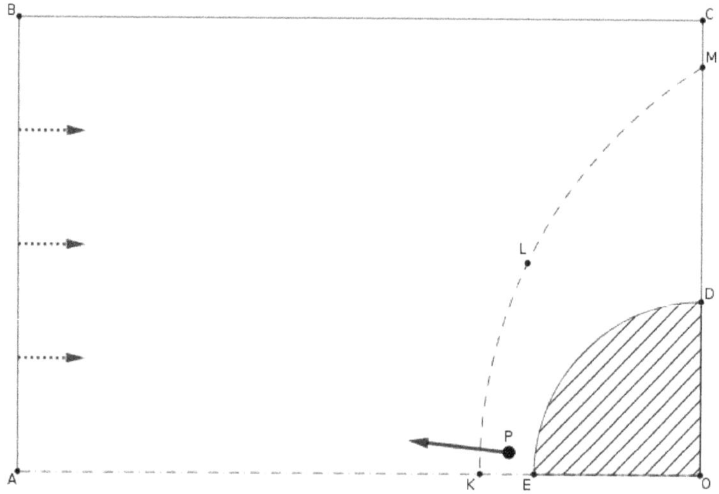

Figure 1. The computational domain: AB is the input boundary, BCD is the output boundary, ODE is a circular cylinder, P is a particle, and KLM is the shock front at the initial moment.

The flow of a viscous compressible gas is described by a system of two-dimensional unsteady Navier–Stokes equations:

$$\frac{\partial \mathbf{q}}{\partial t} + \frac{\partial \mathbf{F}(\mathbf{q})}{\partial x} + \frac{\partial \mathbf{G}(\mathbf{q})}{\partial y} = \frac{\partial \mathbf{F}^v(\mathbf{q})}{\partial x} + \frac{\partial \mathbf{G}^v(\mathbf{q})}{\partial y}$$

$$\mathbf{q} = \begin{pmatrix} \rho \\ \rho u \\ \rho v \\ \rho e \end{pmatrix}, \mathbf{F} = \begin{pmatrix} \rho u \\ \rho u^2 + p \\ \rho u v \\ \rho u H \end{pmatrix}, \mathbf{G} = \begin{pmatrix} \rho v \\ \rho u v \\ \rho v^2 + p \\ \rho v H \end{pmatrix}$$

$$\mathbf{F}^v = \begin{pmatrix} 0 \\ \tau_{xx} \\ \tau_{xy} \\ \tau_{xx} u + \tau_{xy} v - q_x \end{pmatrix}, \mathbf{G}^v = \begin{pmatrix} 0 \\ \tau_{yx} \\ \tau_{yy} \\ \tau_{yx} u + \tau_{yy} v - q_y \end{pmatrix}$$

where t is time, ρ is the gas density, p is pressure, T is temperature, u and v are gas velocity components along the x and y axes, γ is the specific heat ratio, $e = \frac{p}{\rho(\gamma-1)} + \frac{1}{2}(u^2 + v^2)$—total specific energy, $H = e + \frac{p}{\rho}$—total enthalpy, and q_x and q_y are heat flux components.

The equation of the state connecting the gas parameters has the form: $p = \rho RT$.

The viscous stress tensor components are: $\tau_{xx} = \frac{2}{3}\mu\left(2\frac{\partial u}{\partial x} - \frac{\partial v}{\partial y}\right)$, $\tau_{yy} = \frac{2}{3}\mu\left(2\frac{\partial v}{\partial y} - \frac{\partial u}{\partial x}\right)$, $\tau_{xy} = \tau_{yx} = \mu\left(\frac{\partial u}{\partial y} + \frac{\partial v}{\partial x}\right)$.

The viscosity coefficient is calculated using Sutherland's formula, and the thermal conductivity coefficient is defined from the viscosity coefficient and the Prandtl number, which is supposed to be constant and equal to 0.72.

To complete the problem formulation, we set the conditions on the boundaries of the computational domain.

The conditions at the input boundary are the following:

$$\frac{\partial \rho}{\partial \mathbf{n}} = 0,\ u = u_\infty,\ v = 0,\ \frac{\partial p}{\partial \mathbf{n}} = 0,$$

where u_∞ is the free flow velocity and **n**—normal vector to the boundary.

The conditions at the output boundary are:

$$\frac{\partial \rho}{\partial \mathbf{n}} = 0,\ \frac{\partial u}{\partial \mathbf{n}} = 0,\ \frac{\partial v}{\partial \mathbf{n}} = 0,\ \frac{\partial p}{\partial \mathbf{n}} = 0$$

On the cylinder surface, we set standard boundary conditions:

$$\frac{\partial p}{\partial \mathbf{n}} = 0,\ u = 0,\ v = 0,\ T = T_w$$

where T_w is the surface temperature.

The same conditions were set at the particle boundaries.

At the initial moment, the first particle started from the body surface against the incident flow along the normal to the surface; the gas flow pattern at this moment corresponded to the stationary regime of the transverse flow around a circular cylinder. Further, other particles were sequentially launched from close by but at different points on the surface.

The motion of the particles in a gaseous medium is described by the classical dynamic equations:

$$\frac{d\mathbf{r}_p}{dt} = \mathbf{v}_p,\ m_p\frac{d\mathbf{v}_p}{dt} = \mathbf{f}_D$$

where m_p $\mathbf{r}_p$, $\mathbf{v}_p$ are the particle mass, position vector, and velocity, and $\mathbf{f}_D$ is the drag force.

In this work, the drag force was calculated by integrating the gas pressure distribution over the particle surface. A comparison with experimental data [21] showed that, although this approach ignored the friction force, it allowed a much more accurate description of the particle dynamics in comparison with the known criterion dependencies for the drag coefficient.

Thus, the Navier–Stokes equations were solved in a complex region with curvilinear, movable boundaries determined by the motion of particles. This significantly distinguishes this approach from the traditional Euler–Lagrangian approach, where the interphase interaction is taken into account in additional exchange terms.

The Navier–Stokes equations were solved using the TVD—monotonized second-order scheme in combination with the AUSM + (Advection Upstream Splitting Method Plus) method for calculating fluxes through the faces of the computational cell [27–29]. Discretization of the Navier–Stokes equations was performed on a rectangular adaptive grid. The boundary conditions were approximated according to the immersed boundary ghost cell method [7,30–32].

We used sliding grids [33–35] to take into consideration the motion of particles. Along with the primary coordinate system associated with the cylinder, local coordinate systems were introduced that were attached to each moving particle. The gas flow around each object was calculated on a separate "local" computational grid in its coordinate system. The coordinate system associated with the cylinder was considered to be stationary, its computational grid is hereinafter referred to as "primary", and the conditions at its input boundary were determined by the parameters of the incident flow. The boundary conditions for the local grid attached to the particle were determined by the gas parameters obtained on the primary grid depending on the particle position and velocity. At each step of the calculation on the primary grid, the gas parameters in the outer cells of the moving grid were calculated using bilinear interpolation since the centers of the cells of the two grids were usually displaced relative to each other. The gas-dynamic equations were solved in the local coordinate system of a moving particle. Its displacement was calculated, and the obtained gas parameters in the inner region were transferred to the primary grid using the inverse transformation. Figure 2 schematically shows the position of the local computational grid relative to the main one at different points in time.

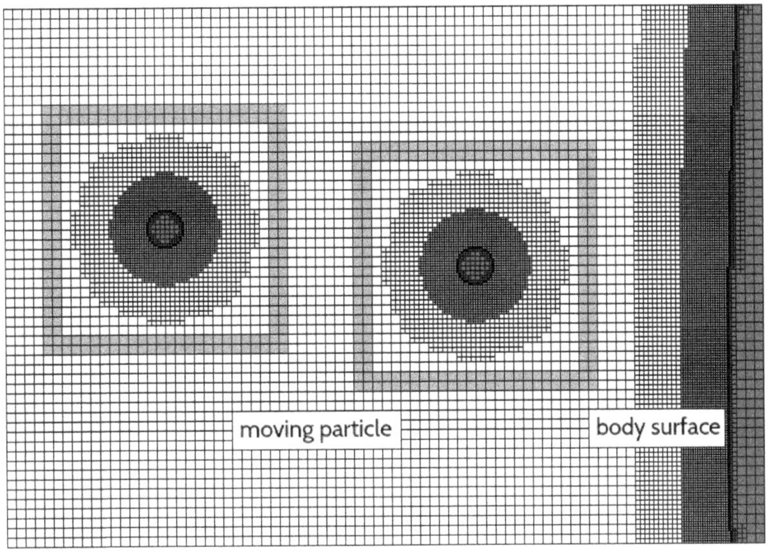

Figure 2. Sliding grids.

3. Results and Discussion

In computational experiments, we simulated a transverse flow of supersonic air around a circular cylinder. The particles were sequentially launched from the cylinder surface. Each particle passed the shock layer, went beyond the bow shock, where it was decelerated by the incident flow, turneds around, and continued to move towards the model.

In Figure 3a the initial, unperturbed state is shown when the particles do not yet affect the shock layer. In the Schlieren image, the detached bow shock wave is clearly visible. In Figure 3b we show the final state when the particles (colored lines show the particle trajectories for the variant with three particles) return to the shock layer, and, despite the presence of local disturbances in their vicinity, they also practically do not affect the flow. The subject of our study is the period between these states. The free flow and particle parameters are given in Table 1.

Table 1. Free flow and particle parameters.

Free Flow Parameters		Particle Parameters	
Mach number	6	Diameter, mm	0.2
Reynolds number	1.09×10^6	Density, kg/m^3	2170
Cylinder diameter, mm	75	Initial velocity, m/s	130–140
Velocity, m/s	1150		
Density, kg/m^3	0.094		
Temperature, K	89.3		

The gas parameters corresponded to the experimental ones [17]. Note that the particles left the surface towards the flow with initial velocities corresponding to the particle reflection from the surface for the case when the particle initially moved in the incoming flow with a horizontal velocity of 880 m/s and reflected from the surface with the recovery coefficient of the normal velocity component equal to 0.15. The magnitude of the initial particle velocity varied depending on the initial vertical displacement of the particle relative to the axis.

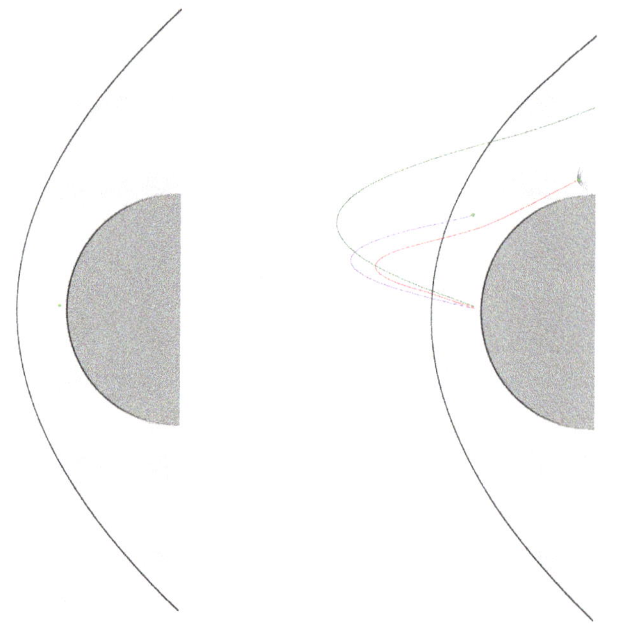

Figure 3. Schlieren images at the starting (**a**) and final (**b**) times.

The computational area was a rectangle 0.125 × 0.2 m divided into 1250 × 2000 large cells. To resolve the boundary layer at the surface, the mesh was refined. As a result, the mesh had cells of six characteristic sizes with sides ranging from 3.125×10^{-6} m to 10^{-4} m. The local grids attached to particles were also adaptive. The total number of cells in the computational grid aws about seven million. The solution to the problem was carried out in the parallelization mode of computations on GPU graphics processors using the OpenCL technology.

Let us first consider the variant of the passage of the bow shock wave by a single particle. Figure 4 shows the evolution of the shock layer. It can be seen that when a particle crosses a shock wave, the stationary shock wave structure is destroyed, and a cone-shaped disturbed region with a vertex moving with the particle is formed. The formation of a complex shock wave and vortex flow structure was analyzed in detail in our previous work [21,22]. From the point of view of the effect of the flow on a body, the fundamental moment is the formation of an impact jet directed towards the surface. In Figure 4a–c it is clearly seen how such a jet is formed in the zone of the lower λ-configuration. In the zone of action of the impact jet, an area of increased pressure is formed on the surface. This is reflected in the intensity of heat transfer. The distributions of the pressure and heat flux along the surface at successive times are shown in Figure 5. Here, all quantities are referred to the values at the critical point for an unperturbed flow. The initial distributions of the pressure and heat flux are shown by curve 1. One can see the appearance of a region of increased heat transfer in the vicinity of the critical point, where the heat flux is more than twice the value in pure gas (curve 2). With time, the increased pressure and heat transfer region shift downstream (Figures 4e,f and 5, curves 3, 4). As a result, the periods of increased heat transfer at a certain position on the surface are changed by periods of a significant decrease in heat flux. Thus, the action of a single particle does not lead to an increase in the integral (over time) heat flux.

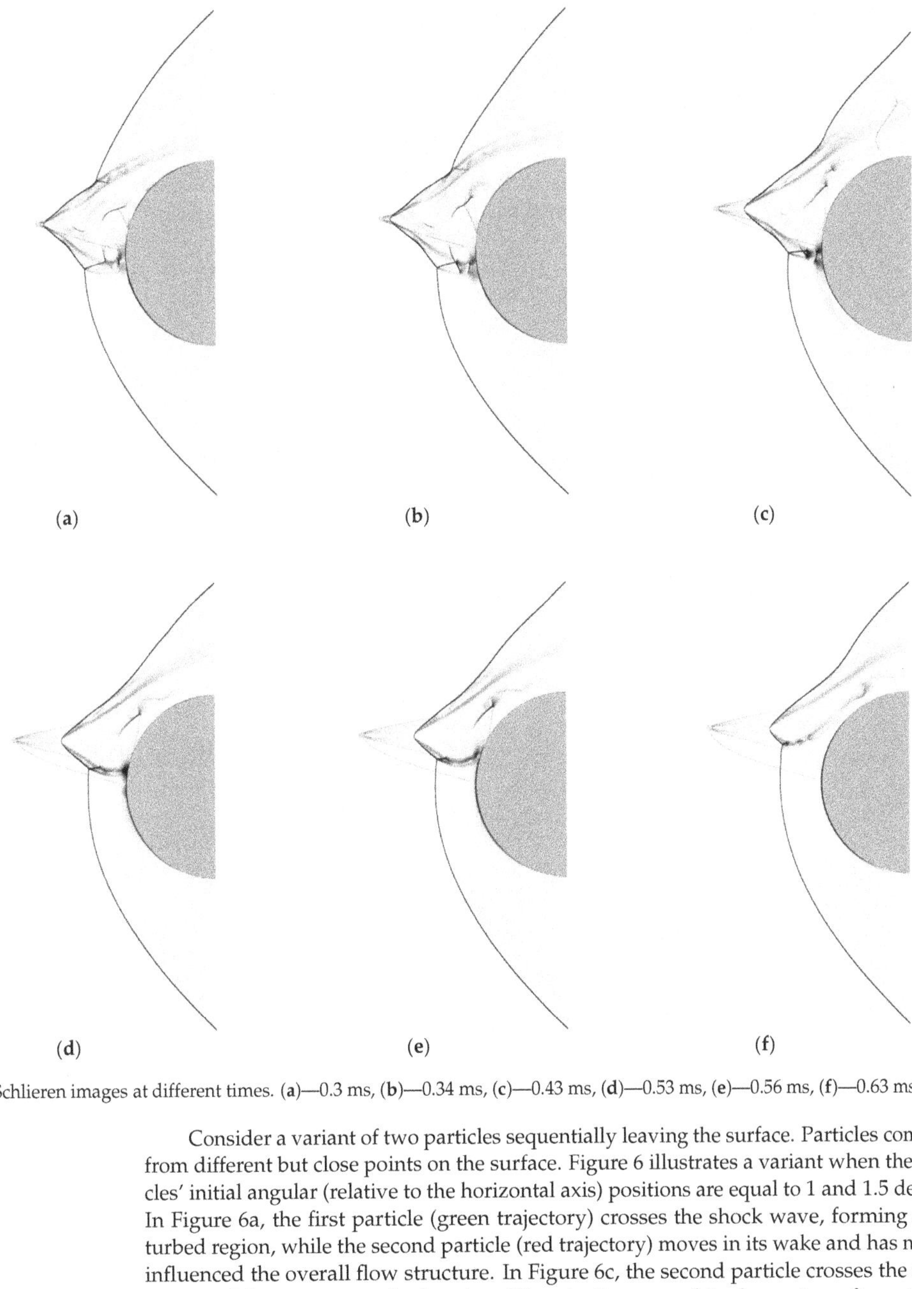

Figure 4. Schlieren images at different times. (**a**)—0.3 ms, (**b**)—0.34 ms, (**c**)—0.43 ms, (**d**)—0.53 ms, (**e**)—0.56 ms, (**f**)—0.63 ms.

Consider a variant of two particles sequentially leaving the surface. Particles come out from different but close points on the surface. Figure 6 illustrates a variant when the particles' initial angular (relative to the horizontal axis) positions are equal to 1 and 1.5 degrees. In Figure 6a, the first particle (green trajectory) crosses the shock wave, forming a perturbed region, while the second particle (red trajectory) moves in its wake and has not yet influenced the overall flow structure. In Figure 6c, the second particle crosses the shock wave and forms its perturbed region. Here, in the zone of the lower λ-configuration, a supersonic jet directed to the surface is clearly visible. Figure 6c,d illustrates the combined hydrodynamic effect of particles on the shock wave structure of the flow. It is characteristic that the zone of action of the impact jet on the surface remains fairly stable during the considered time interval. A similar picture is observed for another variant with two particles, whose initial angular positions are shifted to 2 and 2.5 degrees. This case is shown in Figure 7. Here the second particle moves in the region of intense wave action of the first particle, and, as a consequence, its distance outside the shock wave is less pronounced than the distance of the first particle. This distinguishes this case from the one considered in Figure 6. However, similar to the first variant, there is a stable zone of action of the impact jet on the body surface. This expresses itself in a relatively stable zone of increased pressure and heat transfer in the vicinity of the critical point, which is seen in the graphs of the pressure and heat flux distributions along the surface (Figure 8). Note that the intensification of heat transfer in the case of the two particles is more pronounced than in the case of a single particle. It can be seen that the heat flux increases more than three times.

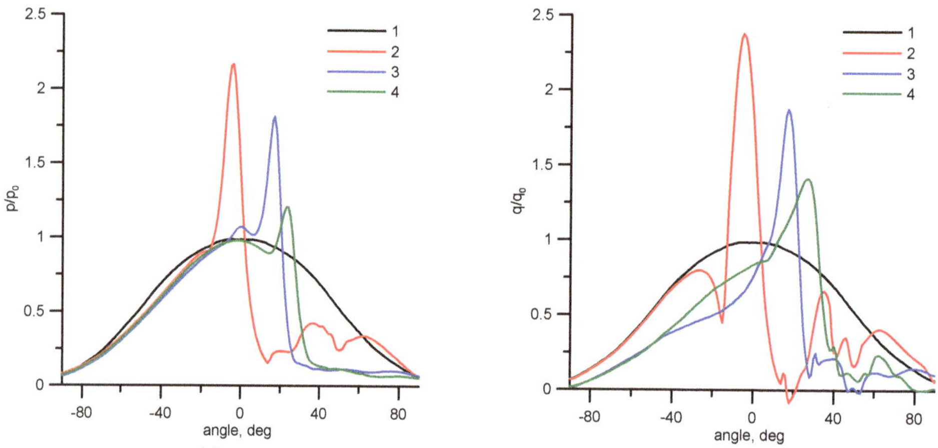

(**a**) (**b**)

Figure 5. Distributions of pressures (**a**) and heat fluxes (**b**) along the surface at successive times for the variant with one particle. 1—initial moment, 2—0.3 ms, 3—0.53 ms, 4—0.63 ms.

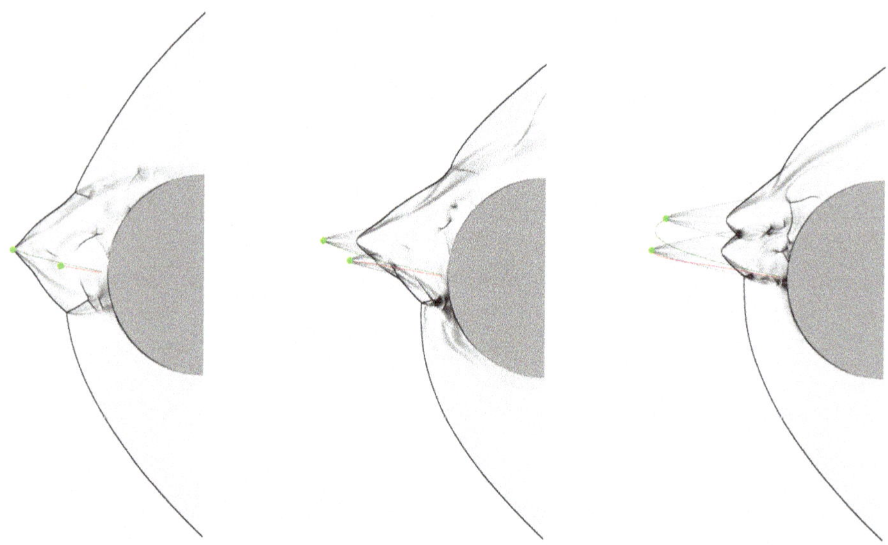

(**a**) (**b**) (**c**)

Figure 6. Schlieren images at successive times. (**a**)—0.3 ms, (**b**)—0.47 ms, (**c**)—0.87 ms. Variant 1 with two particles.

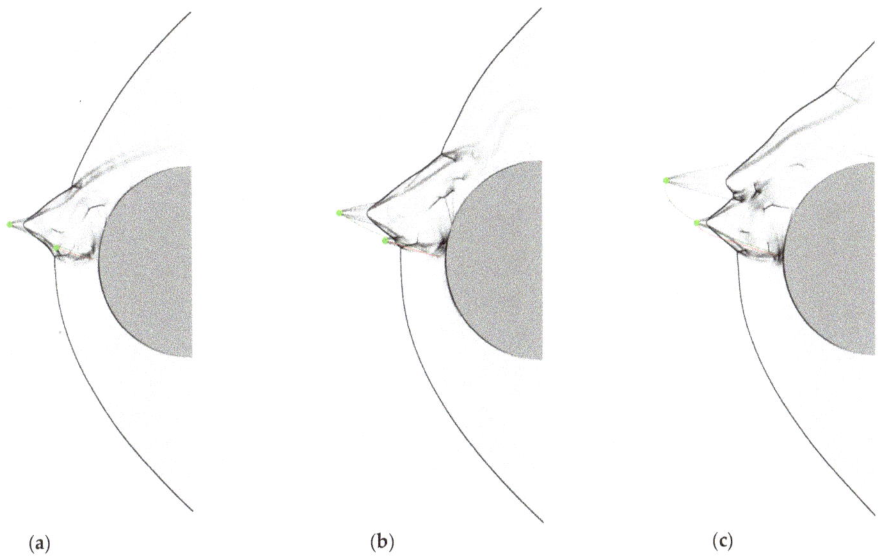

Figure 7. Schlieren images at successive times. (**a**)—0.3 ms, (**b**)—0.38 ms, (**c**)—0.69 ms. Variant 2 with two particles.

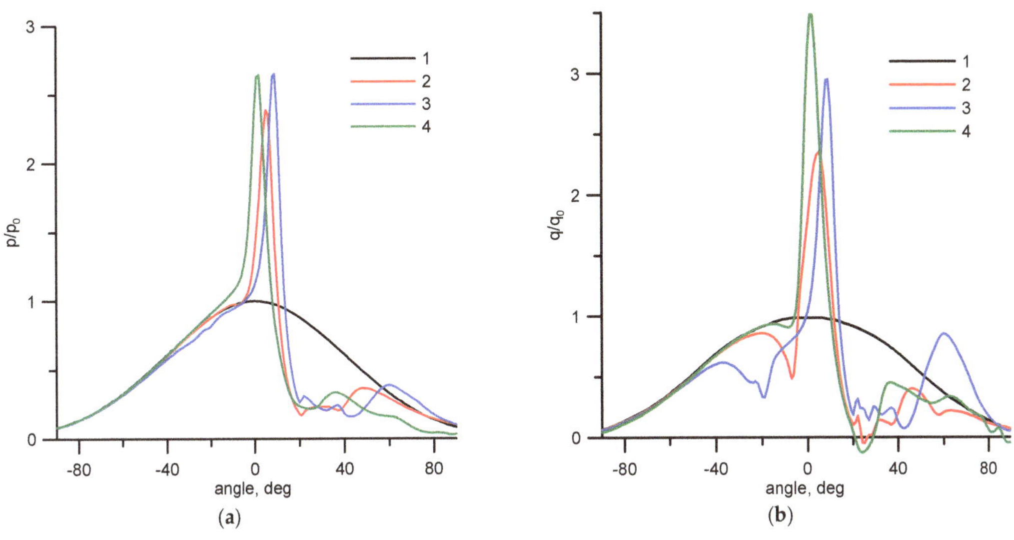

Figure 8. Distributions of pressures (**a**) and heat fluxes (**b**) along the surface at successive times for variant 2 with two particles. 1—initial moment, 2—0.3 ms, 3—0.38 ms, 4—0.69 ms.

Figure 9 shows a variant with three particles sequentially launched from the surface. Figure 9a corresponds to the time when the first two particles leave the shock layer, while the third one is still within the shock layer and practically does not affect the flow structure. The flow pattern is almost identical to that observed in the case of two particles (Figure 7b). In Figure 9b,c, the third particle crosses the bow shock. The picture of the gas-dynamic interaction is more complicated here. However, the tendency towards the formation of a stable region of action of the impact jet takes place, and a zone of increased pressure and heat transfer on the surface is observed (Figure 10).

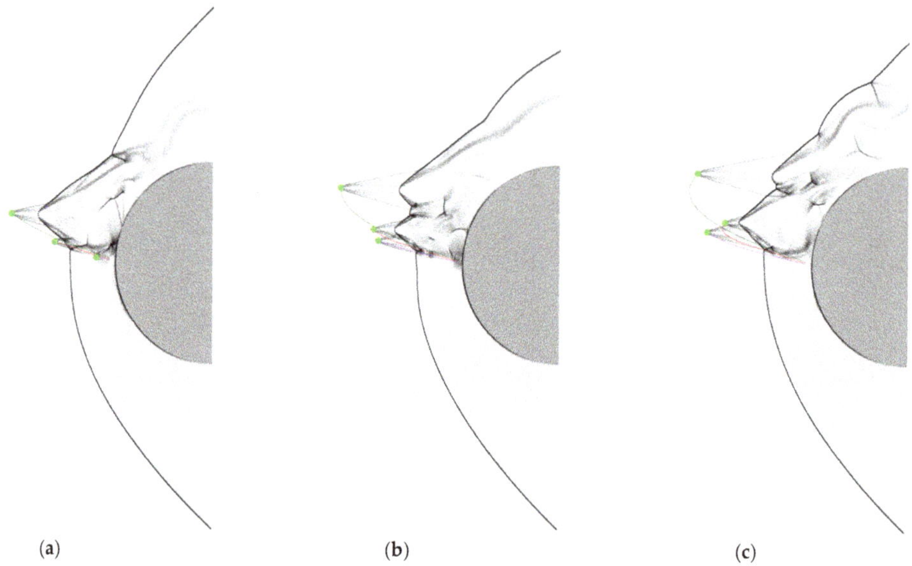

Figure 9. Schlieren images at successive times. (**a**)—0.38 ms, (**b**)—0.63 ms, (**c**)—0.79 ms. A variant with three particles.

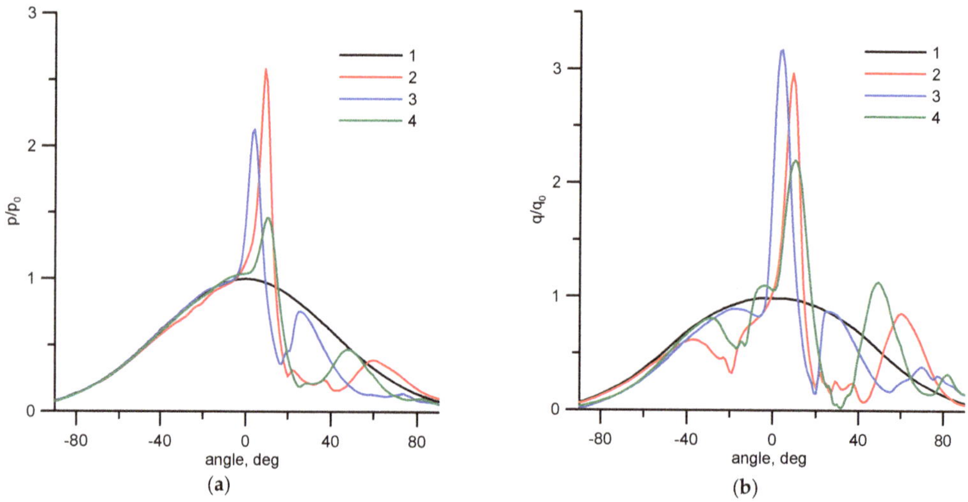

Figure 10. Distributions of pressures (**a**) and heat fluxes (**b**) along the surface at successive times for the variant with three particles. 1—initial moment, 2—0.38 ms, 3—0.63 ms, 4—0.79 ms.

4. Conclusions

A series of computational experiments were carried out aimed at identifying characteristic shock wave structures formed when particles reflected from the surface pass through the bow shock wave. The values of pressure and heat flux obtained in the computations were significantly higher than those in particle-free flow. We show that the key role in the intensification of heat transfer is played by the formation of an impact jet flowing onto the surface. In the zone of action of the impact jet, an area of increased pressure and heat flux is formed. This effect is maintained over time by the successive action of particles. The directions for further research are connected with three-dimensional modeling and the study of complex flow structures and heat transfer enhancement caused by the particle–shock layer interaction.

Author Contributions: Methodology, software, investigation, A.S.; methodology, investigation, validation, D.R. All authors have read and agreed to the published version of the manuscript.

Funding: This research was carried out within the framework of the state assignment issued by the Ministry of Education and Science of Russia, project number FSFF-2020-0013.

Informed Consent Statement: Not applicable.

Conflicts of Interest: The authors declare no conflict of interest.

References

1. Crowe, C.T.; Schwarzkopf, J.D.; Sommerfeld, M.; Tsuji, Y. *Multiphase Flows with Droplets and Particles*, 2nd ed.; CRC Press: New York, NY, USA, 2011.
2. Mikhatulin, D.S.; Polezhaev, Y.V.; Reviznikov, D.L. *Heat Transfer and Destruction of Bodies in a Supersonic Heterogeneous Flow*; Yanus-K: Moscow, Russia, 2007. (In Russian)
3. Tsirkunov, Y.M. Gas-particle flows around bodies–key problems, modeling and numerical analysis. In Proceedings of the Fourth International Conference on Multiphase Flow. New Orleans, LA, USA, 27 May–1 June 2001; Paper No 609. Michaelides, E., Ed.; p. 31.
4. Varaksin, A.Y. Fluid dynamics and thermal physics of two-phase flows: Problems and achievements. *High Temp.* **2013**, *51*, 377–407. [CrossRef]
5. Varaksin, A.Y. Gas-Solid Flows Past Bodies. *High Temp.* **2018**, *56*, 275–295. [CrossRef]
6. Dombrovsky, L.A.; Baillis, D. *Thermal Radiation in Disperse Systems: An Engineering Approach*; Begell House: New York, NY, USA, 2010.
7. Ershova, T.V.; Mikhatulin, D.S.; Reviznikov, D.L.; Sposobin, A.V.; Vinnikov, V.V. Numerical Simulation of Heat and Mass Transfer between Heterogeneous Flow and an Obstacle. *Comput. Therm. Sci.* **2011**, *3*, 15–30. [CrossRef]
8. Vasilevsky, E.B.; Osiptsov, A.N.; Chirikhin, A.V.; Yakovleva, L.V. Heat transfer on the frontal surface of a blunt body in a high-speed flow containing low-inertia particles. *J. Eng. Phys. J.* **2001**, *74*, 1399–1411.
9. Volkov, A.N.; Tsirkunov, Y.M.; Oesterle, B. Numerical simulation of a supersonic gas–solid flow over a blunt body: The role of inter-particle collisions and two-way coupling effects. *Int. J. Multiph. Flow.* **2005**, *31*, 1244–1275. [CrossRef]
10. Oesterlé, B.; Volkov, A.N.; Tsirkunov, Y.M. Numerical investigation of two-phase flow structure and heat transfer in a supersonic dusty gas flow over a blunt body. *Prog. Flight Phys.* **2013**, *5*, 441–456.
11. Osiptsov, A.N.; Egorova, L.A.; Sakharov, V.I.; Wang, B. Heat transfer in supersonic dusty-gas flow past a blunt body with inertial particle deposition effect. *Prog. Nat. Sci.* **2002**, *12*, 887–892.
12. Reviznikov, D.L.; Sposobin, A.V.; Sukharev, T.Y. Numerical simulation of the flow around a blunt body in supersonic polydisperse stream. *High Temp.* **2017**, *55*, 400–406. [CrossRef]
13. Molleson, G.V.; Stasenko, A.L. Acceleration of Microparticles and their Interaction with a Solid Body. *High Temp.* **2017**, *55*, 906–913. [CrossRef]
14. Mironov, A.; Isaev, S.; Skrypnik, A.; Popov, I. Numerical and physical simulation of heat transfer enhancement using oval dimple vortex generators —Review and recommendations. *Energies* **2020**, *13*, 5243. [CrossRef]
15. Isaev, S.A.; Popov, I.A.; Mikheev, N.I.; Guvernyuk, S.V.; Nikushchenko, D.V.; Sudakov, A.G. Heat transfer enhancement by surface vortex generators. New basic mechanisms and industrial technologies. *J. Phys. Conf. Ser.* **2020**, *1683*, 022084. [CrossRef]
16. Fleener, W.A.; Watson, R.H. Convective heating in dust-laden hypersonic flows. *AIAA Pap.* **1973**, 73–761. [CrossRef]
17. Holden, M.; Duryea, G.; Gustafson, G.; Hudack, L. An Experimental Study of Particle-Induced Convective Heating Augmentation. *AIAA Pap.* **1976**, 76–320. [CrossRef]
18. Dunbar, L.E.; Courtney, J.F.; McMillen, L.D. Heating augmentation in Erosive Hypersonic Environments. *AIAA J.* **1975**, *13*, 908–912. [CrossRef]
19. Hove, D.T.; Shih, W.C.L. Re-entry Vehicle Stagnation Region Heat Transfer in Particle Environments. *AIAA J.* **1977**, *15*, 1002–1005. [CrossRef]
20. Vladimirov, A.S.; Ershov, I.V.; Makarevich, G.A.; Khodtsev, A.V. Experimental investigation of the process of interaction between heterogeneous flows and flying bodies. *High Temp.* **2008**, *46*, 512–517. [CrossRef]
21. Reviznikov, D.L.; Sposobin, A.V.; Ivanov, I.E. Change in the Structure of a Flow under the Action of Highly Inertial Particle when a Hypersonic Heterogeneous Flow Passes over a Body. *High Temp.* **2018**, *56*, 884–889. [CrossRef]
22. Reviznikov, D.L.; Sposobin, A.V.; Ivanov, I.E. Oscillatory flow regimes resulting from the shock layer-particle interaction. *High Temp.* **2020**, *58*, 278–283. [CrossRef]
23. Reviznikov, D.L.; Sposobin, A.V.; Ivanov, I.E. Comparative Analysis of Calculated and Experimental Data on an Oscillating Flow Induced by the Gasdynamic Interaction of a Particle with a Shock Layer. *High Temp.* **2020**, *58*, 839–845. [CrossRef]
24. Sposobin, A.V.; Reviznikov, D.L.; Ivanov, I.E.; Kryukov, I.A. Pressure and Heat Flux Oscillations Induced by Gas-Dynamic Interaction between a High Inertia Particle and a Shock Layer. *Russ. Aeronaut.* **2020**, *63*, 677–685. [CrossRef]

25. Holden, M.S. Experimental Studies of Separated Flows at Hypersonic Speeds. Part I: Separated Flows over Axisymmetric Spiked Bodies. *AIAA J.* **1966**, *4*, 591–599. [CrossRef]
26. Zapryagaev, V.I.; Kavun, I.N. Experimental study of the reverse flow in the forward separation region in a pulsating flow around a spiked body. *J. Appl. Mech. Tech. Phys.* **2007**, *48*, 492–500. [CrossRef]
27. Kitamura, K.; Eiji, S. Evaluation of Euler Fluxes for Hypersonic Heating Computations. *AIAA J.* **2010**, *48*, 763–776. [CrossRef]
28. Kitamura, K.; Eiji, S. Towards shock-stable and accurate hypersonic heating computations: A new pressure flux for AUSM-family schemes. *J. Comput. Phys.* **2013**, *245*, 62–83. [CrossRef]
29. Kotov, D.V.; Surzhikov, S.T. Calculation of viscous and inviscid gas flows on unstructured grids using the AUSM scheme. *Fluid Dyn.* **2011**, *46*, 809–825. [CrossRef]
30. Peskin, C.S. The immersed boundary method. *Acta Numer.* **2002**, *11*, 479–517. [CrossRef]
31. Tseng, Y.H.; Ferziger, J.H. A ghost-cell immersed boundary method for flow in complex geometry. *J. Comput. Phys.* **2003**, *192*, 593–623. [CrossRef]
32. Lima, E.; Silva, A.L.F.; Silveira-Neto, A.; Damasceno, J.J.R. Numerical simulation of two-dimensional flows over a circular cylinder using the immersed boundary method. *J. Comput. Phys.* **2003**, *189*, 351–370. [CrossRef]
33. Bakhvalov, P.A.; Bobkov, V.G.; Kozubskaya, T.K. Application of schemes with a quasi-one-dimensional reconstruction of variables for calculations on nonstructured sliding grids. *Math Models Comput. Simul.* **2017**, *9*, 155–168. [CrossRef]
34. Yamakawa, M.; Chikaguchi, S.; Asao, S.; Hamato, S. Multi Axes Sliding Mesh Approach for Compressible Viscous Flows. In *Lecture Notes in Computer Science*; Krzhizhanovskaya, V., Závodszky, G., Lees, M., Dongarra, J., Sloot, P., Brissos, S., Teixeira, J., Eds.; Computational Science–ICCS 2020; Springer: Cham, Switzerland, 2020; Volume 12143. [CrossRef]
35. Dürrwächter, J.; Kurz, M.; Kopper, P.; Kempf, D.; Munz, C.D.; Beck, A. An efficient sliding mesh interface method for high-order discontinuous Galerkin schemes. *Comput. Fluids* **2021**, *217*, 104825. [CrossRef]

fluids MDPI

Article

Modelling Experimental Measurements of Fluid Flow through Railway Ballast

Raed Alrdadi and Michael H. Meylan *

School of Information and Physical Sciences, The University of Newcastle, Callaghan, NSW 2308, Australia; raed.alrdadi@uon.edu.au
* Correspondence: mike.meylan@newcastle.edu.au

Abstract: The flooding of railway ballasts can cause extensive damage. This process has been the subject of several experimental investigations. In the present work, a relatively easy to implement approach to modelling this fluid flow is presented. It is shown that good agreement with the experimental results is obtained. The fluid flow is modelled by Darcy's law, which we extend to the free fluid flowing above the ballast. The main complexity is in determining the free surface position, which is accomplished using an iterative procedure. The equations are solved using the finite element method. The method is illustrated by careful numerical calculations that are carefully compared with the experimental results reported in the literature. The method is then extended to realistic railway ballast, including the effects of ballast fouling. It is shown that when the flow begins to overtop the ballast, the free surface shifts to greatly increase the chance of ballast scouring.

Keywords: railway ballast; free surface; Darcy's law; finite element method; iterative algorithm

Citation: Alrdadi, R.; Meylan, M.H. Modelling Experimental Measurements of Fluid Flow through Railway Ballast. *Fluids* **2022**, *7*, 118. https://doi.org/10.3390/fluids7030118

Academic Editor: Mehrdad Massoudi

Received: 3 February 2022
Accepted: 17 March 2022
Published: 21 March 2022

1. Introduction

One of the critical features of a railway ballast is its high permeability so that water can flow freely through it. In principle, this protects the ballast from damage due to the actions of floodwater. However, under certain conditions, the flow of floodwaters can be sufficiently strong to scour the ballast [1,2]. This scour, in turn, destroys the structural strength of the ballast and leads to a severe accident in the worst-case scenario. Figure 1 shows flood damage to a rail track in the Hunter Valley, NSW, Australia. The porosity of the ballast can change over time, primarily due to ballast fouling, which alters the ballast's ability to resist water damage. The exact process of ballast scouring is not fully understood, and it is the subject of ongoing research.

Figure 1. Scouring of railway ballast during flooding in the Hunter Valley, NSW, Australia, in October 2010. Copyright Australian Rail Track Corporation.

Theoretical and experimental work on the flow of water through the ballast has been reported by [3], who investigated the effect of fouling on the conductivity of railway ballast. This study only calculated the flow through a realistic ballast numerically and did not report any experimental research, although they determined the permeability experimentally. Moreover, their analysis did not account for the free boundary, which is critical to understanding fluid flow. However, it represented the first attempt to model this challenging problem. This work was the starting point for the work of [4], which extended the work of [3] to the case of variable ballast properties and a free surface. However, this work assumed that the free surface was within the ballast itself and could not account for a free surface that extended in any way above the ballast. Such a case occurs both for realistic ballasts and for the experimental studies in the literature. We also note that Darcy's law was assumed in both [3,4], which is a simplification. More complex fluid models have been considered by [5,6]. These more complex models allow for more complicated fluid flows and more complicated porous media laws. However, they require much greater computational power, and they are also time-dependent and not easily adapted to constant flow problems which appear in typical ballast scouring models [3] and the experiments we examine here.

The problem of fluid flow through porous media with a free boundary has been the subject of extensive research. The benchmark solution was reported in [7,8] based on previous results calculated as far back as the 1940s. However, with the advances in computational power, simple numerical methods based on iterative algorithms have become practical, and that is the method used in [4] and the present work. For a general discussion of the numerical method, see [9].

There are very limited experiments that have been conducted on the flow of floodwaters through railway ballast. Experimental investigation of railway ballast scour was conducted by [5,10]. They conducted a series of experiments in a hydraulic laboratory in which the scour of railway ballast was investigated. In particular, they tried to determine the point at which the breaching of the railway ballast occurred. We note that the exact moment of breaching was found by [11] to be a complex property of the ballast structure.

In this study, we will simulate the two experiments of [5,10] numerically. In particular, we will show that we can compute the flow rate and the free surface position. Our method extends the Darcy's law formulation to the fluid that flows above the ballast by making the conductivity in the fluid outside the ballast much higher than that for the fluid in the ballast region. We then use the method we have developed to simulate flow through a realistic railway ballast, including the effect of layers with different conductivity.

2. Numerical Modelling of Flow through Railway Ballast

We assume that the fluid flow is incompressible, an assumption which seems entirely reasonable for the kinds of flows we are modelling, and that it is in steady state. Therefore, by conservation of mass, we can write the continuity equation for incompressible fluid and steady-state as:

$$\nabla.\mathbf{V} = 0 \tag{1}$$

Ref. [12]. $\mathbf{V} = u\hat{i} + w\hat{k}$ is the velocity of a fluid particle, where u and w represent the velocity in x and z directions, respectively. The velocity of seepage by Darcy's law is:

$$\mathbf{V} = \frac{k}{\mu}(-\nabla P + \rho g z), \tag{2}$$

where k is a permeability, which may vary spatially, μ is a viscosity, P is the water pressure, z represents the elevation, ρ is the fluid density, and g is the acceleration due to gravity [13]. We can rewrite Equation (2) as:

$$\mathbf{V} = K_s(\mathbf{x}) \cdot \nabla\phi, \tag{3}$$

where $\phi = -\left(\frac{P-P_{atm}}{\rho g} + z\right)$ is the hydraulic head that has the atmospheric pressure P_{atm} (which we set to be zero) and $K_s(\mathbf{x}) = \frac{k\rho g}{\mu}$ represents the hydraulic conductivity, where $\mathbf{x} = (x,z)$ is a vector. We obtained Equation (3) by letting $u = K_s(\mathbf{x})\frac{\partial\phi}{\partial x}$ and $w = K_s(\mathbf{x})\frac{\partial\phi}{\partial z}$, then from the expression for velocity in Equation (2) in two dimensions, we derive the formula of ϕ. It follows that:

$$V = \frac{k}{\mu}(-\nabla P + \rho g z) \quad \Rightarrow \quad V = \frac{k\rho g}{\mu}\left(\frac{-\nabla P}{\rho g} + z\right),$$

so that $\frac{\partial \phi}{\partial x} = \frac{\partial}{\partial x}\left(\frac{-P}{\rho g}\right)$ and $\frac{\partial \phi}{\partial z} = \frac{\partial}{\partial z}\left(\frac{-P}{\rho g} - 1\right)$ since the minus sign in $\frac{\partial \phi}{\partial z}$ represents the direction of velocity of seepage. Hence, the previous calculations led us to:

$$V = \begin{bmatrix} u \\ \\ v \end{bmatrix} = \begin{bmatrix} K_s(\mathbf{x})\frac{\partial \phi}{\partial x} \\ \\ K_s(\mathbf{x})\frac{\partial \phi}{\partial z} \end{bmatrix} = K_s(\mathbf{x}) \begin{bmatrix} \frac{\partial}{\partial x} \\ \\ \frac{\partial}{\partial z} \end{bmatrix} \phi = K_s(\mathbf{x}) \cdot \nabla \phi.$$

We then obtain the classical flow equation for fluids governed by Darcy's law:

$$\nabla \cdot (K_s(\mathbf{x}) \cdot \nabla \phi) = 0. \tag{4}$$

In the calculations which follow, we will assume that K_s is piecewise constant, but more complicated situations can be considered [4]. The boundary conditions are shown in Figure 2. The equations are solved using the finite element method. Details of the solution method can be found in [4].

The challenge to finding the solution is to calculate the free surface. We use an iterative algorithm here, in which we make an initial guess of the position of the free surface and then compute the solution assuming that $\phi_n = 0$ on the free surface. We then use the condition $\phi = -z$ to calculate a new free surface, and the calculation is repeated until convergence is achieved. Critical to our method is to treat the water outside the ballast using the same system of equations with a significantly reduced porosity (or higher permeability). As we will see, such a method gives good agreement with the experimental results.

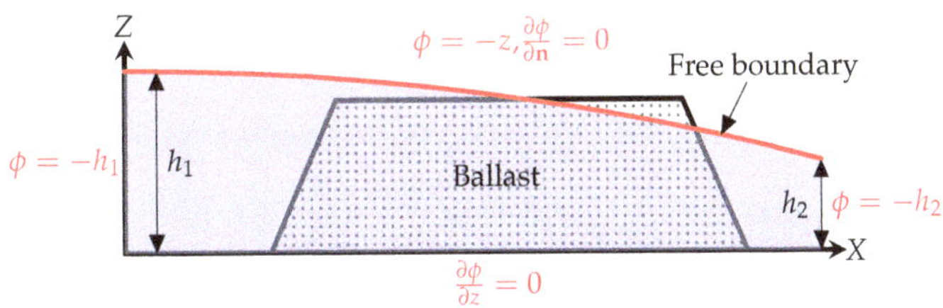

Figure 2. Schematic diagram showing the boundary conditions. The governing equation is $\nabla \cdot (K_s(\mathbf{x}) \cdot \nabla \phi) = 0$, where $K_s(\mathbf{x})$ is variable.

3. Materials and Experiment

We compare our method to two experiments reported in the literature. The Atmojo experiment is reported in [10] and the Tsubaki experiment in [5]. The experiments were different, and we will explain each of them in detail. In particular, the experiments had different dimensions and different ballast properties. Before beginning our comparison, we must determine the porosity of the ballast from the reported properties. To accomplish this, we use the Sieve size chart Figure 3 reported in [14]. This chart allows us to determine the hydraulic conductivity for the experiments. The hydraulic conductivity of clean ballast gradation in Figure 3 is 0.34 m/s, and it is 1.4×10^{-2} m/s for crushed ballast fines [14].

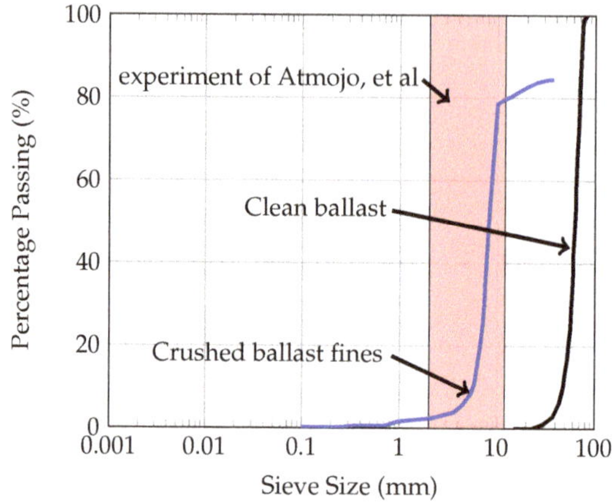

Figure 3. Grain size distribution for ballast as reported in [14]. The red area represents the gradation of the ballast for Atmojo's experiment.

3.1. Atmojo's Experiment

Atmojo and others have performed an experiment that simulated the effect of water flow on railway ballast [10]. A schematic of the experiment is shown in Figure 4. They gradually increased the height of inflow, and their study concluded when breaching of the railway ballast commenced at 15 cm. At this point, the height of inflow was higher than the elevation of the railway ballast. It should be noted that we have exchanged left and right here for consistency with Tsubaki's experiment. The experiment was not conducted on full scale, and the material used for the railway ballast in this experiment was 0.2 to 1.2 cm. We can use this ballast gradation and Figure 3 to determine that the hydraulic conductivity of the railway ballast in Atmojo's experiment Figure 4 was 1.4 cm/s.

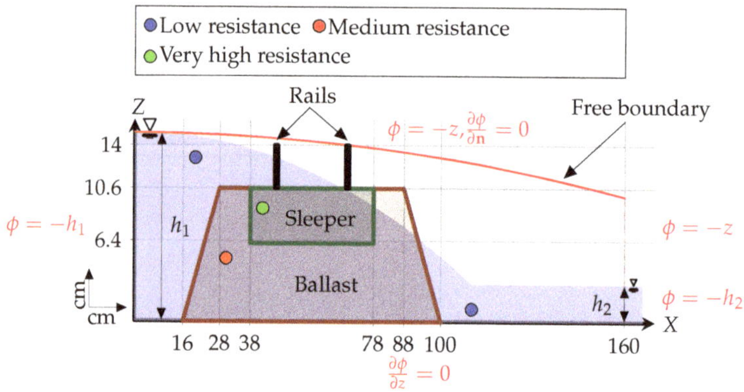

Figure 4. Schematic diagram of the experiment in [10]. The boundaries conditions of the hydraulic head ϕ are for this study. The position of the free boundary is the initial guess above the final water level.

3.2. Tsubaki's Experiment

The Tsubaki's experiment [5] involved many types of railway ballast. We focus on the case of ballast without sleepers and rails since their dimensions were not reported in the study. The study was conducted with a realistic railway ballast, and the gradient of this ballast according to [5] is identical to the distributions of particle sizes as for clean ballast in

Figure 3. Therefore, the hydraulic conductivity equals 0.34 m/s. We present a schematic of the experiment in Figure 5.

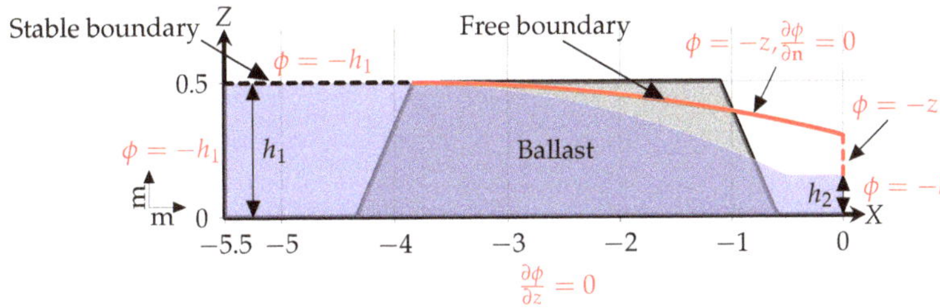

Figure 5. Schematic diagram of the Tsubaki's experiment and the boundaries conditions of the hydraulic head ϕ for this study, where h_2 is assumed to be 0.001.

4. Results

Our focus here is on comparison with the experiments conducted on railway ballast. In the Appendix A we present validation against other results and study convergence. We begin with the Atmojo experiment. We assume the flow site starts at $x = 0$, and we have maintained the distance of the railway ballast and the observations of the free surface as in [10]. We present their observations of the free surface in Table 1. Figure 4 presents the solution where we have calculated the free surface using Equations (1) and (3). This study only covers these initial experiments of [10] because the inflow at 15 cm or greater caused breached the railway ballast. Therefore, we have focused only when the elevation of inflow equals 10, 12, 14, and 15 cm.

Table 1. Free-surface observations as in [10] and the x positions used in this study.

Water Level (h_1)	X	Z			
	cm	10 cm	12 cm	14 cm	15 cm
	0	10	12	14	15
	16	10	12	14	15
	28	9	11.5	13.7	14.5
	46	8.4	10.5	13.4	14.1
	48	8.4	10.5	13	12.2
	69	7	9.5	11.8	12.2
	71	7	9.5	11	10.7
	88	4.2	4.2	7.5	7.5
	100	1.3	1.1	2.7	2
	140	1.3	1.5	2	2
	148	1.3	1.3	1.8	1.9
	157	1.3	1.3	1.8	1.9
	160	1.3	1.3	1.8	1.9

As the properties of the ballast and sleeper required for our study were not reported, we have some freedom in choosing them. The results that best matched the experimental observations were $K_s(\mathbf{x}) = 1.4, 0.7$, and 100 cm/s for the ballast, sleeper, and fluid outside the ballast domain, respectively, as shown in Figures 6–9. The convergence of the iterative method is shown in Figure 6 when the height of the flow was 10 cm. There is a low error between the free surface we calculate and the experiment of [10], and the error rate is less than 4.1%, as shown Table 2. By using the same method for the Tsubaki experiment Figure 5, we obtain approximately the free surface as shown in Figures 10 and 11 for a different flow rate. We have chosen the hydraulic conductivity of the railway ballast equal to 0.34 m/s since it has the same gradations as the clean ballast in Figure 3 and 1 m/s for the flowing outside. The stable boundary in Figure 5 may have affected the result for the free surface in Figure 11, but we can still see convergence if we pay attention to the axis in Figure 11.

Table 2. The area of flow of the experiment in [10] and this study.

h_1	Area of Flow cm^2		Error
	Experiment	**This Study**	
10 cm	830.8500	838.3808	0.8983%
12 cm	1.0101×10^3	1.0444×10^3	3.2783%
14 cm	1.2951×10^3	1.2432×10^3	4.0086%
15 cm	1.3125×10^3	1.3100×10^3	0.1934%

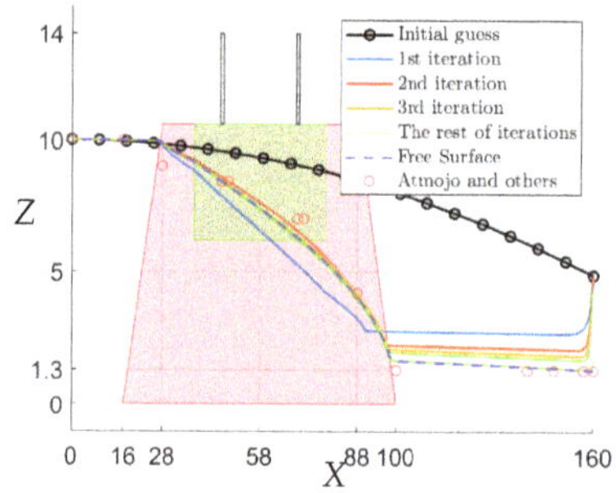

Figure 6. Free-surface of inflow at 10 cm with the convergence of free boundary by iterative algorithm.

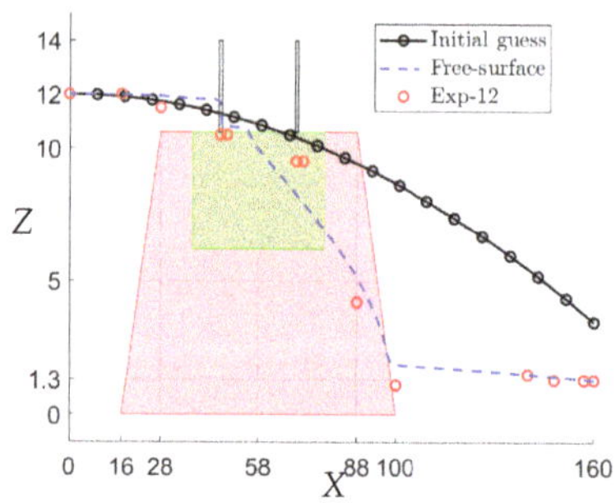

Figure 7. Free-surface of inflow at 12 cm.

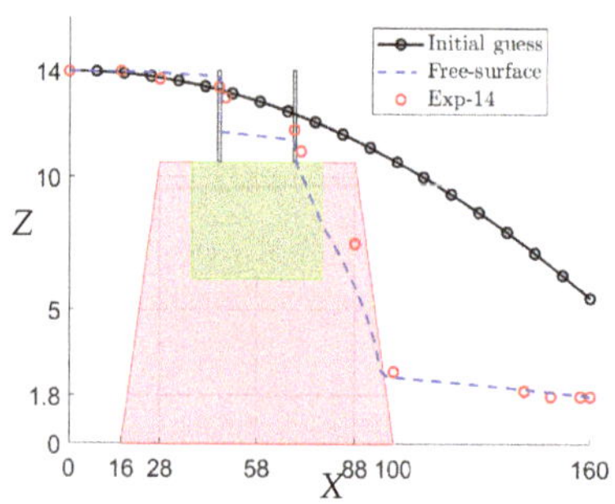

Figure 8. Free-surface of inflow at 14 cm.

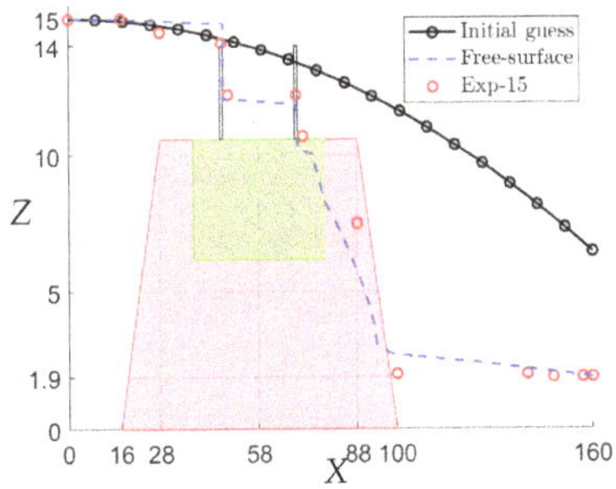

Figure 9. Free-surface of inflow at 15 cm.

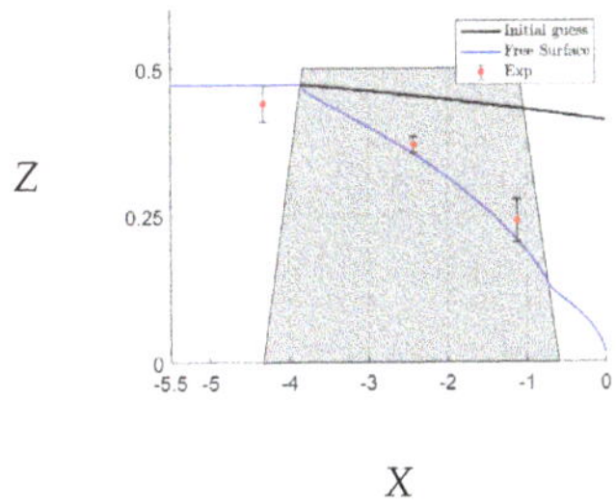

Figure 10. Free surface at flow rate 0.016 m^2/s.

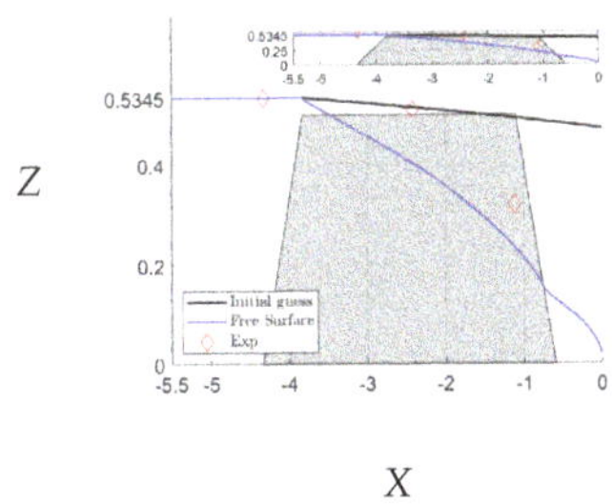

Figure 11. Free surface at flow rate 0.034 m^2/s.

5. Computation for Real Railway Ballast

Having validated our method against experiments, we can now predict the solution for the realistic railway ballast. Our previous results are consistent with both experiments using the hydraulic conductivity of each gradation based on sieve size (Figure 3). If we accept this method, we can predict the free surface of water flow through a realistic railway ballast. According to [3], a typical Australian railway ballast is shown in Figure 12. They suggested different models for realistic railway ballasts with fouling, as shown in Figure 13. These show the fouling ratio in the railway ballast depicted in Figure 14. We can calculate the effect of the fouling ratio by following [3]. Further discussion of fouling can be found in [15]. We use the following equation to calculate the hydraulic conductivity:

$$K_s = \frac{k_b \times k_f}{k_f + (VCI/100) \times (k_b - k_f)},$$

where (VCI) is the void contaminant index, k_b is the speed of hydraulic conductivity for a clean ballast, and k_f is the speed of hydraulic conductivity for crushed ballast fines (as fouling material), which is 1.4×10^{-2} m/s. The hydraulic conductivity K_s for the ballast is 0.34 m/s based on measurements for a clean ballast, and for the flow outside the ballast region, we assume 1 m/s. The flow starts at $x = 0$ to $x = 5.6$, and the ballast's width is $0.5 < x < 5.1$. We examine different heights for the flow. The free surface for the railway ballast is shown in Figures 15–17, where $h_2 = 0.001$ for all of them. We can see the influence of the VCI on model 1 Figure 13 The affect on the free surface is shown in Figure 18. We have increased the height of h_2 to 0.1 m to obtain the convergence for the 40% VCI.

These figures show that even a small overtopping flow greatly alters the flow and the free surface moves close to the top far corner of the ballast. This may increase the chance of ballast scouring as the upper layer of the ballast may be less well supported. We can also see that, as expected, the presence of ballast fouling pushes the free surface of the flow further to the right, again increasing the change in the scour. All of these match the experimental results previously reported in the literature.

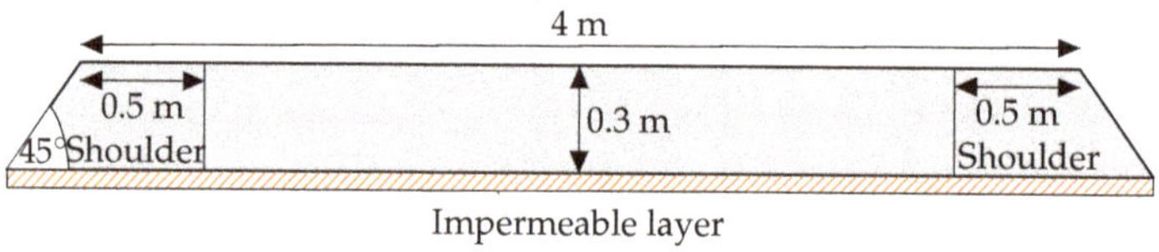

Figure 12. The railway ballast geometry following the Australian track dimensions [3].

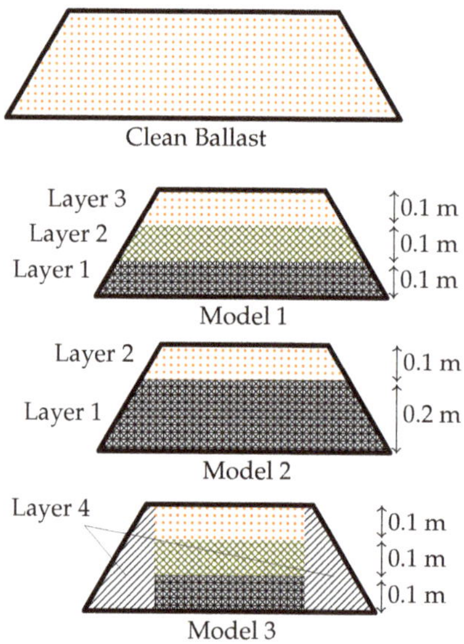

Figure 13. Models of realistic railway ballast including fouling following [3]. The layers of colour coding correspond to different values for the conductivity, and they are given in Figure 14.

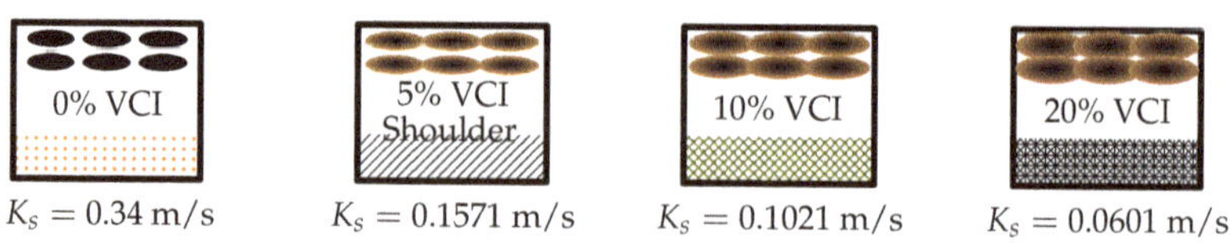

Figure 14. Fouling ratio or VCI and the equivalent hydraulic conductivity K_s. The lower portion of these figures shows the corresponding colouring used in Figure 13.

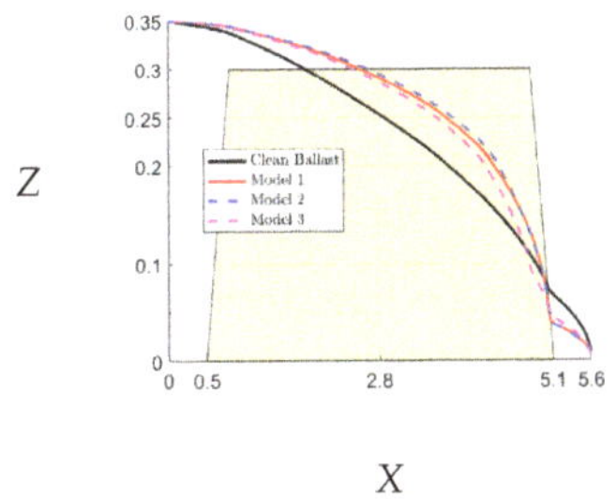

Figure 15. Free-surface for realistic railway ballast at height $h_1 = 0.35$.

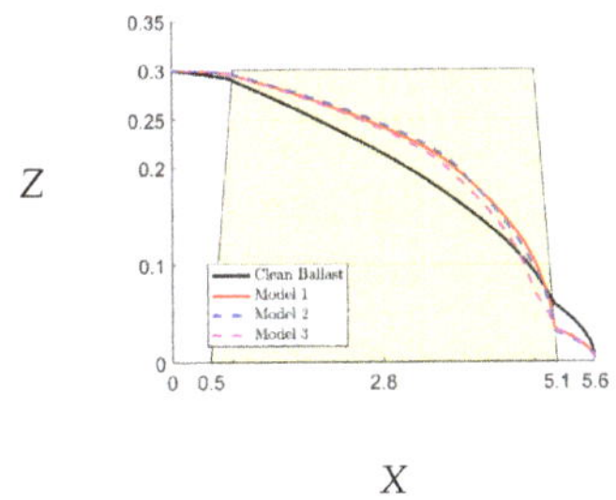

Figure 16. Free-surface for realistic railway ballast at height $h_1 = 0.3$.

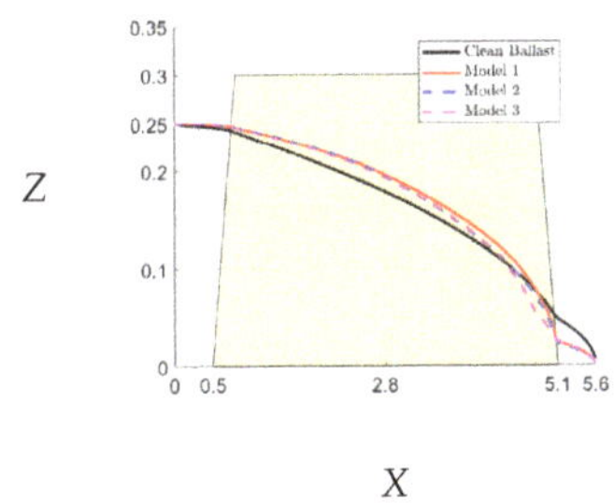

Figure 17. Free-surface for realistic railway ballast at height $h_1 = 0.25$.

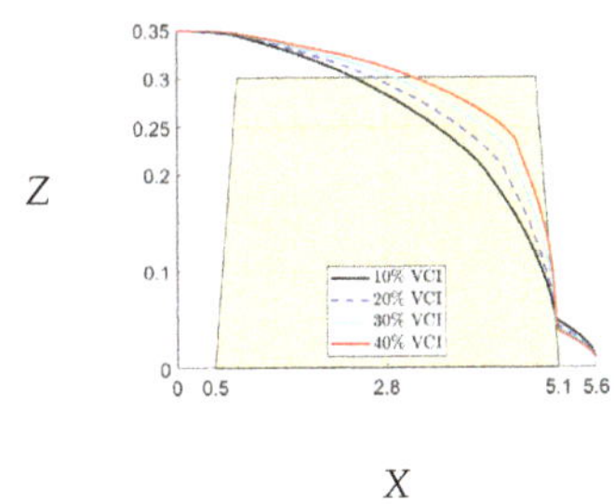

Figure 18. The influence of the fouling ratio on the railway ballast (Model 2) when the layer 1 has different percentages of VCI where $h_2 = 0.1$.

6. Conclusions

We have shown that we can modify the method to solve for flow through porous media to investigate more complex flows in which the free surface extends beyond the ballast. We have made a detailed comparison with the two previous experiments reported in the literature and have shown that we can obtain a good agreement. Such a comparison has not been reported in the literature previously. Finally, we have presented some preliminary calculations for a realistic rail track. These show that the fluid overtopping the ballast greatly shifts the free surface of the flow to the discharge end of the ballast. Any fouling of the ballast increases this shift in the position of the free surface. This fits with the reported experimental results. The model is relatively straightforward to implement using standard finite element code and can be used to make more detailed predictions and to aid in the prediction of potential flood damage to rail track.

Author Contributions: Conceptualisation, M.H.M. and R.A.; methodology, M.H.M. and R.A.; software, R.A.; validation, R.A.; formal analysis, R.A.; investigation, R.A.; writing—original draft preparation, R.A.; writing—review and editing, M.H.M. and R.A.; visualisation, R.A. All authors have read and agreed to the published version of the manuscript.

Funding: This research was supported under Australian Research Council's Industrial Transformation Training Centres Scheme (ARC Training Centre for Advanced Technologies in Rail Track Infrastructure; IC170100006).

Institutional Review Board Statement: Not applicable.

Informed Consent Statement: Not applicable.

Data Availability Statement: Not applicable.

Conflicts of Interest: The authors declare no conflict of interest.

Appendix A. Further Validation and Convergence Study

Although our method is designed to solve for railway ballast, it is generally enough to compare it to other problems. We can compare our solution with the well-known test case reported in [8], which in turn follows from the analytical solution of [7]. This problem is shown in Figure A3. The iterative solution procedure is shown in Figure A2, and the final free surface comparison is shown in Figure A3. Further validation is shown in Figure A4 based on the previously reported numerical solution for a trapezoidal shape, and finally, we can compare with other experimental results Figure A5. The robustness and validity of our method are again confirmed.

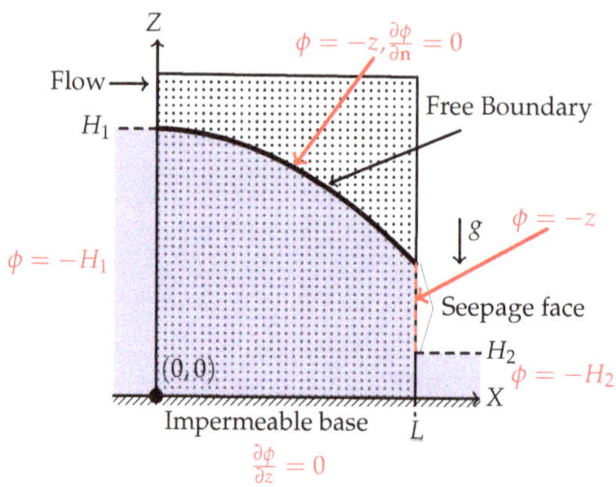

Figure A1. Schematic diagram of the test problem where $H_1 = 1$ and $H_2 = \frac{1}{6}$ [8].

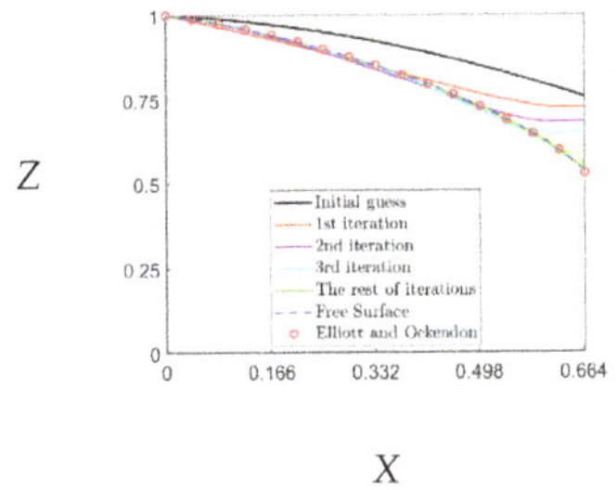

Figure A2. The iterations of free boundary to obtain the free surface and the comparison for the results reported in [8].

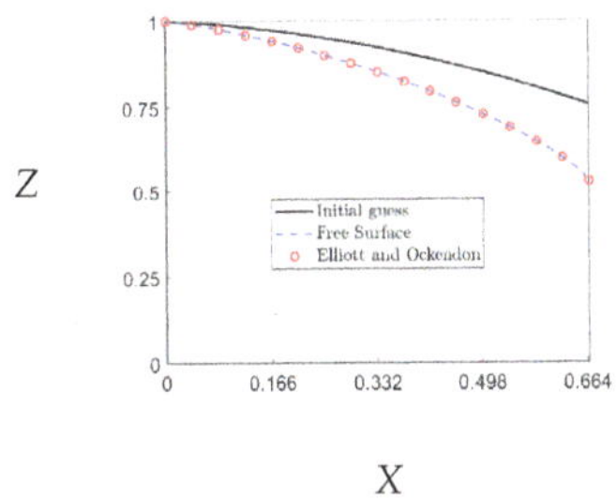

Figure A3. Position of the free surface from our calculations and from [8] showing the excellent agreement.

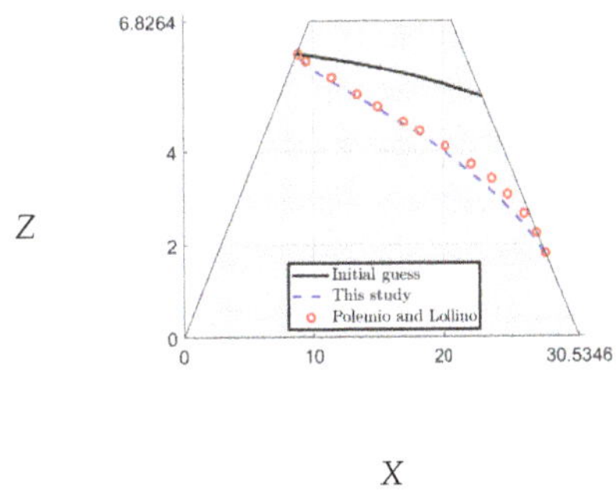

Figure A4. Comparing our results for free surface for a trapezoid dam with the numerical solution of [11].

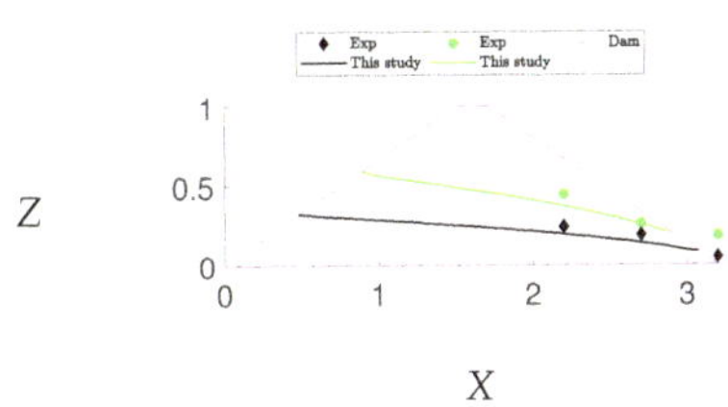

Figure A5. Validating our results for the free surface for a trapezoid dam with the experiments of [16].

References

1. Tsubaki, R.; Bricker, J.D.; Ichii, K.; Kawahara, Y. Development of Fragility Curves for Railway Embankment and Ballast Scour due to Overtopping Flood Flow. *Nat. Hazard. Earth Syst.* **2016**, *16*, 2455–2472. [CrossRef]
2. Johnston, I.; Murphy, W.; Holden, J. A Review of Floodwater Impacts on the Stability of Transportation Embankments. *Earth-Sci. Rev.* **2021**, *215*, 103553. [CrossRef]
3. Tennakoon, N.; Indraratna, B.; Rujikiatkamjorn, C.; Nimbalkar, S.; Neville, T. The Role of Ballast-Fouling Characteristics on the Drainage Cpacity of Rail Substructure. *Geotech. Test. J.* **2012**, *35*, 629–640. [CrossRef]
4. Alrdadi, R.; Meylan, M.H. Modelling Water Flow through Railway Ballast with Random Permeability and a Free Boundary. *Appl. Math. Model.* **2021**, *103*, 36–50. [CrossRef]
5. Tsubaki, R.; Kawah, Y.; Ueda, Y. Railway Embankment Failure due to Ballast Layer Breach Caused by Inundation Flows. *Nat. Hazards* **2017**, *87*, 717–738. [CrossRef]
6. Kalyanaraman, B.; Meylan, M.H.; Lamichhane, B. Coupled Brinkman and Kozeny–Carman Model for Railway Ballast Washout using the Finite Element Method. *J. R. Soc. N. Z.* **2021**, *51*, 375–388. [CrossRef]
7. Kochina, P. *Theory of Ground Water Movement*; Princeton University Press: Princeton, NJ, USA, 1962.
8. Elliott, C.M.; Ockendon, J.R. *Weak and Variational Methods for Moving Boundary Problems*; Pitman: Boston, MA, USA, 1982.
9. Xiao, J.E.; Ku, C.Y.; Liu, C.Y.; Fan, C.M.; Yeih, W. On Solving Free Surface Problems in Layered Soil using the Method of Fundamental Solutions. *Eng. Anal. Bound. Elem.* **2017**, *83*, 96–106. [CrossRef]
10. Atmojo, P.S.; Sachro, S.S.; Edhisono, S.; Hadihardaja, I.K. Simulation of the Effectiveness of the Scouring Prevention Structure at the External Rail Ballast using Physical Model. *Int. J. GEOMATE* **2018**, *15*, 178–185. [CrossRef]
11. Polemio, M.; Lollino, P. Failure of Infrastructure Embankments Induced by Flooding and Seepage: A Neglected Source of Hazard. *Nat. Hazards Earth Syst.* **2011**, *11*, 3383–3396. [CrossRef]
12. Pritchard, P.J.; Mitchell, J.W. *Fox and McDonald's Introduction to Fluid Mechanics*; John Wiley & Sons: Hoboken, NJ, USA, 2016.
13. Vafai, K. (Ed.) *Handbook of Porous Media*; CRC Press: Boca Raton, FL, USA, 2015.
14. Danquah, W.O.; Ghataora, G.S.; Burrow, M.N. The Effect of Ballast Fouling on the Hydraulic Conductivity of the Rail Track Substructure. In Proceedings of the XV Danube-European Conference on Geotechnical Engineering (DECGE), Austrian Society for Soil Mechanics and Geotechnical Engineering, Vienna, Austria, 9–11 September 2014.
15. Koohmishi, M.; Palassi, M. Effect of Gradation of Aggregate and Size of Fouling Materials on Hydraulic Conductivity of Sand-Fouled Railway Ballast. *Constr. Build. Mater.* **2018**, *167*, 514–523. [CrossRef]
16. Larese A.N.; Rossi R.; Onate E.; Idelsohn S. A Coupled PFEM–Eulerian Approach for the Solution of Porous FSI Problems. *Comput. Mech.* **2012**, *50*, 805–819. [CrossRef]

MDPI
St. Alban-Anlage 66
4052 Basel
Switzerland
www.mdpi.com

Fluids Editorial Office
E-mail: fluids@mdpi.com
www.mdpi.com/journal/fluids

www.ingramcontent.com/pod-product-compliance
Lightning Source LLC
Chambersburg PA
CBHW040829250926
51850CB00007BA/1293
9783725806850